Fundamentals of Process Control

Principles and Concepts

Fundamentals of Process Control

Principles and Concepts

By Patrick F. Bouwman

Notice

The information presented in this publication is for the general education of the reader. Because neither the author nor the publisher has any control over the use of the information by the reader, both the author and the publisher disclaim any and all liability of any kind arising out of such use. The reader is expected to exercise sound professional judgment in using any of the information presented in a particular application.

Additionally, neither the author nor the publisher has investigated or considered the effect of any patents on the ability of the reader to use any of the information in a particular application. The reader is responsible for reviewing any possible patents that may affect any particular use of the information presented.

Any references to commercial products in the work are cited as examples only. Neither the author nor the publisher endorses any referenced commercial product. Any trademarks or tradenames referenced in this publication, even without specific indication thereof, belong to the respective owner of the mark or name and are protected by law. Neither the author nor the publisher makes any representation regarding the availability of any referenced commercial product at any time. The manufacturer's instructions on the use of any commercial product must be followed at all times, even if in conflict with the information in this publication.

Printed in the United States of America.
Version: 1.0

ISBN-13: 978-1-64331-139-5 (print)
ISBN-13: 978-1-64331-140-1 (ePub)
ISBN-13: 978-1-64331-141-8 (Kindle)

ISA
67 T. W. Alexander Drive
P.O. Box 12277
Research Triangle Park, NC 27709

Library of Congress Cataloging-in-Publication Data in process

SKY10138735_012626

This book is dedicated to the late Paul W. Murrill, who wrote the original Fundamentals of Process Control Theory *and whose material greatly inspired this edition,*

and

to all my students and their many questions.

Contents

Preface

The evolution of instrumentation and industrial process control has led to today's intricacies of computers, microcontrollers, distributed control systems, and sophisticated networks of intelligent devices. They are all connected to human-machine interfaces, either directly or through wireless networks and handheld devices. The fundamental scientific principles on which process control is based are unchanged. Today, they are better understood in the academic and scientific communities, but they have been neglected in the study and application of process control in pursuit of applying the latest technologies.

There is considerable variation in the hardware and software available from vendors and in the application and practice of these scientific principles from one industry to another. The material presented in this book is generally oriented toward the automation practitioner in such processing industries as petroleum, petrochemical, chemical, pulp and paper, mining, power, drug manufacturing, food processing, and water/wastewater treatment.

This book is designed to teach the fundamental principles, concepts, and theory of process control and to provide detailed examples of how they are applied in controlling industrial processes.

It is designed for both those who want to study independently and those who are studying at a college or university. It also serves as a reference resource for engineers, senior technicians, technologists, and first-line supervisors who are working in or involved in the field of process control.

Much of the book's content is based on my many years of teaching process control at the college level. Therefore, each chapter can be considered a lecture or presentation of a specific process control topic. By reading each chapter, students can deepen their understanding of the subject. However, for those who have some prior knowledge of process control, the chapters can be read and studied independently or used as a reference for the topic covered. This modular approach also allows college and university instructors to select specific chapters and the appropriate sequence to tailor their course material.

There are no specific prerequisites for this study, though an appreciation for industrial concerns and their philosophies will be helpful. It is inevitable that parts of the text will involve mathematics. However, the student does not need to be intimately familiar with such mathematics beforehand to appreciate the fundamental principles and concepts that are presented and applied here. Often, mathematics becomes one of the barriers that prevent people from understanding and using fundamental principles. I have attempted to minimize such barriers by providing detailed worked examples to help students understand the mathematics used. Engineering students will recognize these examples and appreciate why mathematics is part of the engineering curriculum.

Organization and Sequence

The text has been divided into three sections reflecting Principles and Concepts (Chapters 1 through 10), Control Strategies (Chapters 11 through 17), and Advanced Topics (Chapters 18 through 22). This allows the text to be used as a two- or three-semester course at a technical college or university.

The book is also divided into 22 chapters, each of which builds on the content of the previous chapter to expand the reader's understanding of the subject.

- Chapter 1 – A broad introduction to process control, its intent, and why it is important to understand and study process control
- Chapters 2 through 4 – The most fundamental concepts encountered in process control
- Chapter 5 – Process dynamics
- Chapter 6 – Process characteristics
- Chapter 7 – Sensors and transmitters, and the importance of the accuracy, precision, sensitivity, and validity of measurements
- Chapter 8 – Final control elements

- Chapter 9 – The process controller
- Chapter 10 – The procedure for tuning a process control loop
- Chapters 11 through 14 – Cascade control, feedforward control, multivariable control, and ratio control, respectively
- Chapter 15 – Special control strategies such as three-element boiler control, selective control, override control, duplex and split-range control, and auto-selector or cutback control
- Chapters 16 and 17 – Statistical process control and fuzzy control, respectively
- Chapter 18 – Process modeling and simulation
- Chapter 19 – Model-based control and the concept of controlling dead time
- Chapter 20 – Nonlinear and adaptive control
- Chapter 21 – A historical view of distributed control, and control system architecture
- Chapter 22 – Digital systems, specifically digital signal processing and computer implementation of control strategies

Each chapter includes a set of specific learning objectives at the beginning of the chapter. Some chapters contain example problems to illustrate specific concepts, and at the end of most chapters, you will find exercises to test and solidify your understanding of the material. The solutions to selected exercises are given in Appendix D.

A concerted effort was made to ensure that this text is consistent with standards relating to terminology and symbols, specifically:

- *Process Instrumentation Terminology*, ISA-51.1-1979 (R 1993)
- *Instrumentation Symbols and Identification*, ISA 5.1-2009

A basic premise of self-study is that students learn best if they proceed at their own comfortable pace. As a result, there will be a significant variation in the time individual students take to complete this book. Most students will complete the material presented in 40 to 50 hours, though previous experience with the material and personal capabilities will do much to vary this time.

My special thanks go to all who made this book possible. It is my hope that the text answers the many questions that my students have asked.

I would also like to thank Héctor Torres, Ahmad Rad, and Maxym Lachance for their many constructive suggestions in reviewing the book.

Lastly, I would like to give special thanks to the International Society of Automation for giving me this opportunity and to my editor, Liegh Elrod, for her diligence. Without her patience, this book would not have been possible. Her many suggestions and corrections made this book a reality and a better final product.

I hope this text proves useful to you, the student.

Patrick F. Bouwman
March 2022

About the Author

Patrick F. Bouwman is a semiretired electrical engineer with over 40 years of industrial experience, including 18 years in the design and development of military electronics, instrumentation, and biomedical equipment. For more than 15 years, Bouwman operated his own company specializing in research, design, and development of data acquisition and signal processing systems for medical research applications and the instrumentation and process industry. He has provided design and consulting services to the process and automation industry and participated in research conducted at McGill University by custom designing data acquisition equipment and instrumentation for medical research applications.

In addition, Bouwman taught industrial electronics at the collegiate level for over 20 years and was department chair for 16 years, a role in which he was actively involved in all facets of curriculum development, laboratory design, industrial training, and continuing education. He has written numerous instructional and lab manuals on topics including analog and digital electronics, signal processing, discrete automation, instrumentation, process control, PLC programming, ISA-5.1 technical documentation, robotics, computer automation, distributed automation, and medical and military electronics.

He has also provided technical training and instructional material to the Canadian Armed Forces; conducted seminars in data acquisition, signal processing, instrumentation, and control for the manufacturing industry; and prepared and conducted seminars/workgroups in instrumentation, process control, and ISA-5.1 for the International Society of Automation (ISA) Montreal Section.

Bouwman is a senior member of ISA, and he served on the Montreal Section board of directors.

He received a B.Sc. in electrical and electronic engineering from Concordia University in Montreal in 1976, and he currently lives in Montreal, Canada.

Section 1

Principles and Concepts

This section (Chapters 1 through 10) is intended to teach the student the fundamental principles and concepts in feedback control and the functional methods that are used in the process control industry.

Chapter 1 gives a broad introduction to what process control is, its intent, and why it is important to study process control. It also outlines some of the scientific and mathematical principles that are fundamental to understanding process control. The chapter ends with a discussion of process variables, specifically temperature, pressure, and flow.

Chapters 2 through 4 introduce the most fundamental concepts encountered in automated process control, namely the control loop and basic terminology. The concept and mathematics of block diagrams and the transfer functions are introduced to describe and analyze the functional structure of a feedback control loop. Combining transfer functions with block diagrams provides the tools needed to understand more complex systems.

System dynamics are the ways a system reacts to a change in set point, load, or disturbances. Chapter 5 describes many of the general concepts and principles of system dynamics and the associated terminology.

Chapter 6 describes the characteristics of process and the relationship between the manipulated variable (input) and the controlled variable (output). This chapter also lays the foundation for Chapter 18, "Process Modeling and Simulation," in Section 3.

One of the most critical aspects in designing and installing a feedback control loop is specifying the measuring system that will obtain a measurement of the controlled variable. Chapter 7, "Sensors and Transmitters," defines and explains accuracy, precision, sensitivity, and the validity of measurements and transmission systems used in process control.

Final control elements have a direct influence on the process by converting the control signal generated by the process controller into the action required to correctly manipulate the process to meet its desired value. Chapter 8 covers some of the fundamental principles of final control elements and actuators, such as control valves, motors, pumps, and solenoids.

Chapter 9 discusses process controllers, the element of the process control loop that processes the information needed to perform the decision-making function. A clear understanding of the computational aspect of the controller is necessary for proper use and tuning of the process control loop.

A properly tuned control loop maintains its process value at its desired set point, regardless of external disturbances. Chapter 10 discusses the process of tuning control loops by setting the proportional, integral, and derivative values to get the best possible control.

1 Introduction and Overview

Welcome to *Fundamentals of Process Control*. The purpose of this book is to provide the reader with an introduction to the fundamental principles, concepts, and techniques of process control. It is designed specifically for students in technical schools who are studying automation; however, students in related fields can benefit from understanding the theories and practices that form the body of knowledge that is process control.

The field of process control encompasses many diverse disciplines, scientific principles, and mathematical concepts. This book touches on each of these in order to help the reader develop a basic understanding of automated process control.

Learning Objectives

- Have an appreciation of process control history
- Understand the meaning of process control
- Recognize the importance of process control
- Become familiar with process control terminology
- Recognize the fundamental concept of differential equations
- Become familiar with the principle of thermodynamics

1.1 History

Automated control systems date back thousands of years. The first feedback control device on record is thought to be the ancient water clock of Ktesibios in Alexandria Egypt around the third century BC. It kept time by regulating the water level in a vessel and, consequently, the water flow from that vessel.

The first well-defined use of feedback control seems to have been James Watt's application of the flyball governor to the steam engine in about 1775. Broader use of automatic control began in the late 1920s, and the first general theoretical treatment of automatic control was written by Harry Nyquist in 1932. The growth in industrial use of automated process control has been steady and consistent.

As new scientific principles were discovered and new technologies implemented, many of these new principles and technologies were applied to process control hardware and software. The industrial use of automation technology has developed and advanced over the past decades and continues to advance into the digital age.

The underlying principles of automated control have kept pace and also developed rapidly. A constructive and broad basis of understanding has been formed, and new ways of controlling processes have been developed. Today's applications are based on this foundation. New approaches to automation and control, such as distributed control, statistical process control, fuzzy logic control, and internet applications such as the Industrial Internet of Things (IIoT) and artificial intelligence (AI), are becoming more entrenched in the industry.

Two important examples of new approaches are the application of analog electronics to replace mechanical controllers, and then the introduction of the digital computer and microprocessor to process control. As a result, automation and control has received a significant and special boost from technology.

The underlying principles and theory of process control have also developed steadily, and a firm and broad foundation of understanding has been created. Today's applications are based on this foundation.

1.2 What Is Process Control?

The best way to describe process control is through an example. Consider a simple situation: keeping a room at a comfortable temperature using a manually controlled heater. When the temperature of the room is too low, we turn the heater on until the temperature reaches the desired level. When this temperature is reached, we turn the heater off so the room will not overheat. As the room cools and is no longer comfortable, we repeat the process. We continue this process to keep the room at a desired comfort level.

In this example, the comfort level is the desired temperature of the room. This is a typical example of controlling a process, namely, the temperature of the room.

Now suppose you are driving a car. The process of driving a car requires that you are looking out the windshield, that you are aware and observe your surroundings, and that you have a direction toward which the car is directed. See Figure 1-1. In this example, the *process* is ensuring that the car is following the intended direction of the driver, while avoiding many of the disturbances and obstacles on the road. These disturbances, such as curves in the road, other cars on the road, people crossing the street, and paying attention to the road conditions, must be observed to drive safely without accident. All these factors and more must be observed by the driver. To safely accomplish these tasks, the driver must engage all his or her senses—sight, hearing, touch, and smell. That is, the driver must have a heightened sense of awareness of his or her situation and be able to respond to typical road disturbances with corrective action.

The entire task of driving is a *process*, controlled by the driver. The driver sees, hears, and senses all factors that will help in driving safely in the desired direction, and the driver makes mental calculations and decisions that determine any corrections that may be required. Finally, the driver applies the necessary actions. This is quite a daunting task for an unskilled driver. In today's automated world, the technology to automate driving is progressing at a rapid rate, but much still must be done.

Process control is sensing, comparing, and taking corrective action. This idea is further discussed in the chapters that follow.

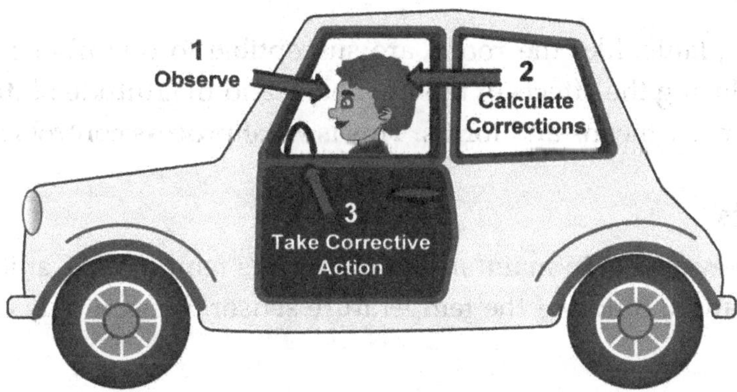

Figure 1-1. The process of driving a car.

Definition

> **Process control** – An operation that regulates parameters by observing the parameter, comparing it with some desired value, and taking some action to bring the parameter as close as possible to its desired value.[1]

1.3. Process Intend

Refining, combining, handling, and otherwise manipulating materials and energy to produce end products profitably and safely can be a precise, demanding, and potentially hazardous process. Small changes in a process can have a large impact on the end result. Variations in proportions, temperature, flow, pressure, turbulence, and many other factors must be carefully measured and consistently controlled to produce the desired end product with a minimum of raw materials and energy. Process control technology is the tool that enables manufacturers to keep their operations running within specified limits and to set more precise limits to maximize profitability while ensuring quality and safety.

A key factor in the engineering design of a process, so that it can be well controlled, is to understand exactly what must be controlled and why it needs to be controlled. Consider a common process such as heating a room on a cold day (see Figure 1-2). The room temperature in this example is the process, and we must answer the following questions: What temperature should the room be to ensure comfort? Will the furnace be able to deliver the required heat rapidly enough considering all the disturbances that may occur?

By *rapidly enough,* we mean whether the control system (the thermostat) and furnace are responsive to disturbances such as opening the room's door. Will the room temperature be brought back to the desired temperature if the door is opened and heat leaves the room?

Most industrial plants, like the room, are susceptible to disturbances that must be minimized. Reducing the effects of the frequency and magnitude of disturbances can be achieved by many means and forms. This is what process control is about.

Measurements

As expected, a key aspect in maintaining the room's temperature at its desired value is the selection and location of the temperature sensor. We can only control what we

1. ISA, *The Automation, Systems, and Instrumentation Dictionary* (Research Triangle Park, NC: ISA [International Society of Automation]), 2003.

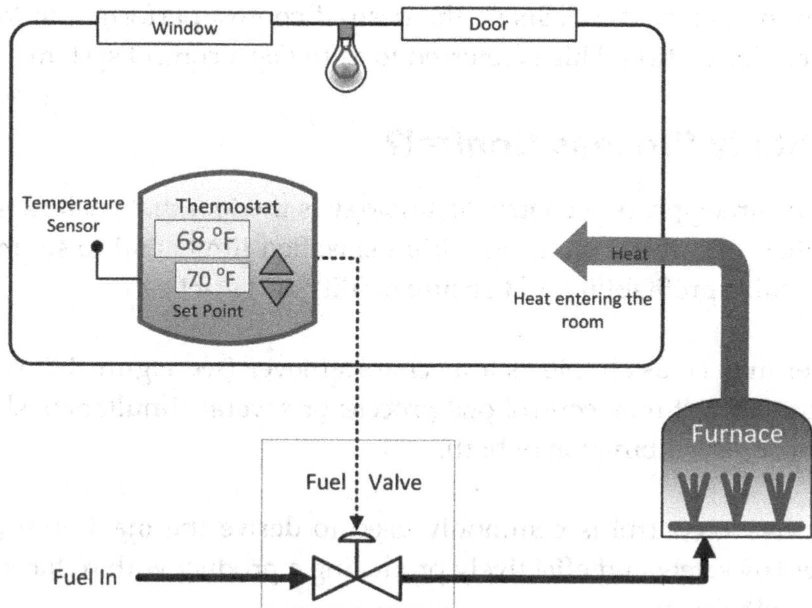

Figure 1-2. Typical room heating process.

are able to measure. The selected temperature sensor that measures the temperature should accurately and rapidly respond to changes in room temperature.

Final Control Elements

The system must also provide proper equipment that can handle the required action to affect the changes required. This is usually referred to as the *manipulated variable* and controls a final element. In the room heating example, the manipulated variable is the fuel valve controlled by the thermostat. If there were no valve on the fuel line, it would not be possible to control the furnace and consequently the room temperature.

Control Arrangement

In the design, some critical issues must be considered, namely, what decisions must be made to apply the necessary corrective action. For example, in the room scenario, how much should the valve be opened or closed to maintain the desired room temperature? It is self-evident that when the room is warm, the heater needs less fuel, and when the room is cold, the heater needs more fuel. This is referred to as *feedback control.*

Control Calculations

After the variables and control arrangement have been selected, equations are selected that use the measurement and desired values in calculating the required fuel. As will be discussed, only a few equations are necessary to provide good control for many types of processes. After the control equations are defined, parameters that appear in the equations are adjusted to achieve the desired control performance for a specific process control application. This is referred to as *tuning* a control system.

1.4 Why Study Process Control?

As mentioned earlier, process control technology is the tool that enables manufacturers to keep their operations running within specified limits and to set more precise limits to maximize profitability and ensure quality and safety.

The controller may be as simple as a mechanical lever (see Figure 1-3) or as complicated as a computer. It may control one process or several simultaneously. It may be analog, digital, or a combination of both.

Automated process control is commonly used to derive the maximum profitability from a process by safely and effectively producing a product with reduced variability and increased efficiency.

In the previous sections there was an implicit assumption that we knew the desired values for the controlled quantities. Once these desired values are known, automation techniques are applied to achieve and/or maintain these desired values, referred to as *set points*. On reflection, however, some of the most significant questions associated

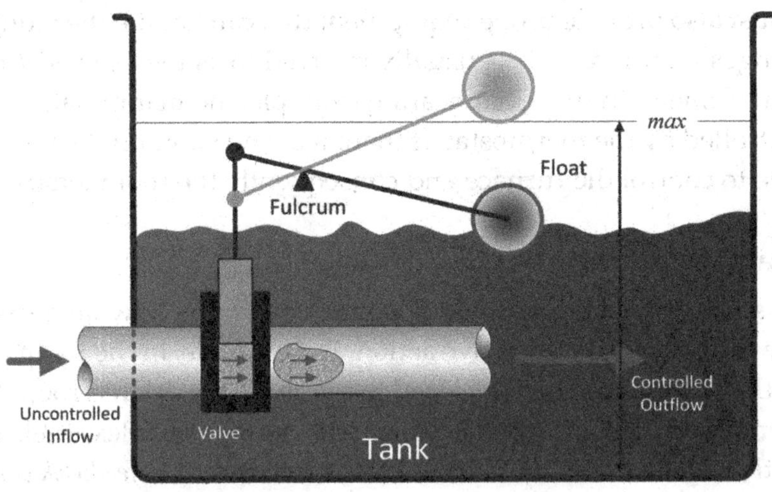

How it works

As the level in the tank increases, the float rises, causing the pivot arm to rise on the float side and descend on the valve side. The lowering of the pivot arm on the valve side causes the valve to close an amount proportional to the rise of the liquid in the tank. The closing of the valve reduces flow into the tank, which reduces the rate of rise of the liquid level. If the level goes down, the reverse takes place.

Figure 1-3. A simple mechanical lever control system.

with the profitability, quality, and safety of a process are those that must be answered to determine the desired values. This is basically a supervisory or management function, and quite often it is left for the operator to determine. But, in recent years, with significant advances in automation and control, many of these supervisory and management functions have themselves become automated, and the ability to achieve technological solutions and hardware answers for such management questions is a significant part of modern control.

As the level of automation increases, more of the automation is associated with process management. The combination of these two phenomena—process control and process management—must be reflected in our overall understanding and appreciation of automated process control.

1.5 Control Strategies

Automatic control systems may be open-loop or closed-loop. The general term *feedback control* implies that the system is closed-loop; in such a system, feedback is used to compare the controlled quantity against a reference quantity, as we saw in the room heating system of Figure 1-2. Regardless of the type of variable (temperature, pressure, fluid flow, speed, torque, or mechanical position), if the controlled output is not at its desired value, then the device used for comparison creates an error signal that feeds an error-correcting device of some form, which then drives the system to restore the variable to the desired value. In a servo system, a motor is used as the error-correcting device, and in many process control applications, valves are used. After the system has been brought within its tolerance band or accuracy limits, it will stay that way until some outside disturbance or a change in desired value (set point) causes an error signal to be created, telling the system to correct itself again.

There are many control strategies besides feedback control, such as feedforward, cascade, ratio, selective, duplex or split-range, auto-selective or cutback, sequential, supervisory, distributed, dead time, adaptive, multivariable, and optimal, just to name the more well-understood strategies. Many will be covered throughout the book.

In feedforward control, for example, corrective action is taken by measuring a disturbance before it can affect the process. In the room heating system, for example, by measuring temperature near the door, the thermostat can pre-empt the room from becoming cold by increasing the fuel, heating the room to compensate for the heat loss caused by opening the door. Ratio control tries to maintain a proper ratio of two or more ingredients or flows. The room heating system example is rather simplified. In most furnaces, not only the fuel must be controlled, but also the ratio of fuel to air (oxygen).

1.6 Process Design

Process design usually focuses on the steady-state operation of the process. Engineering companies tend to design primarily for the steady state and have little experience with variability or dynamics, at least until the design is in operation. The emphasis on minimizing changes in design to minimize risk and maintain the budget and schedule often results in a plant built just like the last one. Not surprisingly, plant variability audits show that most plants have similar problems.

Process control deals with change. If there were no variations in raw materials, utilities, process and ambient conditions, equipment performance, production rates, or desired operating points, there would be little need for control. In actual plant operations, nothing is completely constant. Thus, most of the opportunities for process control deal with change.

1.7 Conservation of Mass and Energy

Manufacturing processes often deal with chemical transformations and are based on the laws of conservation of mass and energy. For that reason, one must consider the importance of these conservation laws when dealing with process control. These laws are also used to develop dynamic models that mathematically represent the process behavior. Models help us greatly in understanding and predicting how a feedback system behaves when changes occur.

Conservation of Mass

The law of conservation of mass states that mass is neither created nor destroyed by chemical reactions or physical transformations. According to the law, the mass of the products in a chemical reaction must equal the mass of the reactants. We can state this mathematically as follows:

$$\textit{rate of mass accumulated} = \textit{rate of mass in} - \textit{rate of mass out} \tag{1-1}$$

Conservation of Energy

The first law of thermodynamics states that energy cannot be created or destroyed; it can only change forms. Work and heat are the forms energy can take to be transferred.

We can state this mathematically as follows:

$$\textit{energy entering} - \textit{energy leaving} = \textit{change of energy} \tag{1-2}$$

The laws of conservation of mass and conservation of energy are useful for several calculations. They can be used to solve for unknown masses, such as the amount of gas consumed or produced during a reaction.

1.8 Concepts of Calculus

Process control, as you might expect, involves many different disciplines, including mathematics and science. It is therefore appropriate to learn the explicit principles of calculations to understand the goals of process control and how it applies to other aspects of science and engineering technology. We will give emphasis to mathematics as a practical tool and language in describing and solving problems. Rather than treating it as a theoretical discipline, as is done in many math courses, this book treats mathematics as the language of science.

The word *calculus* often scares people away from science and technology. Calculus is simply the mathematics of how things change. It provides a method and structure for describing systems in which there is change, and a way to predict and understand change.

The development of calculus and its applications to science and technology is one of the most significant contributions to modern science, because it gave science and technology a language in which to develop. Mathematicians, scientists, and engineers use the concepts of calculus in all sorts of contexts, and its language and notations describe those contexts. Without some knowledge of the language of calculus, it would be difficult to fully comprehend process control. This brief preface of calculus is aimed at giving you the "mathematical language" to relate to more advanced studies in process control or, for that matter, in science and engineering in general.

Functions

Functions are used to describe things. The simplest definition of a function is:

> A sequence of ordered pairs of numbers, with the property that the first members of the pairs are all different from one another.

Here is an example of a function:

0, 21	1, 22	2, 21	3, 23	4, 22.5	5, 21	6, 21

This function consists of seven pairs (0 through 6). It is conventional to give functions names, like f or g, and if we call this function T, we use the following mathematical notation:

$$T(0) = 21, \quad T(1) = 22, \quad T(2) = 21, \ldots \text{and so forth.}$$

The first numbers of the pairs are called *arguments*, and the whole set of them is called the *domain* of the function. Therefore, the arguments of T are 0, 1, 2, 3, and so on, and the set consisting of these seven numbers is its domain, also referred to as the *independent variable*. The second numbers of the pairs are called the *values* of the functions, and the set of these numbers is called the *range* of the function or *dependent value* (they depend on the independent variable). Many functions are expressed in the form of an equation or are represented graphically. Often in process control, the independent variable is time, and the dependent variable is the process variable such as temperature or pressure.

The term *variable* simply describes how a parameter such as temperature or pressure is expressed as a function of time. Time is the independent variable, meaning it does not depend on other variables. Of course, temperature does not only occur at the specified values of the independent variable but often is continuous. These functions are typically referred to as *signals*. See Figure 1-4.

Derivatives

The derivative of a function or of a signal is the *rate of change* of the function. In other words, how fast is the signal (temperature) changing as a function of time? Looking at the function in Figure 1-4, we see that during the time interval from 2 to 3 s, the

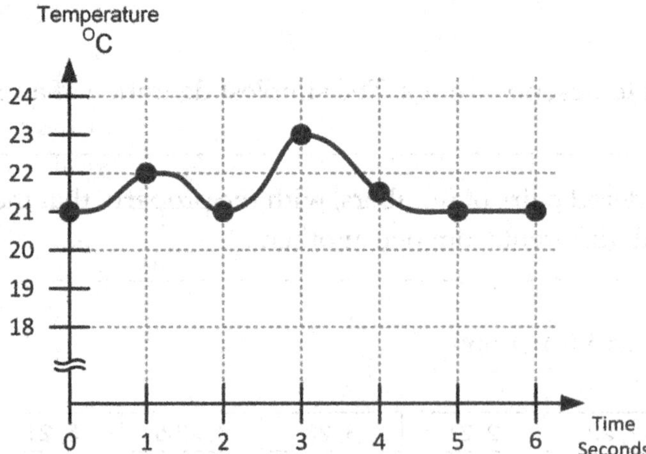

Figure 1-4. Graphical representation of a function.

average increase in temperature is 2°C/s (23 – 21 = 2). We can denote this using the following mathematical notation:

$$rate\,of\,change = \frac{\Delta T}{\Delta t} \tag{1-3}$$

where

ΔT = the change in temperature (dependent value)
Δt = the change in time (independent variable)
$\frac{\Delta T}{\Delta t}$ = the average rate of change

When dealing with continuous functions, and as the time interval (Δt) in Equation 1-3 diminishes to an instant, the average rate of change of temperature becomes the instantaneous rate of change, and we denoted this as: $\frac{dT}{dt}$.

In mathematical notation:

$$\text{As} \quad \Delta t \to 0 \quad \text{then} \quad \frac{\Delta T}{\Delta t} \to \frac{dT}{dt}$$

Effectively, the derivative of a function is a measure of the slope of the function at any point on the function. When a signal is increasing, its derivative is positive; when it is decreasing, its derivative is negative. When there is no change, the derivative is zero. As shown in Figure 1-4, between 2 s and 3 s the derivative is positive, and between 3 and 4 s the derivative is negative. Also, between 5 and 6 s the derivative is zero because the temperature is neither increasing nor decreasing.

Integration

Mathematically, integration is the reverse of differentiation. However, if a function is represented in graphical form, the integral of the function effectively becomes the area under the curve. See Figure 1-5. For each point, we can estimate the area under the curve by multiplying the value by its duration. For the function shown, if we calculate the area between 4 s and 10 s, we get:

$$area\ from\ 4\ to\ 10 = 50 \times 2 + 60 \times 2 + 60 \times 2 = 340$$

This is mathematically denoted as:

$$area = \sum_{4}^{10} Q \times \Delta t \tag{1-4}$$

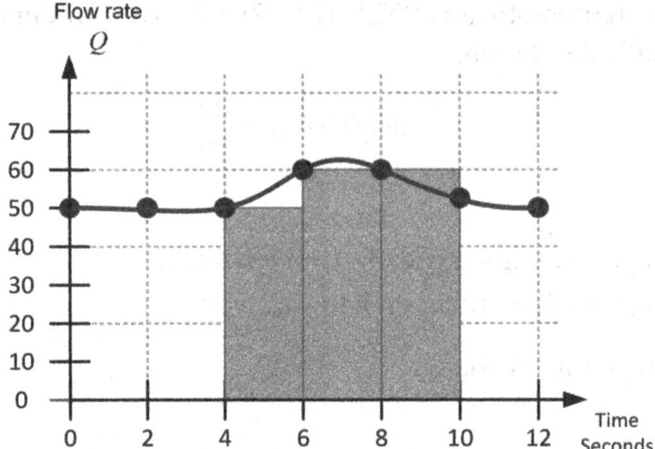

Figure 1-5. Graphical representation of integration.

where

Q = flow rate
Σ = summation
Δt = duration of time

The calculation of each area is an approximation. However, as Δt gets smaller, the calculation becomes more accurate.

Again, when dealing with continuous functions, and as the time interval (Δt) in the above equation diminishes to an instant, the equation is denoted as:

$$\text{As } \Delta t \to 0 \quad \text{then} \quad \sum Q\,\Delta t \to \int Q\,dt$$

For example, if the function represents a constant flow rate of liquid (q_{in}) into a container, then the integral of the flow rate will provide the amount of liquid in the container at any moment in time. See Figure 1-6.

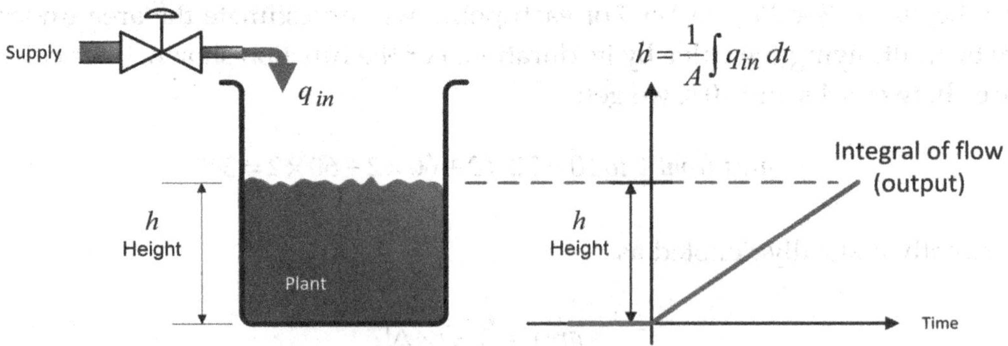

Figure 1-6. Graphical illustration of integration.

Differential Equations

It should be clear that differentiation and integration are the mathematical descriptions of change. A differential equation expresses how a variable changes as a function of other variables. The study of differential equations and the means for obtaining solutions are both best done by the mathematician. In the study of the fundamentals of process control, one is rarely, if ever, called on to actually solve a differential equation. However, it is beneficial to understand how a differential equation describes a system.

Once the system can be described using differential equations, we can use software to simulate the system's behavior and make appropriate predictions.

The Laplace Transform

In solving differential equations, automation professionals use the concept of a *transfer function*. A transfer function is an algebraic equation derived from a differential equation using the Laplace transform. Essentially, the Laplace transform replaces the differential notation $\left(\frac{d}{dt}\right)$ with the Laplace s (or sometimes p) operator.

where: s implies differentiation and $\frac{1}{s}$ integration. In other words,

$$\frac{d}{dt} \rightarrow s \quad \text{and} \quad \int \rightarrow \frac{1}{s}$$

These terms are again taken from standard calculus, but the student who has not been introduced to such mathematics need not worry. The mathematics here is only used for notation. Those who are interested in learning more about the Laplace transform can refer to Appendix C.

Much of this discussion of calculus will become clearer in the chapters that follow.

1.9 Process Documentation

Process control, like most disciplines, has its own language and terminology. Information is often conveyed in the form of technical diagrams using special symbols in addition to terminology, both of which are used in the process control industry. Technical diagrams are important in the design, development, and documentation of complex process control systems.

Figure 1-7 shows a typical diagram of a small portion of a process.

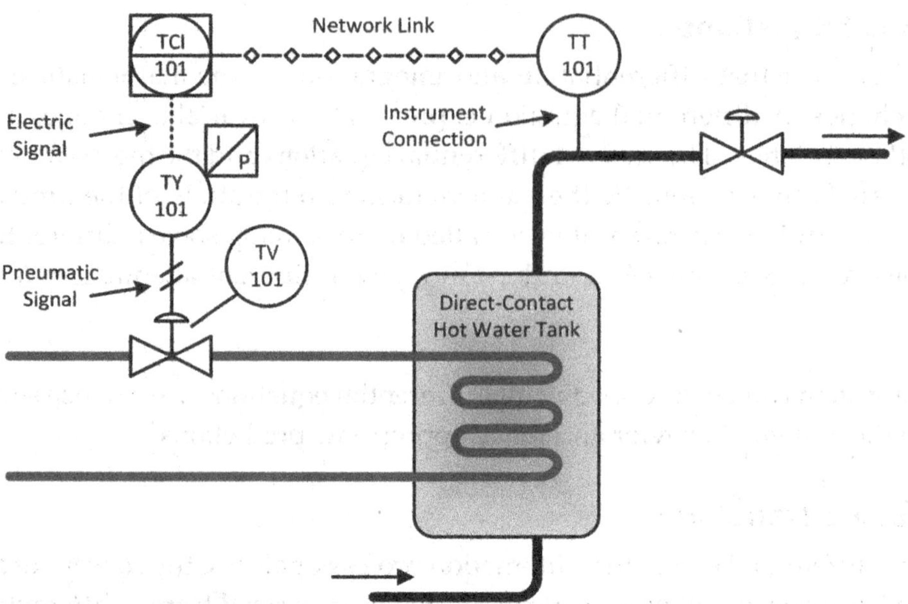

Figure 1-7. Example of a P&ID.

Figure 1-7 depicts a heat exchanger that consists of a temperature sensor (TT 101), a controller (TCI 101), and a valve (TV 101). However, specific details can be extracted, such as that the connection between the sensor (TT 101) and the controller (TCI 101) is connected via a digital bus, or that the connection between the controller and the control valve (TV 101) is electrical and goes through a current-to-pressure converter. These types of details are understood by automation professionals. Process control professionals extract this information from knowing and understanding a *piping and instrumentation diagram* usually abbreviated as *P&ID*.

Being able to clearly understand how a product is manufactured enables those working in the process control industry to communicate precisely what must be done. The major types of drawings that are used in process control are:

- Process flow diagrams (PFDs)
- P&IDs
- Instrumentation loop diagrams

PFDs

A PFD shows all major equipment and process flows, typically in a single drawing. Only major control variables are identified. The drawing may include a material balance (and sometimes an energy balance), which is a list of typical or design flow rates into and out of the process, key stream compositions, and the pressures and temperatures of the key stream and equipment. The drawing is usually a first look at what is being manufactured and how it is manufactured.

P&IDs

These drawings provide specific details of all piping and instruments, including valves (automated and manual), bypass lines, pipe sizes, and sample points.

Detailed symbolism and a more complete identification are used to describe the control systems, the type and location of hardware, and the signals used. All instrumentation is shown. A moderate level of mechanical detail is provided for process equipment so that the piping and instrumentation can be precisely documented. Symbols are used for piping, equipment, and instruments. These symbols, usually called *ISA symbols*, are defined by ISA-5.1-2009.[2]

Instrumentation Loop Diagrams

An extension of the P&ID is the instrumentation loop diagram, which shows the individual components and accessories of each specific control loop, how they are interconnected, and their locations.

More detail about documenting process control is provided in Appendix B. Students are strongly encouraged to consult this appendix while reading this book whenever they encounter diagrams with symbols that are unfamiliar.

1.10 Units

The study of process control is deeply related to instrumentation. We cannot control a variable if we cannot quantify it. A brief review of standards of measurement is helpful because accurate measurements are the basis of all science.

There are six base units of measurement:

1. Length
2. Mass
3. Temperature

---

2. ISA-5.1-2009, *Instrumentation Symbols and Identification* (Research Triangle Park, NC: ISA [International Society of Automation]).

4. Electric current
5. Light
6. Time

All other units of measurement are expressed in terms of the six fundamental units and are known as *derived units of measurement.*

Throughout the text, measurements are expressed in US customary, metric, and SI[3] units. More details on units can be found in Appendix A.

1.11 Process Variables

We end this chapter by providing some insight into the process variable that we will be controlling by going back to some basic ideas and making sure we understand the variable itself. The following sections explain the basics of temperature, pressure, and flow.

Temperature

Temperature, unlike most other process quantities such as level, mass, pressure, and flow, is an abstract quantity, and its scale is defined in terms of the behavior of materials. These behaviors—such as change in volume of a liquid, length of a bar, electrical resistance of a wire, pressure of a gas at constant volume, and color of a lamp filament—are used to indicate temperature.

A change in temperature of a substance (or body) is an indication of the amount of energy a body absorbs (increase in temperature) or the amount of energy a body releases (decrease in temperature). Temperature is related to the kinetic energy of the molecules at a localized region in a body. This kinetic energy cannot be measured directly, so its temperature cannot be inferred. To overcome this difficulty, international temperature scales have been defined in terms of the behavior of several materials at thermodynamic fixed points. Figure 1-8 compares the different temperature scales used: Fahrenheit (°F), Rankin (°R), Celsius (°C), and Kelvin (K).

Energy enables a body to do work. Energy can exist in two states: potential and kinetic. Potential energy is the energy a body possesses due to its position or its specific physical or chemical state. Kinetic energy is the energy a body possesses due to its motion.

3. Système International, or International System of Units.

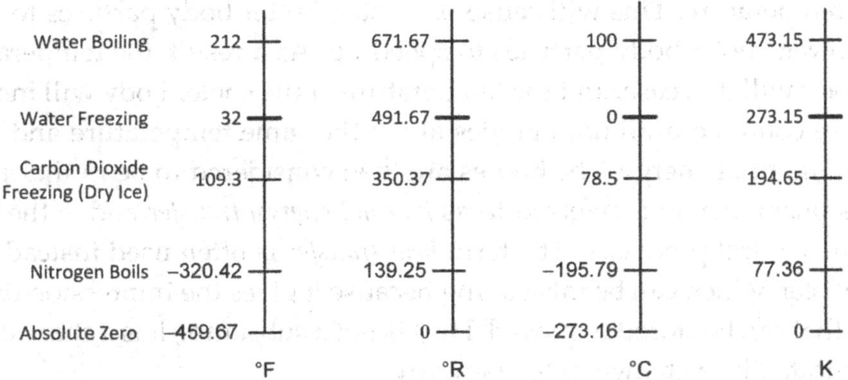

Figure 1-8. Common temperature scales.

Depending on how energy is derived, energy can be described as nuclear, gravitational, mechanical, electrical, thermal, or chemical. In all cases, it is either potential, kinetic, or both. Thermal energy is made up of both potential and kinetic components (see Table 1-1).

Thermal energy is associated with the random or disorganized motion of the particles that make up a body. In ideal fluids (liquid or gas), the particles move randomly, and thermal energy is all kinetic energy. In solids, the particles move randomly around an equilibrium position. The energy of the particles constantly changes between kinetic and potential. Thermal energy is then a combination of both kinetic and potential energy. When we speak of heat, we are referring to a body's thermal energy.

Temperature is then the measure of the average kinetic energy of the particles that make up the body. The greater the kinetic energy, the higher the temperature. When two bodies of different temperature are brought together and the bodies' particles collide, energy transfer will take place due to random movement of the particles. The particles of the body with a higher temperature will transfer energy to the body with

Table 1-1. Energy exists in two forms.

Type of Energy	Form
Nuclear	Potential
Gravitational	Potential
Mechanical	Potential (if at rest) Kinetic (if in motion)
Electrical	Potential (when not used) Kinetic (when used)
Thermal	Potential and Kinetic
Chemical	Potential

the lower temperature. This will cause the faster, hotter body particles to slow down and the slower, cooler body particles to speed up. As a result, the temperature of the warmer body will decrease, and the temperature of the cooler body will increase. This transfer will continue until both bodies are at the same temperature and have equal amounts of thermal energy. The bodies are then considered to be in thermal equilibrium. This phenomenon is referred to as *thermal energy transfer* and is the basis of all temperature control processes. The term *heat transfer* is often used instead of thermal energy transfer, which can be misleading because it gives the impression that heat is a substance that can be stored or moved. Heat is not a substance; it is a change in thermal energy indicated by a change in temperature.

The rate at which thermal energy transfer occurs is directly proportional to the thermal conductivity of the bodies, the surface area where thermal energy transfer is occurring, and the difference in the temperatures of the bodies. Thermal energy transfer occurs via three mechanisms: conduction, convection, and radiation.

Conduction

Conduction is the transfer of thermal energy by direct contact of two bodies of different temperature. The transfer of thermal energy by conduction is most effective for metals, which are good thermal conductors, and less effective for gases, which are poor thermal conductors.

Convection

Convection occurs when thermal energy is transferred through the particles of a fluid and is either natural or forced. In natural convection, the warmer fluid becomes lighter and rises above a cooler, denser region. One example is warm air rising to the ceiling of a room (air is classified as a gas, which is considered to be a fluid). Forced convection occurs when the fluid is circulated by means of a mechanical device, such as a fan, which increases the rate of thermal transfer. This type of convection is present in home heating and air-conditioning systems.

Radiation

Radiation is heat transfer by electromagnetic waves. For example, when your face warms by exposure to the sun, radiation is the cause. In this case, the medium through which the waves travel has not absorbed all the thermal energy. In this case the medium can be vacuum, gas, or liquid.

Changes in Phase

The relationship between thermal energy (heat) and temperature seems to indicate that an increase in thermal energy will result in an increase in temperature, and a

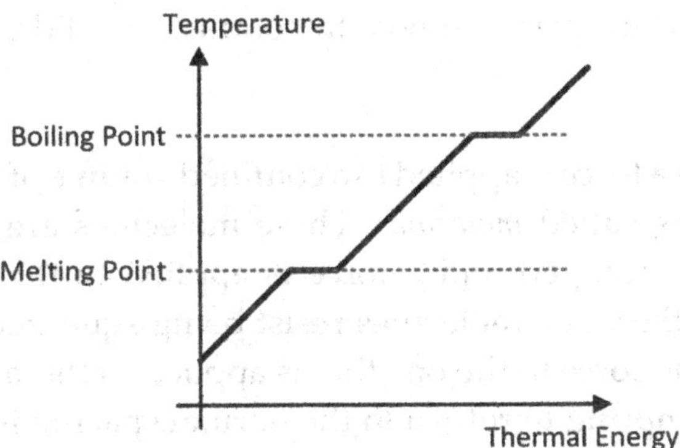

Figure 1-9. Thermal energy and phase change.

decrease in thermal energy will result in a decrease in temperature. However, there is an important exception to this rule: when a substance changes phase from solid to liquid or liquid to gas. In each case, thermal energy is added with no increase in temperature (see Figure 1-9). The reverse happens when a substance is cooled and undergoes a phase change.

The thermal energy gained or lost during the change in phase is often called *latent energy* (or *latent heat*) because it does not produce a change in the temperature of the body. This reiterates the point that thermal energy must not be confused with temperature.

Pressure

The control of fluid pressure is required in many industrial applications, such as distillation control, boiler control, evaporator control, extrusion, and gas pipelines in chemical reactors. As an example, the operation of gas pipelines can be achieved by controlling pressure at the end of the pipeline or by controlling the reaction rate of a reactant.

Pressure is defined as the *force per unit area* that a fluid (liquid or gas) exerts on its surroundings. For a gas, the pressure is uniform on all the walls of its enclosure. For a liquid, the pressure varies, being greatest on the bottom of its enclosure and zero at the top of the enclosure. These statements are explicitly true for a fluid that is not moving. In this case, the pressure is referred to as *static pressure.* For a fluid in motion, the pressure that is exerted on its surroundings depends on the motion and is known as *dynamic pressure.*

A fluid is a substance that tends to move or flow in response to a force. When a fluid is confined to a closed vessel, the fluid conforms to the shape of the vessel. Fluids include gases, liquids, and certain plastic solids. In contrast to solids, fluids cannot resist a deformation force.

Pressure occurs when a force is applied to a confined volume of fluid. Fluids are made of many tiny particles called *molecules*. These molecules are spaced relatively far apart. When they are trapped and a force is applied to them, the molecules are squeezed closer together. The molecules resist being squeezed together and create an equal and opposing force to the one that is applied to them. The force exerted by the molecules in attempting to return to the normal spacing is called *pressure.*

Figure 1-10a shows a container filled with a fluid. When a force is applied to the top of the stopper, the fluid inside the container resists compression by pushing back in all directions, causing pressure to occur inside the container. Because the fluid is confined, this pressure is transmitted equally throughout the entire volume of fluid. This concept is known as Pascal's law and was first discovered by the French scientist Blaise Pascal in the seventeenth century.

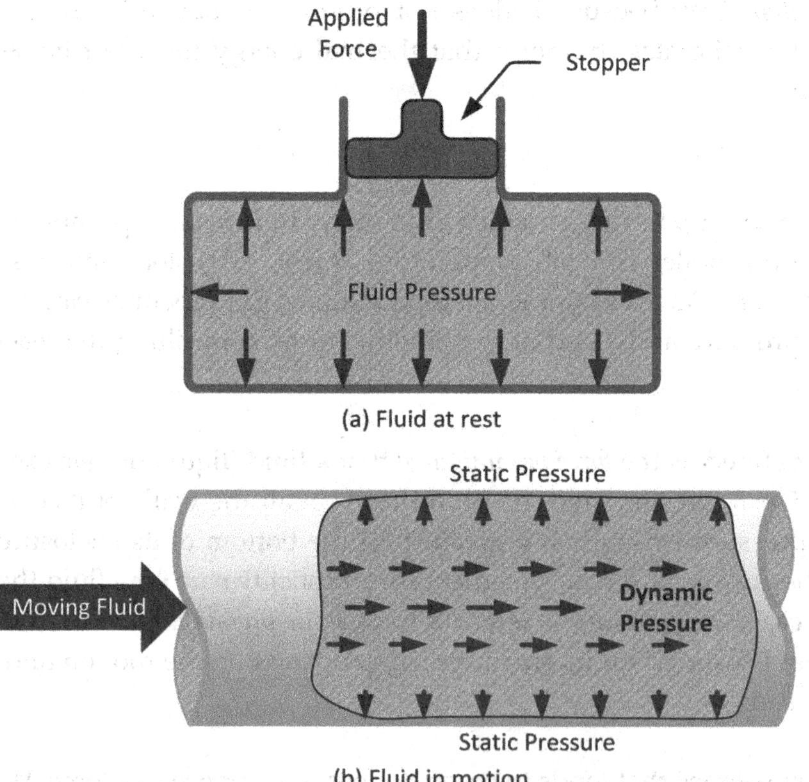

Figure 1-10. Fluid at rest and in motion.

All fluids exert pressure, whether moving or at rest. The pressure exerted by a fluid consists of two parts: the static pressure and the dynamic pressure.

1. Static pressure is related to the potential energy of the fluid and occurs whether the fluid is moving or at rest.
2. Dynamic pressure is related to the velocity of the fluid and is directly related to its kinetic energy. As mentioned earlier, dynamic pressure occurs when the fluid is moving.

In Figure 1-10a the fluid is at rest, so the exerted pressure is static only. In Figure 1-10b, however, the fluid is moving through a pipe, so both static and dynamic pressures occur within the pipe. In this case, the static pressure is exerted perpendicular to the wall of the pipe, whereas the dynamic pressure is exerted in the direction of fluid flow. The total pressure exerted by the fluid is equal to the sum of the static and dynamic pressures.

Units

The units used to describe pressure are directly based on its definition, force per unit area. The SI unit is newton per square meter (N/m^2). This unit is better known as the *pascal* (Pa), and this means that $1 \text{ Pa} = 1 \text{ N/m}^2$. A more practical unit is the kPa, where the prefix *k* stands for kilo or times 1000. In the US customary system of units, the most common unit of measuring pressure is pound per square inch, lb/in^2. This is usually written as *psi*. The conversion is that 1 psi is approximately equal to 6.895 kPa. Other units of pressure are the *atmosphere* (at), which is 101.325 kPa or approximately 14.7 psi, and the *bar*, which is 100 kPa.

The centimeter (cm), millimeter (mm), or inches of water (cm H_2O, mm H_2O, or in H_2O), and mm of mercury (mm Hg) are also commonly used. Table 1-2 shows common measurements for pressure and their conversion.

Table 1-2. Conversion table for common units of pressure.

From/To	psi	in H_2O	cm H_2O	mm Hg	atm	KPa	millibar
1 psi =	1.000	27.6807	70.308	51.7148	0.068046	6.89473	68.9473
1 in H_2O =	3.6127×10^{-2}	1.000	2.54	1.86826	0.002458	0.249081	2.49081
1 cm H_2O =	1.4223×10^{-2}	0.3937	1.000	0.7355	9.678×10^{-4}	9.806×10^{-2}	0.9806
1 mm Hg =	1.9337×10^{-3}	0.53526	1.3595	1.000	0.001358	0.133322	1.33322
1 atm =	14.696	406.795	1033.26	760.000	1.000	101.325	1013.25
1 kPa =	0.145038	4.01475	1019.75	7.50062	9.869×10^{-3}	1.000	10.000

Pressure Types

Pressure can be calculated by dividing the magnitude of a force by the area over which the force is acting. Mathematically:

$$P = \frac{F}{A} \tag{1-5}$$

where

P = pressure	in pascals (Pa) or pounds per square inch (psi)
F = force	in newtons (N) or pounds (lb)
A = area	in square meters (m^2) or square inches (in^2)

Atmospheric Pressure

The weight of the air on the surface of the earth causes a pressure on objects. This pressure is referred to as the *atmospheric pressure* and varies (decreases) with the distance above the earth's surface. At sea level, this pressure is usually given as 14.7 psi (but also depends on weather conditions), and it drops off to 12.2 psi at an elevation of 5000 ft.

Absolute Pressure

Pressure is most often measured from a point of reference. If this point is zero pressure, it is called *absolute pressure.* Atmospheric pressure is usually measured and stated in terms of absolute pressure.

Gauge Pressure

Another common measurement for pressure is *gauge pressure.* Gauge pressure is measured using atmospheric pressure as the point of reference. Gauge pressure is really a differential pressure where the value is given as positive for a pressure above atmospheric pressure and negative for a pressure below atmospheric pressure.

Differential Pressure

In situations when the pressure between two points is of interest, the pressure is known as *differential pressure.*

Figure 1-11 illustrates the different pressure relationships.

Head

For liquid fluids at rest, the pressure gradient in the vertical direction depends only on the specific weight of the fluid. Based on this, a pressure can then be expressed as a vertical height of a column of liquid that can be supported by this pressure. When expressed as a height of a liquid column, pressure is called *head.* Head is often used to

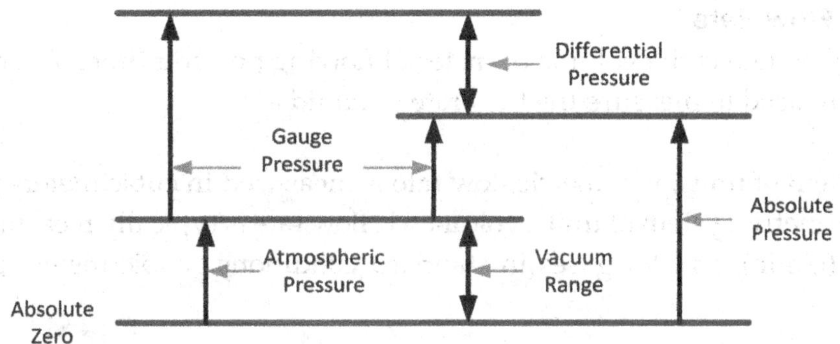

Figure 1-11. Relationship between absolute, atmospheric, and gauge pressure.

measure low-pressure levels. Head is usually measured in meters (m) or centimeters (cm) of water or mercury in the SI system of units, and it is measured in feet (ft) or inches (in) of water or mercury in the US customary system of units.

A pressure of 1 kPa corresponds to a head of 10.2 cm of water at 15.5°C. Consequently, applying a pressure of 1 kPa to a column of water at 15.5°C will cause the water to rise by 10.2 cm or 0.102 m within the column. Similarly, a pressure of 1 psi corresponding to a head of 2.31 ft or 27.7 in at 60°C will cause the water to rise by 2.31 ft or 27.7 in within the column.

Flow

The control of flow rate can be critical in industrial process control. In steam turbines, controlling the flow rate of steam into the turbine controls the rotational speed of the turbine. In chemical reactors, controlling the flow rate of the reactant controls the changing of the reactants in a predetermined ratio. In steam boilers, controlling the flow rate of the combustion air controls the proper and complete combustion of the fuel.

Flow or *flow rate* is defined as the amount of material flowing past a point during an explicit period. Flow rate can be expressed either as *mass* flow rate or *volumetric* flow rate.

Mass Flow Rate

Mass flow rate is the mass of the material flowing per unit time. Mass flow rate is usually used to measure the flow rate of solids and slurries.

In the SI units, mass flow rate is measured in kilograms per second (kg/s). In the metric system of units, we typically use kilograms per hour (kg/h).

In the US customary system of units, mass flow rate is measured in pound-mass per hour (lbm/h), pound-mass per minute (lbm/min), or pound-mass per second (lbm/s).

Volumetric Flow Rate

Volumetric flow rate is the volume of material flowing per unit time. Volumetric flow rate is usually used to measure the flow rate of liquids.

In the SI system of units, volumetric flow rate is measured in cubic meters per second (m^3/s). In the metric system of units, volumetric flow rate is typically measured in liters per minute (l/min), and for gases in standard conditions,[4] cubic meters per minute (m^3/min).

In the US customary system of units, volumetric flow rate is measured in US gallons per minute (gpm) for liquids and standard cubic feet per minute (ft^3/min) for gases.

Note that the abbreviation scfm (standard cubic feet per minute) is often used erroneously instead of ft^3/min to measure flow rate of gases.

Flow rate for liquids is often determined using the following definition:

$$Q = \overline{v} \times A \tag{1-6}$$

where

Q = liquid flow	in cubic meters per second (m^3/s)
$\overline{v}$ = average flow velocity	in meters per second (m/s)
A = pipe cross-sectional area	in square meters (m^2)

Note: In the $\overline{v}$ symbol, the bar indicates average.

Nature of Fluids

A fluid is a substance that tends to move or flow in response to a force. When confined in a closed vessel, the fluid will alter its shape to conform to the outline of the vessel. Fluids include gases, liquids, and certain plastic solids. In contrast to solids, fluids cannot resist a deformation force.

Compressibility of Fluids

All fluids can be characterized by their compressibility. Compressibility is defined as the capability of a fluid to decrease in volume when subjected to pressure.

Liquids are relatively incompressible, whereas gases are considered to be very compressible. The near incompressibility of liquids implies that their pressure will increase

4. Standard conditions are usually defined as 101.325 kPa, 15.5°C, and a relative humidity of 36%.

rapidly when they are confined and pushed on. For this reason, liquids are used in hydraulic systems to develop very high pressure and transmit tremendous amounts of power.

Four important characteristics of fluids are mass density, specific gravity, dynamic viscosity, and vapor pressure. We will describe each briefly.

Mass Density

Mass density is defined as the mass of fluid contained in a unit volume of the fluid. Mass density is measured in kilograms per cubic meter (kg/m^3) in the metric system, or in the pound-mass per cubic foot (lbm/ft^3) in the US customary system.

Compression of fluids causes an increase in mass density of the fluid. This occurs because the mass remains the same, but the volume containing this mass is decreased. Because gases are very compressible, their mass density is significantly affected by an increase in pressure. In contrast, liquids are considered to be incompressible, so their mass density is relatively unaffected by an increase in pressure, except at very high pressures.

Specific Gravity

Specific gravity, sometimes referred to as *relative density*, is the ratio of the mass density of a liquid to the mass density of an equal volume of a reference fluid. The reference fluid is usually water at standard conditions. Because a ratio of identical units is a dimensionless number, specific gravity has no units.

When two liquids that do not mix are in the same container, the one with the lower specific gravity will float on top. An example is when oil floats on top of water.

Viscosity

Viscosity is an internal property of a fluid that resists flow. For example, water moves through a pipe much more easily than motor oil does. In fact, one of the differences between types of oil is its viscosity.

Dynamic or absolute viscosity is a measure of how thick a fluid is and how easily it flows. Dynamic viscosity is measured in pascal second (Pa·s) in the SI system of units, and in pound-force per foot second (lbf/ft·s) in the US customary system of units. Also, dynamic viscosity is often measured in units of poises (P)[5] in the centimeter-gram-second (CGS) system of units. In that system, 1 P is equal to 0.1 Pa·s or 0.067 lbf/ft·s.

5. Named after Jean Louis Marie Poiseuille.

Viscosity is more commonly expressed in centipoise (cP). The centipoise is used because water has a viscosity of 1.0020 cP at 20°C; the closeness to 1 is a convenient coincidence.

In flow situations, we are more concerned with the ratio of the viscous force to the inertial force, characterized by the fluid density. This ratio is better known as *kinematic viscosity* and is defined as:

$$v = \frac{\eta}{\rho} \tag{1-7}$$

where

v = kinematic viscosity	in square meters/second (m^2/s)
η = viscosity (absolute)	in newtons second per square meter ($N\ s/m^2$)
ρ = density	in kilograms per cubic meter (kg/m^3)

The SI unit for kinematic viscosity is m^2/s. The CGS unit for kinematic viscosity is the *stokes* (abbreviated S or St).[6] It is sometimes expressed in terms of *centistokes* (cS or cSt).

Temperature has a significant effect on viscosity. Liquids decrease in viscosity (get thinner) when their temperature is increased. In contrast, gases increase in viscosity when their temperature is increased. Table 1-3 shows typical liquid viscosities.

Vapor Pressure

Vapor pressure is a measure of how volatile a fluid is. Vapor pressure corresponds to the pressure developed by a fluid when confined in a closed vessel, for a given temperature and with the fluid at equilibrium. Vapor pressure is usually measured in

Table 1-3. Typical liquid viscosities.*

	Specific Gravity	Absolute Viscosity (centipoise)	Kinematic Viscosity (centistokes)
Water	1.0020	0.98	0.978
Gasoline	0.71	0.48	0.676
Acetone	0.79	0.33	0.42
Corn Oil	0.93	26.5 (130°F)	28.7 (130°F)
Freon	1.33	0.20	0.15

Note: *At room temperature unless otherwise stated.

6. Named after George Gabriel Stokes.

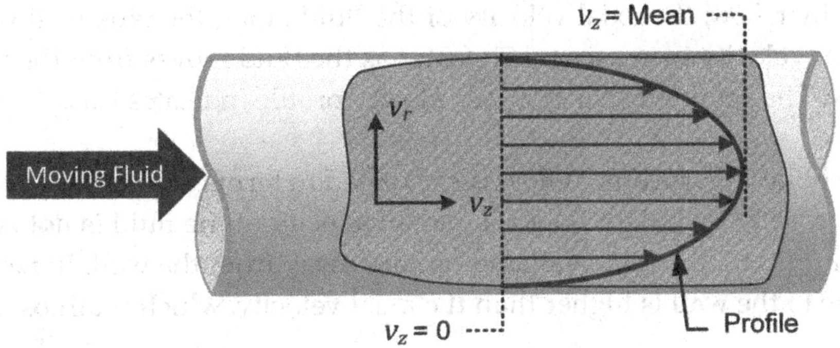

Figure 1-12. Velocity profile for laminar flow.

kilopascals absolute (kPa, absolute) in the SI system of units and in pounds per square inch absolute (psia) in the US customary system of units.

There are three different types of fluid flow in a component such as a pipe: laminar, turbulent, or transitional. They are described in the sections that follow.

Laminar Flow

In laminar flow, the fluid particles move smoothly and orderly along parallel paths without "eddying or turbulence," as shown in Figure 1-12.

The velocity profile for laminar flow has a parabolic shape as shown. The axial velocity (v_z) of the fluid adjacent to the pipe wall is zero. A short distance from the pipe wall, the axial velocity increases, and it reaches a maximum value at the center of the pipe. Note that the radial velocity (v_r) of the fluid is zero in laminar flow.

Turbulent Flow

In turbulent flow, the particles move along irregular paths causing the flow to be completely random, as shown in Figure 1-13.

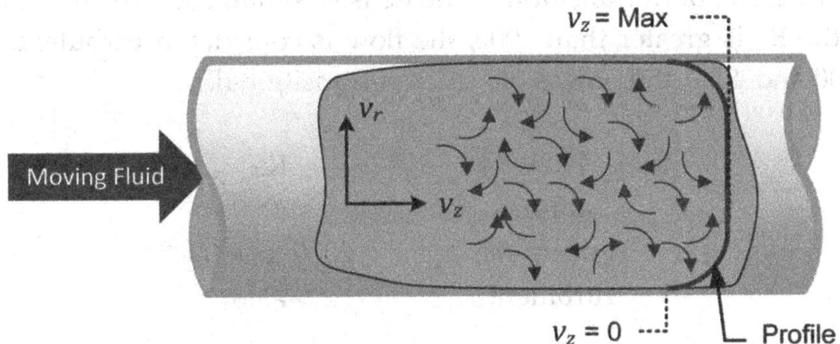

Figure 1-13. Velocity profile for turbulent flow.

As with laminar flow, the axial velocity of the fluid along the pipe wall is zero; however, the axial velocity increases much faster as the fluid moves from the wall than in laminar flow. The rounded shape of the velocity profile indicates this.

The velocity of fluid in turbulent flow is expressed in terms of a *mean value* rather than a *maximum value.* In turbulent flow, the radial velocity of the fluid is not zero because molecules of fluid hit the pipe wall and bounce away from the wall. In fact, the radial velocity close to the wall is higher than the axial velocity, which is almost zero.

Transitional Flow

In transitional flow, the flow changes from laminar to turbulent. The particles move from one layer to another in a direction transverse to the direction of the main flow. The type of flow that occurs depends on the component, specifically the component's orifice diameter, and the characteristics of the fluid including specific gravity, viscosity, and velocity. These factors can be combined to form a dimensionless number called the *Reynolds number* (Re).

Re is calculated as follows:

$$Re = \frac{u \times \rho \times D}{\mu} \tag{1-8}$$

where

Re = Reynolds number viscosity	(dimensionless)
u = mean velocity of fluid	(m/s or ft/s)
ρ = mass density of fluid	(kg/m^3 or lbm/ft^3)
D = orifice diameter of component	(m or ft)
μ = dynamic viscosity of fluid	(Pa · s or lbf/ft · s)

Re can be used to determine whether the fluid flowing through a specific component is laminar, turbulent, or in transition. If the Re is less than 2000, the flow is considered laminar. If the Re is greater than 4000, the flow is considered turbulent. If the Re is between 2000 and 4000, the flow is considered transitional.

Flow	Re
Laminar	<2000
Transitional	2000 to 4000
Turbulent	>4000

Table 1-4. Conversion table for common units of flow.

From/To	gpm	gph	l/min	m^3/h	cm^3/min	ft^3/min
1 gpm =	1.000	60.00	3.785	0.2271	3785	0.1337
1 gph =	0.01667	1.000	0.6309	0.003785	63.09	8.022
1 l/min =	0.2642	15.85	1.000	0.600	1000	0.0453
1 m^3/h =	4.403	264.2	16.67	1.000	16.667	0.5886
1 cm^3/min =	2.642×10^{-4}	0.01585	0.001	0.00006×10^{-5}	1.000	3.53×10^{-5}
1 ft^3/min =	7.479	0.1247	28.31	1.699	28.312	1.000

Turbulent flow is more likely to occur in processes where the pipe diameter is relatively small, the fluid viscosity is relatively low, and the fluid velocity (proportional to flow rate) is high. Any restriction or irregularity in the flow path may cause laminar flow to go temporally into a turbulent condition. Table 1-4 shows common units for flow and their conversion.

1.12 Summary

This chapter discussed process control history, and the meaning and the importance of process control. It also introduced fundamental terminology and some scientific and mathematical concepts.

Table 1-4. Conversion table for common units – Flow

From/to	[illegible]	[illegible]	[illegible]	[illegible]	[illegible]	[illegible]
1 gpm	[illegible]	[illegible]	[illegible]	[illegible]	[illegible]	[illegible]
1 gph	[illegible]	[illegible]	[illegible]	[illegible]	[illegible]	[illegible]
1 [illegible]	[illegible]	[illegible]	[illegible]	[illegible]	[illegible]	[illegible]
1 [illegible]	[illegible]	[illegible]	[illegible]	[illegible]	[illegible]	[illegible]
[illegible]	[illegible]	[illegible]	[illegible]	[illegible]	[illegible]	[illegible]
1 [illegible]	[illegible]	[illegible]	24	[illegible]	29.5[illegible]	1,000

Turbulent flow [illegible] increases when the pipe diameter is relatively small, the fluid [illegible] and the fluid velocity is proportional to [illegible] the [illegible] [illegible] units [illegible] common units for [illegible] their conversion.

1.12 Summary

This chapter discussed [illegible] of process control [illegible] and maintenance [illegible].

2

Basic Control Concepts

This chapter introduces fundamental concepts and terminology encountered in automated process control, namely, the concept of the control loop.

Learning Objectives

- Define the following terms:
  - Controlled (process) variable
  - Set point
  - Disturbances
  - Manipulated variable
- Understand the basic concept of feedback control
- Understand the basic concept of a control loop
- Have a general overview of process automation

2.1 Process and Process Control

Process as used in the terms *process control* and *process industry* refers to the methods of changing or refining raw materials to create end products. The raw materials, which either pass through or remain in a liquid, gaseous, or slurry (a mix of solids and liquids) state during the process, are transformed, transferred, measured, mixed, heated or cooled, filtered, stored, or handled in some way to produce an intermediate or end product. Process industries include the chemical, oil and gas, food and beverage, pharmaceutical, water treatment, and power industries, among others.

Process control refers to the methods used to control process variables when manufacturing a product. For example, factors such as the proportion of one ingredient to another, the temperature of the materials, how well the ingredients are mixed, and the pressure under which the materials are held all can significantly impact the quality of a product. Manufacturers control the production processes for three main reasons, to:

1. Reduce variability
2. Increase efficiency
3. Ensure safety

Reduce Variability

Process control can reduce variability in a product to ensure a consistently high-quality product. Manufacturers can also reduce cost by reducing variability. For example, in a gasoline blending process, 12 or more different components may be blended to make a specific grade of gasoline. If the refinery does not have precise control over the flow of the separate components, the gasoline may get too much of the high-octane components. As a result, customers would receive a higher grade and more expensive gasoline than they paid for, and the refinery would lose money. The opposite situation would be customers receiving a lower grade at a higher price.

Reducing variability can also lower cost by reducing the need for activities such as product padding to meet required product specifications. *Padding* refers to the method of making a product higher quality than it must be to meet specifications. When there is variability in the product (i.e., when process control is poor), manufacturers are forced to pad the product to ensure that specifications are met, which adds to the cost. With accurate, dependable process control, the *set point* (desired or most favorable point) can be moved closer to the actual product specification and thus reduce manufacturing cost.

Increase Efficiency

Some processes must be maintained at a specific point to maximize efficiency. For example, a control point might be the temperature at which a chemical reaction takes place. Accurate temperature control ensures process effectiveness and efficiency. Manufacturers save money by minimizing the resources required to produce the product.

Ensure Safety

Process control is often implemented to maintain substances under high temperature and pressure. A runaway process, such as an out-of-control chemical reaction, may result if manufacturers do not maintain precise control of all the process variables. The consequences of a runaway process can be and have been catastrophic.

Precise process control is required to ensure safe operation of the manufacturing process. For example, maintaining proper boiler pressure by controlling the inflow of air used in combustion and the outflow of exhaust gases is crucial in preventing a boiler explosion that can clearly threaten the safety of workers.

2.2 The Process

To understand process control, we first must understand the process and its terminology. A *process* in process control can be defined as a succession of actions taken to achieve a specific result, such as maintaining temperature at a specific value. Four important terms are associated with any process and must be fully understood:

1. Controlled variable (process variable)
2. Manipulated variable
3. Set point or desired value
4. Disturbances

The *process variable* (PV) is the end result of the process and is the variable that is normally measured. It is the quantity that we desire to control or maintain at some desired level, such as temperature, level, pressure, flow, or other process variable. The process variable is often referred to as the *controlled variable* (CV). The desired level or reference value is referred to as the *set point* (SP). This is the value set by an operator that determines the reference value that is desired of the controlled variable.

Figure 2-1 shows the relationship of controlled variable, the manipulated variable, and the disturbances related to a process. The controlled variable is the quantity or conditions that the process must maintain at some desired level. These may be flow rates, levels, pressures, temperatures, compositions, or other such variables. For each controlled variable, the process also has a desired set point or reference input.

Often the term *process variable* is preferred to *controlled variable* to emphasize that it is the process (temperature, pressure, etc.) that we are controlling. However, both are in common use.

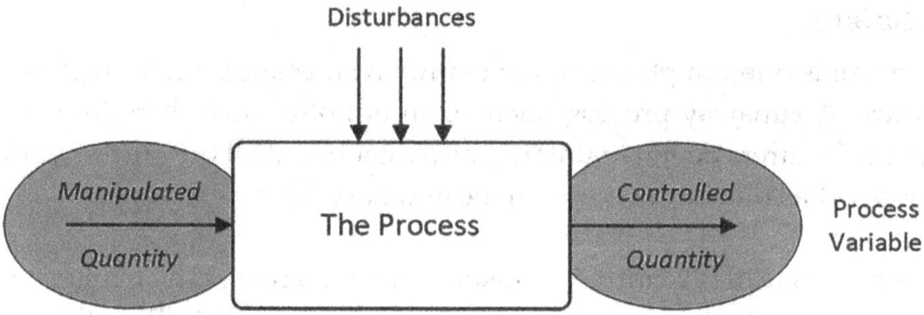

Figure 2-1. The variables involved.

For each controlled quantity, there is an associated manipulated quantity or *manipulated variable.* The manipulated variable is the quantity that drives the process to maintain a constant process variable by adjusting the controlled variable. In process control, this is often a flowing stream, and in such cases the flow rate of the stream is often manipulated using a control valve or other actuator and final control element.

Disturbances enter the process and tend to drive the CV away from the desired reference or set point conditions. The automatic control system must then adjust the manipulated variable so that the set point value of the controlled quantity is maintained regardless of the effects of the disturbances. Also, the set point may be changed by an operator, in which case the manipulated variables must be changed to adjust the controlled quantity to the new desired set point.

Typical PVs are pressure, temperature, mass, flow, density, position, level, liquid interface, position, conductivity, speed, composition, and pH (acidic or alkalinity).

Figure 2-2 shows a typical home heating system. In this system, the process variable is room temperature. The intent, of course, is to maintain the room at a comfortable level. The room temperature is maintained by controlling a variable that can be measured easily, such as temperature. Several types of disturbances cause room temperature to vary, for example, outside ambient temperature, the number of people in the room, the type of activity taking place in the room, heat loss through doors and windows, heat generated by light fixtures, and so forth. The control system is designed to manipulate the fuel flow to the furnace to maintain room temperature at the desired value or set point despite the various disturbances.

This example clearly identifies the four terms associated with process control: the process variable or controlled variable, which is temperature; the manipulated variable, which is the gas flow; the disturbances; and the required set point.

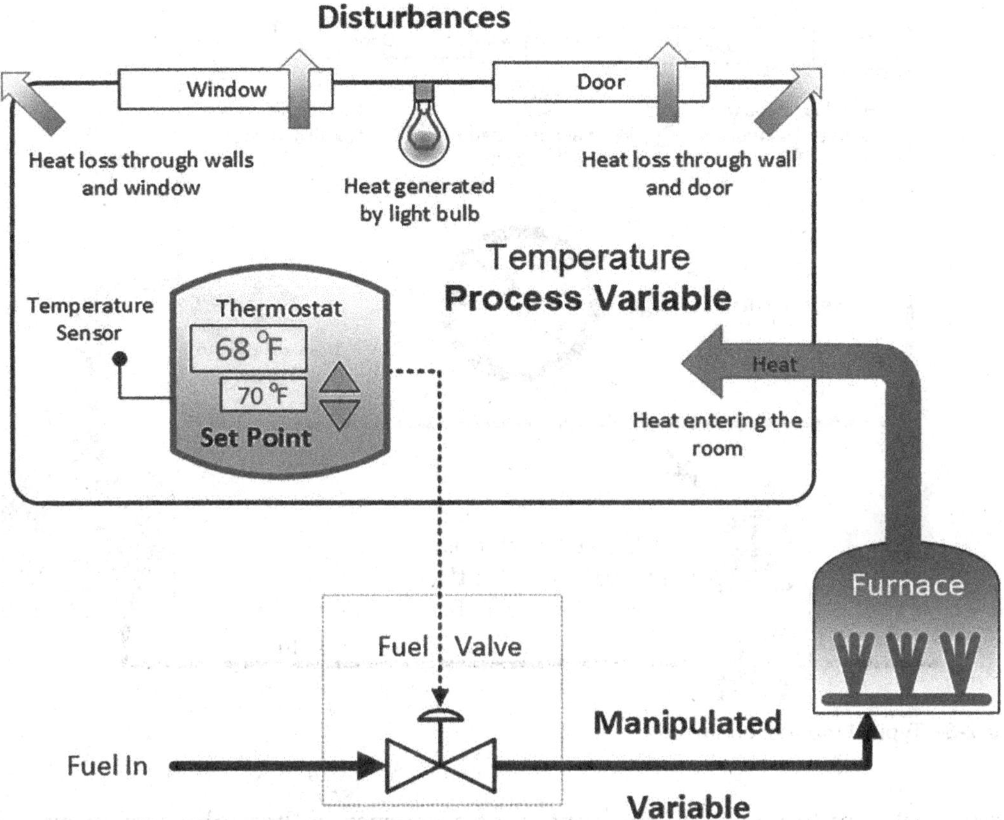

Figure 2-2. Typical home heating system.

2.3 The Control Loop Concept

Before studying automated process control, it is helpful to spend a moment reviewing a typical manual process control procedure. This is illustrated in Figure 2-3, which shows a process with one manipulated quantity, namely, fuel. The process variable or controlled variable is again temperature, and there is a temperature gauge that displays the current actual temperature in the room—the process variable—to an operator. The operator can inspect this indicator visually and, as a result, manipulate the fuel flow into the furnace to achieve some desired value or set point of the process variable. The set point is, of course, in the operator's mind, and the operator makes all the control decisions.

This is an example of a *control loop.* In the control loop, if the CV (temperature) falls below the set point, the operator will take action to bring the process back to the desired condition by opening the fuel valve and increasing the manipulated variable, thereby adding fuel to the furnace. The control loop will now remain stagnant until the temperature again rises above or falls below the desired set point.

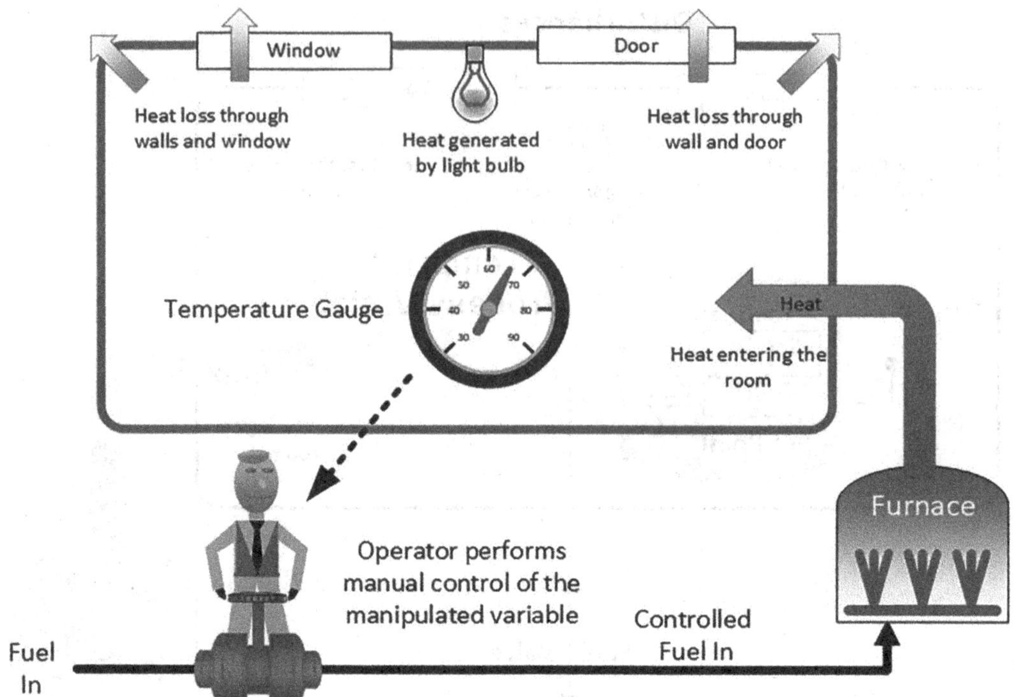

Figure 2-3. Typical manual control.

The simplest way to automate the control of a process is through conventional feedback control. This widely used concept is illustrated in Figure 2-4, based on the home heating system discussed earlier. Sensors or measuring devices are installed to measure the actual values of the process (temperature). These measured values are then transmitted to feedback control hardware, and this hardware makes a comparison between the set point (desired values) of the process variables and the measured values of these same variables. Based on the differences ("errors") between the actual values and the desired values of the process variables, the feedback control hardware calculates the required control action needed by the manipulated variable to minimize the error. This control action is then transmitted to the adjusting device (control valves)

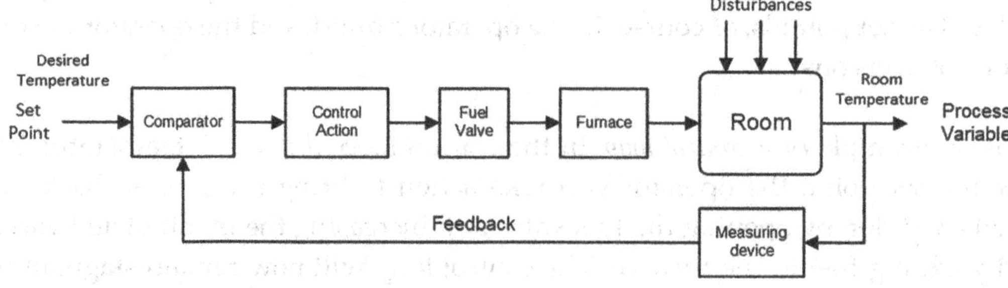

Figure 2-4. Feedback control concept.

that manipulates the inputs to the process. Referring to Figure 2-2, we see a thermostat in the room that measures the room temperature and is set to a specific set point. The thermostat compares the temperature reading with the set point and sends a control signal to the fuel valve, providing the required fuel to the furnace.

This configuration is shown in block diagram form in Figure 2-4. This arrangement of control hardware is used in a standard format, and most feedback control loops tend to reflect this general conceptual framework. To a significant extent, this configuration exists regardless of the specific nature of the process or the process variables involved.

The advantage of feedback control is that the system does not need to know exactly what disturbances will affect the process, and, consequently, the system designer does not need to know the specific quantitative relationships between the disturbances or their eventual effects on the controlled variables.

Three Tasks

Most feedback control loops in the process control industry work in the same way. They require three tasks to occur:

1. Measurement
2. Comparison
3. Adjustment or manipulation

Referring to Figure 2-5, the room example redrawn, a temperature element (TE) measures the temperature in the room and transmits it to a temperature-indicating controller (TIC). The controller compares the reading to a desired temperature value (set point). The controller then sends a signal to the fuel control valve either increasing or decreasing the amount of flow, depending on the error signal from the comparator in the controller.

The next chapter will discuss this topic in much more detail.

The symbols used in Figure 2-5 are standard process control symbols according to the ISA-5.1 standard.[1] See Appendix B.

---

1. ANSI/ISA-5.1-2009, *Instrumentation Symbols and Identification.*

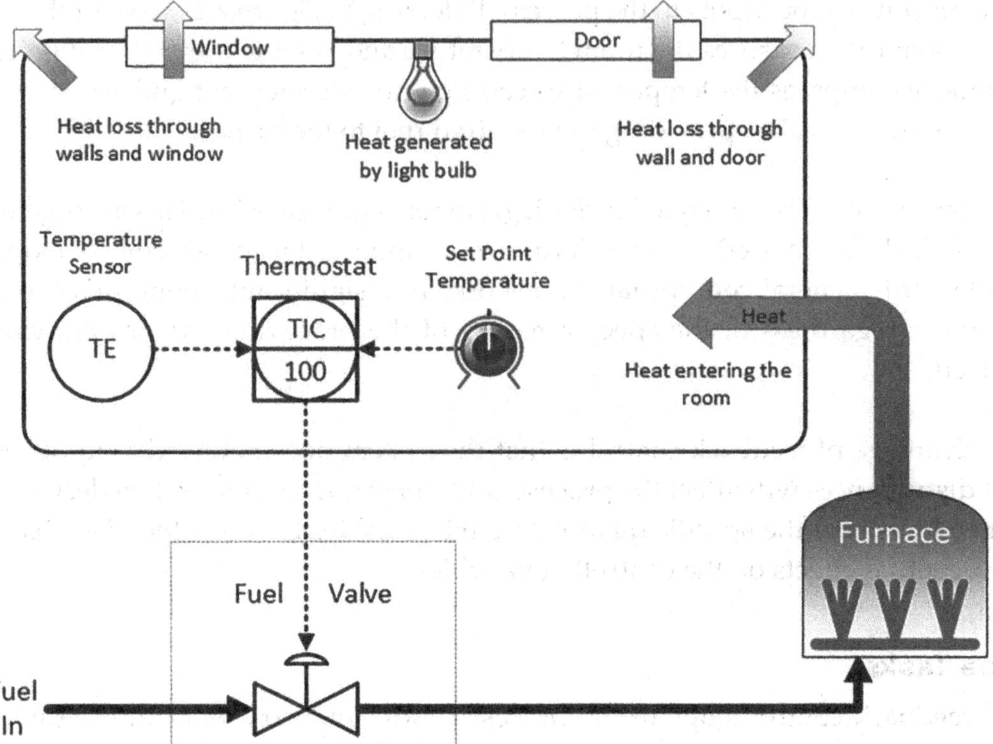

Figure 2-5. Home heating system illustrated using ISA-5.1 symbols.

2.4 Summary

This chapter introduced the fundamental concept of feedback control encountered in the automated process control industry, along with its terminology. Understanding the terminology and this fundamental concept of process control will help in understanding subsequent chapters.

Exercises

2.1 Consider an electric oven in a typical modern kitchen. Identify the controlled variable (CV), the manipulated variable, and the disturbances.

2.2 Consider an automatic, gas-fired, home hot-water tank. Identify the controlled variable, the manipulated variable, and the disturbances.

2.3 Imagine you own a backyard swimming pool. Describe a manual control system to measure the pH of the pool water and to add an acidic solution to adjust the pH. Define the controlled variable, the manipulated variable, and the disturbances.

2.4 Describe how to automate the control of your swimming pool. Assume you have a tank of acid solution to pump into your pool to control pH; use feedback control.

2.5 The cruise control feature used to control speed in an automobile is a good example of feedback control. Outline its operation in terms of feedback control.

3

The Control Loop

The concept of a feedback control loop was presented in Chapter 2. This chapter expands on this concept by examining the functioning layout of a typical feedback control loop.

Learning Objectives

- Understand the functional layout for a typical feedback control loop
- Understand process control terminology
- Explain the components of a feedback control loop in terms of a block diagram

3.1 A Typical Feedback Control Loop

Any given process may have more than one controlled variable (CV), and for each controlled variable an associated manipulated variable. In practice, a specific controlled variable is paired to a specific manipulated variable through appropriate feedback control hardware. This is illustrated in Figure 3-1.

The CV is measured with appropriate instrumentation, such as a sensor and its associated electronics. This measured value of the controlled variable is then transmitted to the controller and compared to the desired value of the controlled variable (the set point). The difference between these two (the error) is used to produce the control action required by the manipulator, which in this example is the valve.

Based on the error or deviation, the controller calculates an output signal to adjust the manipulated variable. Usually, the manipulated variable is flow, and the output of the

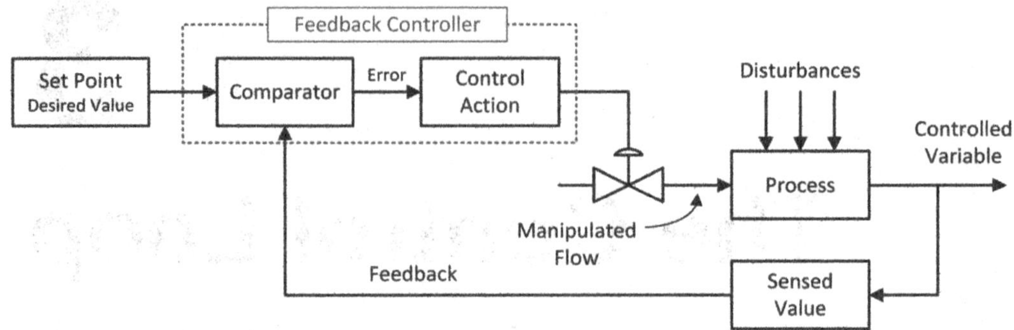

Figure 3-1. A single feedback control loop.

feedback controller is a signal to a control valve, as illustrated in Figure 3-1. While all this is happening in a continuous fashion, disturbances may enter the process and tend to drive the controlled variable in one direction or another. The single manipulated variable is used to compensate for all such changes produced by the various disturbances. In addition, if there are changes in the set point, the manipulated variable is also changed accordingly to produce the needed change in the controlled variable.

In a functional sense, all feedback control loops operate as illustrated in Figure 3-1.

3.2 The Three Tasks

The control loop performs three specific functions

1. measurement,
2. comparison, and
3. adjustment or manipulation.

Figure 3-2a shows a typical level control system. A level measuring device referred to as a *level transmitter* (LT1) measures the level in a tank and transmits the measured value (MV) associated with the level reading to a level controller (LC1). The controller compares the measured value to a desired set point (SP), in this case, a level established by the plant operator. In the meantime, there is a demand, adjusted by the hand valve (LV2) that removes liquid from the tank. This is detected by LT1, and then the controller sends a signal to the proportional control valve (LV1) bringing the level back to its required level. Note that all three devices—LT1, LC1, and LV1—have the first letter *L*, indicating they are part of the single level control loop as specified by ISA-5.1.[1]

1. ANSI/ISA-5.1, *Graphic Symbols for Process Measurement and Control* (Research Triangle Park, NC: ISA [International Society of Automation]).

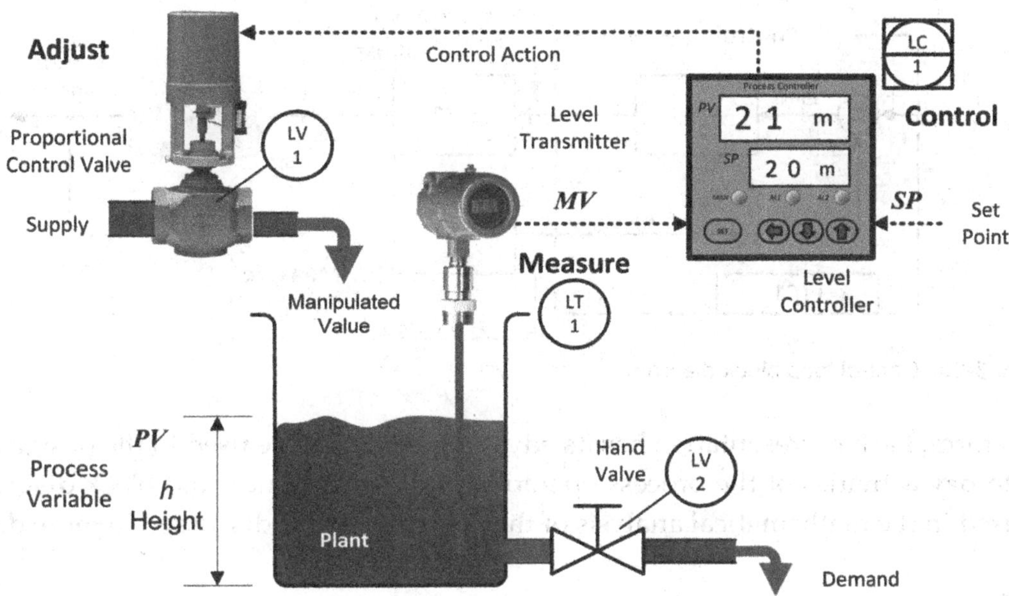

Figure 3-2a. Level control loop.

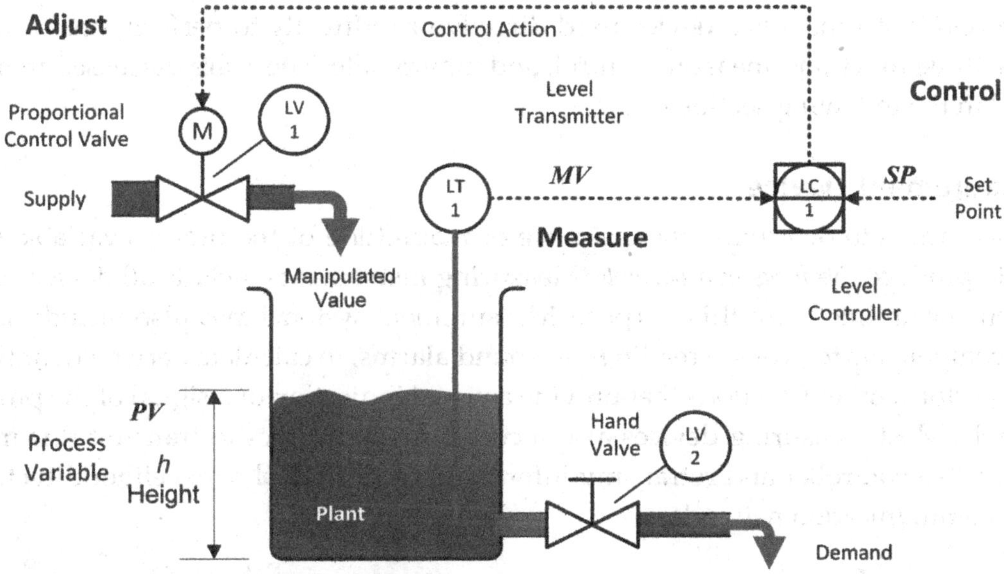

Figure 3-2b. Control loop schematic (ISA-5.1).

Control loops may use different instruments and control devices (e.g., transmitters, sensors, controllers, valves, and pumps), but the three tasks of measurement, comparison, and adjustment are always present. Figure 3-2b shows the same arrangements using ISA-5.1 schematic symbols and is referred to as a *process and instrumentation diagram* (P&ID).[2] Figure 3-2c shows the same arrangements, but this time in block dia-

2. P&IDs are discussed in detail in Appendix B.

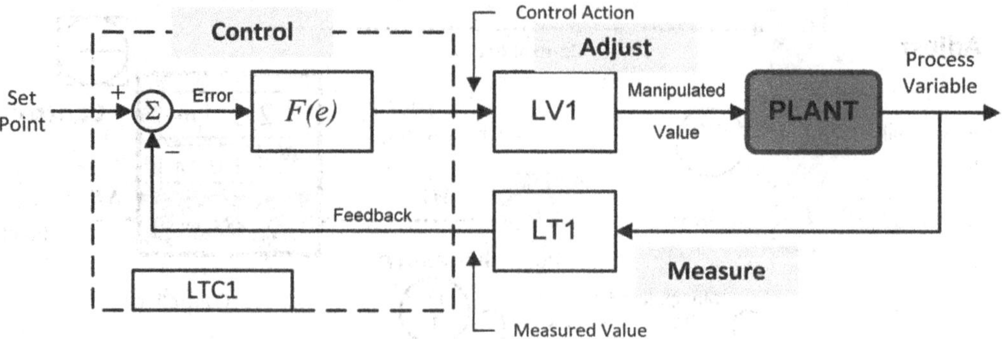

Figure 3-2c. Control loop block diagram.

gram form. Each representation has its advantages. P&IDs are used in the design and day-to-day activities of the process operation and maintenance, and block diagrams are used in the mathematical analysis of the process. Each is discussed in more detail later.

3.3 Process Instruments

A process instrument is a device used directly or indirectly to perform one or more of the three functions: measure, control, and manipulate. These are discussed in more detail in the following sections.

Measurement Device

To measure is to determine the existence or magnitude of the process variable (PV) and to produce the *measured variable.* Measuring instruments include all devices used directly or indirectly for this purpose. Measurement systems may also include auxiliary components to provide readings, to sound alarms, to calculate derived quantities, or to perform other functions that provide a visual indication or a signal of the process variable. Most measuring devices also incorporate electronics to transmit this information to a controller and retransmit information to a control panel either directly or via a communication network.

The *measurement device* is the "eye" of the system. It senses the process variable and produces an output signal that represents the status of the controlled variable. Examples in a process control system include a thermistor to measure temperature and a humidity detector to measure moisture. Examples in a motion control system include an optical device to measure presence or position and a tachometer to measure rotational speed. Other terms for a measurement device are *detector, transducer,* and *sensor.*

In the level control loop example, the measured variable is level, which must be held close to its set point. In this example, and in most instances, the measured variable is also the process variable. The measured variable is the condition of the process fluid that must be kept at the designated set point.

Sometimes the measured variable is not the same as the process variable. For example, it is common to measure the level in a tank by measuring pressure at the bottom of the tank. In this setup, pressure is actually measured but mathematically converted to measure level, and the process fluid level is the process variable. Chapter 7 discusses measuring devices in greater detail.

Controller

To control a process is to cause the process variable, to be maintained at a specified value or within specified limits, or to be altered in a specified manner, by a set point change. A controller requires both a set point and the measured value as inputs, and commands another device, which is usually a *final control element* or *manipulator*. The controller calculates what output to generate to the actuator. This output is calculated from its provided set point and measured value. Many controllers have additional functionality similar to measuring devices. For example, most temperature controllers include all the functionality available in measuring devices. Such controllers only need an additional sensor connected to the controller.

The *controller* is then the computational part of the process control loop. It receives as its input the measured value, determines the amount of deviation from the set point (error signal for closed-loop control), and develops an output signal that is expected to cause the controlled variable to reach the value specified by the set point. Most controllers are operated electronically, although some process control systems use air pressure in pneumatic devices for safety reasons.

Most modern controllers consist of a microcontroller and software as well as analog circuitry. Some older and specialized controllers are pure analog.

The controller usually produces a voltage or current signal that may need to be conditioned or modified before it is sent to the next element. For example, it must be amplified if it is applied to an electrical motor or converted to a proportional air pressure if it is applied to a pneumatic positioner or a control valve. The controller function can also be performed by programmable logic controllers (PLCs) and panel-mounted microprocessor controllers.

Control Algorithm

All controllers perform a *control algorithm,* which is a mathematical procedure or expression of a control function. Using the level control loop example, if *V* represents the linear control valve position, and *e* is the error, the relationship in a control algorithm can be expressed as

$$V = f(e)$$

The linear control valve position (*V*) is a function (*f*) (positive or negative) of the error (*e*).

Control algorithms are used to calculate the requirements of both simple and much more complex control loops than the one described here. In more complex control loops, questions such as "How far should the valve be opened or closed in response to a given change in the set point?" and "How long should the valve be held in the new position after the process variable moves back toward the set point?" must be answered. Controllers determine their corrective action based on the value of the error, positive or negative, on its past value and its predicted future value.

Controllers and their algorithms are discussed in detail in Chapter 9, along with their different control modes.

Manipulation (Adjustment)

To adjust or manipulate is to cause a final control element to directly affect a change to achieve control of the process variable. For example, in Figure 3-2 the controller manipulates the flow valve. The *final control element* is the valve that carries out the controller's commands.

The actuator is the part of the system that affects the change in the process variable (sometime referred to as the "muscle" of the system). It is a device that physically alters some type of energy or fuel supply, causing the controlled variable to match the desired set point. Examples of energy or fuel are the flow of steam, water, air, gas, or electrical current. A practical application is a commercial bakery where the objective might be to keep the temperature in an oven at 375°. The temperature is the controlled variable and is determined by how much gas is fed to the oven burner. A valve in the gas line controls the flow by the amount it opens or closes. The valve is the actuator in the system. In a typical liquid reservoir system, the actuator is the flow control valve, connected to the inlet pipe. Other terms for actuator are the *final control element,* or *final correcting device.* Common types of actuators are louvers, hydraulic cylinders, pumps, and motors.

Final control elements and actuators are discussed in more detail in Chapter 8.

3.4 Elements of a Closed-Loop System

A more detailed block diagram of a closed-loop control system is shown in Figure 3-3. Each block shows an element of the system that performs a significant function in the operation. The lines between the blocks show the input and output signals of each element, and the arrowheads indicate the direction in which they flow.

Provision is made for either the operator or some hardware to provide a set point to the control loop. This set point is the desired value of the controlled variable and will have the same units of measurement as the controlled variable; for example, if the controlled variable is gallons per minute then the set point also will be gallons per minute. The input elements provide a functional conversion of the input set point signal into the operating units of the controller; for example, millivolts (mV), milliamps (mA), or pounds per square inch gauge (psig) air pressure. Often both the set point and the controlled variable are expressed in the percentage of full range.

The controlled variable is measured by a sensor, and the measured value of the controlled variable is transmitted back to the controller using a transmitter. The transmission element inside the controller ensures that the set point and measured value have similar ranges. Inside the controller is the comparator. This important device calculates the controlled variable by comparing (taking the algebraic difference between) the value of the set point (after conversion by the input elements) and the measured value of the process variable transmitted back into the controller. The comparator, or error detector, is common to all feedback control systems. Note that this is negative feedback control, that is, the signal fed back to the comparator is subtracted from the signal that represents the set point. All applied feedback control is negative feedback control. (Positive feedback control is inherently unstable.)

The error signal, that is, the output of the comparator, becomes the input to the feedback controller. Proper, formal process instrumentation terminology refers to this

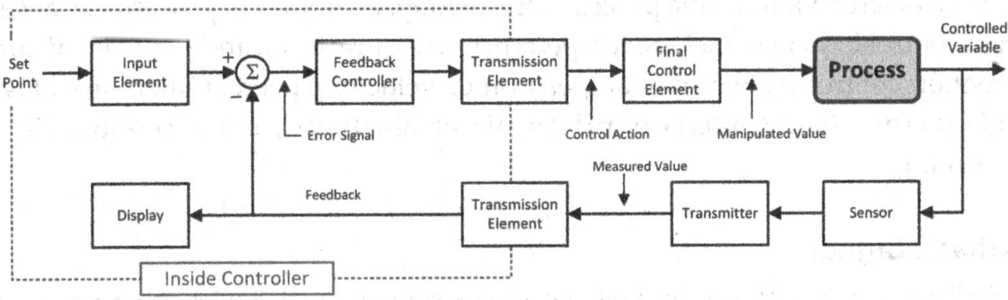

Figure 3-3. Elements of a closed-loop system.

error signal as the *actuating error signal* or the *system deviation.* Based on the error, the controller calculates a signal to send to the final control element, typically a control valve, which applies the signal—the manipulated variable—to the process.

Also shown with the feedback control loop in Figure 3-3 is a display (for the controlled variable), which is optional. In practice, many people refer to all the items contained within the controller as the feedback controller. Although common, this broad usage is not necessarily precise and will not be used in this book. Instead, it is more desirable to make functional distinctions between the various elements that perform the functions of the feedback control loop and are within the controller case.

Because electronics, microprocessors, and software have all had a major impact on process control equipment, the loop as described here has become smaller and smaller, shrinking into a single electronic unit with lines of code. This in itself leads to a "black box" view of feedback control. All the component parts are shown in the figure as individual items to better illustrate each function and make them easier to visualize.

3.5 Process Control Terms

As is true in any field, process control has its own set of terms and technical jargon. Now that you have a better understanding of process control and the process control loop, review the following sections to familiarize yourself with the terms commonly used in discussing process control applications and technology.

Controlled Variables

The *controlled variable,* also referred to as the *process variable,* is the parameter of the process that can change the desired condition of a manufacturing process in some way. In the example of the level tank in Figure 3-2, the process variable is level.

The controlled variable is the actual variable being measured, monitored, and maintained at a desired value in the process control loop. Examples of process variables in a process control loop may include temperature, pressure, level, and flow rate. Examples in a motion control system may be position or velocity (speed). Other variables may include density, liquid interface, pH (acidic or alkalinity), mass, conductivity, and composition.

Feedback Signal

The *feedback signal* is the output of the measurement device. Other terms used are *measured value* and *measurement signal.* Feedback is an essential part of any automated control system because it provides the controller with the information to produce corrective action to the process. It is the feedback signal that closes the control loop.

Set Point

The set point is the value of the process variable that is desired. It is important that the set point and process variable have the same units. Consider our level control example; if the process level must be kept within ±5 m of 100 m, then the set point should be set at 100 m. A level sensor can be used to help maintain the level at the set point. The sensor is inserted into the process, and the controller compares the level reading from the sensor to the adjusted set point. If the level reading is 110 m, then the controller determines that the process is above the set point and signals the control valve to close slightly until the process is brought back to 100 m.

The set point may be manually set by a human operator, automatically set by an electronic device, or programmed into a computer. Other terms for set point are *command* or *reference*.

Error

Error is the difference between the measured variable and the set point and can be either positive or negative. In the level control loop example, the error is the difference between the 110 m measured variable and the 100 m set point; that is, (100–110) m or –10 m. The error is calculated by an *error detector.* The error detector compares the set point to the feedback signal. It then produces an output signal that is proportional to the difference between them.

The *error signal* is the output of the error detector. If the set point and the feedback signal are not equal, an error signal proportional to their difference develops. When the feedback and set point signals are equal, the error signal goes to zero, and the feedback controller provides no update to the output. Other terms used are *comparator* or *comparer* and *summing junction.*

The objective of any control scheme is to minimize or eliminate error. Therefore, it is important that the error is well understood. Any error has three major components:

1. Magnitude (±)
2. Duration
3. Rate of change (±)

These three components are shown in Figure 3-4.

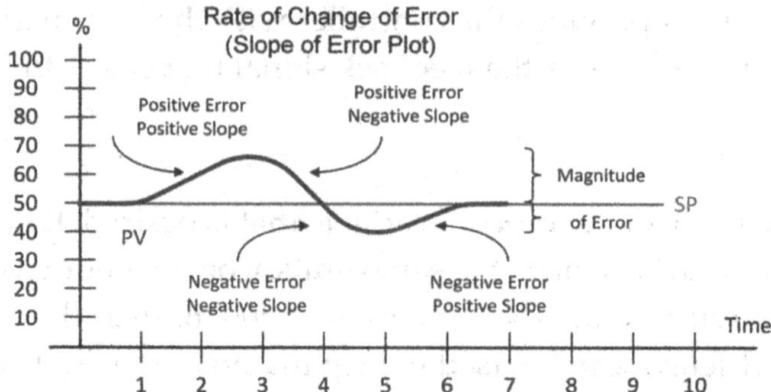

Figure 3-4. Error signal expressed in percentage.

Depending on the error's magnitude, duration, and rate of change, the controller will determine appropriate corrective action.

Magnitude

The *magnitude of the error* is simply the deviation (mathematical difference) between the values of the set point and the process variable as measured. The magnitude of error at any point in time compared to the previous error provides the basis for determining the change in error.

Duration

Duration refers to the length of time that a particular error condition exists. For example, how long has the error been positive or negative?

Rate of Change (Slope)

The *slope* or *rate of change* of the error is an indication of how fast (or slow) the error increases or decreases. This can be very important, because it is part of determining how fast or slow corrective action is required.

Offset

Offset is a sustained deviation of the process variable from the required set point. In the level control loop example, if the control system held the process fluid at 95 m consistently, even though the set point is set at 100 m, then an offset of 5 m exists.

Load Disturbance

A *load disturbance* is a change in factors affecting the process variable. In the level control loop example, increasing the removal of process fluid from the vessel would be a load disturbance because it would decrease the level of the tank. The controller then calculates a new manipulated value for the control valve.

Manual and Automatic Control

Before automation and control, people, rather than machines, performed many of the process control tasks. For example, an operator might be watching a level gauge and close a valve when the level reaches the required set point. Control operations that involve human action to make the required adjustment are called *manual control* systems. Conversely, control systems in which operations are performed with no human intervention, such as an automatic valve actuator that responds to a level controller, are called *automatic control* systems.

Closed-Loop and Open-Loop Control

A *closed-loop control* exists where a process variable is measured, compared to a set point, and action is taken to correct any deviation from the set point. An *open-loop control* exists where the process variable is not compared, and action is taken not in response to feedback on the condition of the process variable but is instead taken without regard to process variable conditions. For example, a water valve may be opened to add cooling water to a process to prevent the process fluid from getting too hot based on a preset time interval, regardless of the actual temperature of the process fluid.

Manipulated Variable

The fuel, energy, or material that is physically altered by the actuator is referred to as the *manipulated variable.* The amount at which the manipulated variable is changed by the actuator affects the condition of the controlled variable. In the commercial oven example, the gas is the manipulated variable, and the temperature is the controlled variable. In the liquid reservoir system of Figure 3.2, the inflow is the manipulated variable. The flow rate is altered by the control valve (actuator), which affects the condition of the CV.

Manufacturing Process

The *manufacturing process* is the operation performed by the actuator to control a physical variable, such as the motion of a machine or the processing of a liquid.

Disturbance

A *disturbance* is a factor that upsets the manufacturing process being performed, causing a change in the CV. In the reservoir system shown in Figure 3-2, the disturbances are the changes in demand that alter the water level. In most manufacturing processes, there are predictable as well as unpredictable disturbances. For example, a set point change is a predictable disturbance.

3.6 Combining Functions in a Single Instrument

In some cases, we may need only one instrument to control a process variable, but often we need a group of instruments that work together.

A typical household electric heater may be controlled by a single thermostat. The thermostat senses the current temperature, indicates the desired temperature, and adjusts the electric heater directly affecting the temperature; it is a complete measuring system, a complete loop.

A common industrial control loop may use two or more instruments to provide a temperature measurement to be read at distant locations. Frequently, the level sensor and level transmitter shown in Figure 3-2 are separate instruments, but they can be joined in a single instrument. There are often many different combinations of sensing, transmitting, controlling, automatic switching, indicating, and other functions in single instruments, such as temperature-recording controllers, flow-indicating switches, and many others.

The individual functions are not always obvious when their hardware is in one physical assembly, yet the functions are there. By understanding the basic functions, we can better understand how such assemblies work. As an example, consider a household iron, which is a single device that is a complete control loop in itself. The iron is a regulator that maintains a constant temperature that can be adjusted for ironing cotton, nylon, or other fabrics. Its use requires human guidance and human energy, but these are separate from its built-in function of controlling temperature automatically. It does this by measuring its own temperature and providing a corrective command, if the temperature is not what the operator has set it to be, by closing or opening an electric switch. On low temperature, the switch closes, electricity flows through an electrical heating element, and the temperature rises. When the temperature gets too high, the switch opens and the flow of electricity stops, the temperature drops, and the on-off cycle starts again.

3.7 Summary

The general concept of a feedback control loop was developed. This concept of a single feedback control loop was broken down into specific functions to better understand the functional layout of a typical feedback control loop.

Exercise

3.1 Develop a functional layout of the basic feedback control loop for the simple level control system shown.

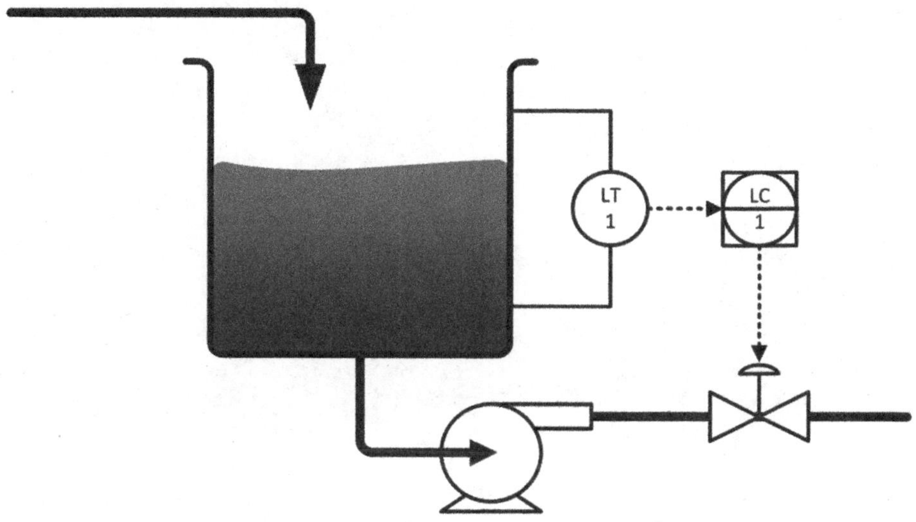

4

Functional Structure of Feedback Control

This chapter further develops the concept of block diagrams and introduces the concept of transfer functions to describe and analyze the functional structure of an automated process control loop. Combining transfer functions with block diagrams provides a powerful tool and method for simplifying and analyzing complex systems.

Learning Objectives

- Understand the functional layout for a single feedback control loop
- Explain the components of block diagrams
- Understand the meaning of a transfer function
- Draw and analyze block diagrams
- Appreciate the mathematical arrangement of a single feedback control loop
- Perform simple block diagram reduction and analysis

4.1 The Transfer Function

A transfer function is a mathematical representation of a component, system, or process. Specifically, it is the ratio of the output divided by the input of the component, system, or process. Usually, transfer functions are represented with capital letters; popular letters are *H*, *G*, and *K*.

Mathematically, a transfer function can be represented as:

$$\text{transfer function} = TF = \frac{\text{output}}{\text{input}} \tag{4-1}$$

This can also be represented graphically in block diagram form as follows:

Because the transfer function is a ratio, its dimension is the output units divided by its input units. When both are of the same dimension, the transfer function is dimensionless. In process control, all elements and instruments can be expressed using transfer functions. Complex transfer functions are usually expressed in functions that describe an element or instrument's dynamic behavior. *Dynamic* means how the output changes relative to the input. The block represents the change that the input undergoes to produce an output. This means the output is equal to the input multiplied by the block transfer function.

$$\text{output} = TF \times \text{input} \tag{4-2}$$

4.2 Block Diagrams

Block diagrams provide a consistent way of representing control systems pictorially. It is useful to take advantage of block diagrams because they are simple, symbolic, graphical tools commonly used in automatic control. Block diagrams have three basic symbols:

1. Block
2. Summing junction
3. Takeoff point

The Block

The block represents the transfer function of the system element.

Input → H → *Output* $= H \times$ *Input*

Blocks have one arrow entering and one arrow leaving. This is the way the algebraic operations of multiplication and division are symbolically represented. The output of the block is merely equal to whatever is contained within the block multiplied by the input. The block represents the equation shown.

Summing Junction

The summing junction (the circle with the summation symbol) represents the place where addition or subtraction of different signals takes place or is done.

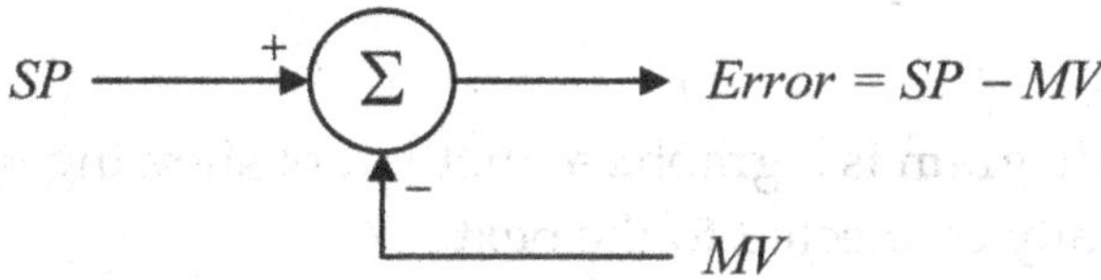

The arrows entering and leaving the circle are not vectors. They do, however, represent variables and the flow of information. The head of each arrow has an algebraic sign associated with it, either plus or minus. If no sign is present, an implied plus is intended. The small circle is a simple way to represent algebraic addition or subtraction. The symbol in the diagram represents the algebraic equation shown leaving the circle. Alternate symbols are shown below.

Takeoff Point

The takeoff points represent the places where more than one input or output signal is used.

The directional arrows represent the flow of the signal.

A constant block (the square with the number 4) can also be used in conjunction with the summing junction to add/subtract constants.

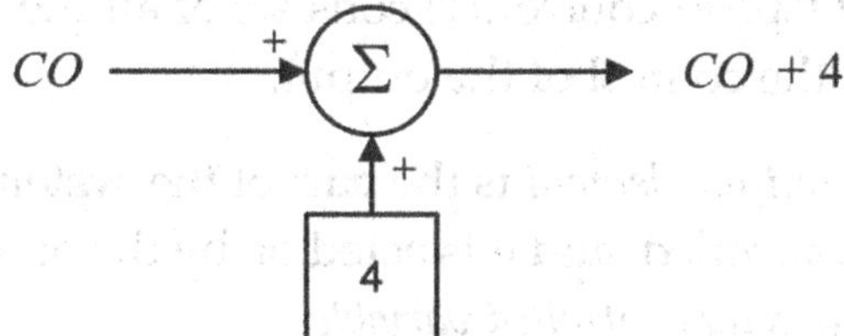

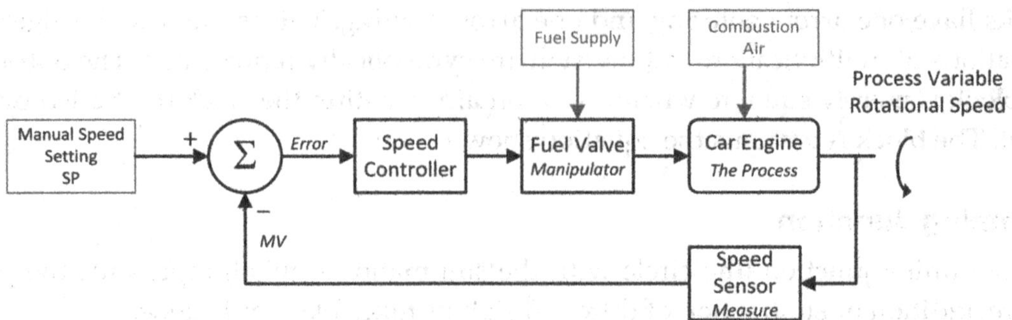

Figure 4-1. Block diagram of a single control loop.

In summary, a block diagram is a graphical method of showing how one element of a system is mathematically connected to the next.

Block diagram symbols may be combined to form systems. Figure 4-1 shows a simplified block diagram of a car's cruise control using a negative feedback control loop; it is "negative" because of the subtraction of the feedback signal being fed back. This figure essentially represents a combination of the three symbols shown earlier.

In this block diagram of a control loop, the speed of an automobile is maintained at a constant speed; the control system is known as *cruise control.*

The block diagram in Figure 4-1 represents the following elements.

- The *input element* (manual speed setting) is the desired speed, or set point (*SP*), of the car; it provides the *reference input* signal that initiates and governs the response of the system. This is the reference against which the actual system performance is compared. It is sometimes called the *command.* In most cars equipped with cruise control, it is the initial position of the gas pedal.
- The *error-measuring element* (comparator) compares the reference input with the measured value (*MV*) output of the system (speed) and generates an error signal.

$$error = SP - MV \tag{4-3}$$

- The *control element* (speed controller) consists of all the parts of the system that have any effect on the control of the output.
- The *manipulator* or *output element* is the part of the system that actually produces the output being controlled, and it is acted on by the other components. It is also referred to as the *directly controlled variable.*

- The *feedback element* is the device that converts the controlled output into a feedback signal, which is a representation of the output. This signal must be of the same form and scale as the reference signal. The difference between these two signals is the *error signal.*

Note that the fuel supply and combustion air are not part of the signal flow, but they are required to make the system function.

Block diagrams will be used throughout this book to pictorially represent automatic control principles and applications.

4.3 The Functional Layout of a Feedback Control Loop

The general layout and structure of the single feedback process control loop can now be examined in much more detail, using what we have learned about block diagram mathematics and the transfer function. Such a functional analysis is illustrated in Figure 4-2. This figure is separated into two broad parts: the functional objectives that are accomplished inside the controller case and the balance of the process control loop that is, of course, external to the controller case.

Each block can be assigned an appropriate transfer function, and then using block diagram mathematics, the response of the control loop can be analyzed and predicted. This is exactly what process simulation software does. Inside the controller, all signals are electrical and are within a limited range of 0 to 5 V or some other electrically convenient range. Outside the controller, signals can be in many different ranges, such as 4 to 20 mA or 3 to 15 psi.

Determining the actual transfer functions of each block is easier said than done. Chapters 5 and 6 will discuss some of the fundamental aspects of deriving transfer functions. An important point is that transfer functions only deal with changes,

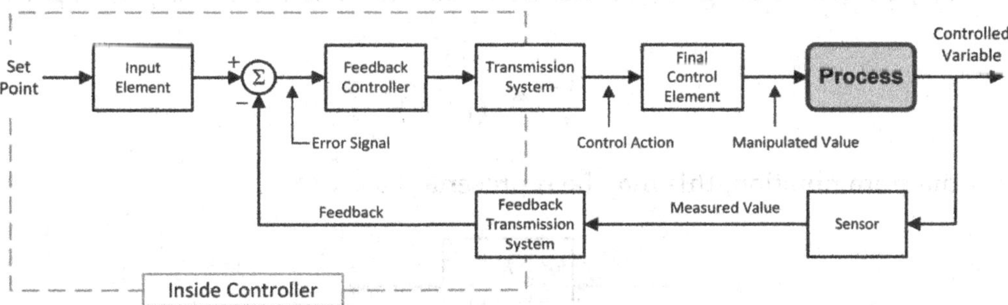

Figure 4-2. The functional layout of a feedback control loop.

referred to as *dynamic behavior.* When constants are introduced, they can either be ignored or dealt with separately.

The purpose of a block diagram in any system, whether it is electrical, mechanical, hydraulic, pneumatic, chemical, or something else, is to simplify the knowledge of the function of the total system by conveying a better understanding of each element that makes up the system, and the relationship of the various elements that make up the system. Only necessary details are shown to illustrate the functional operation of all elements within the total system. The block diagram is combined with the transfer function of each element in the system to give a clear picture of the relationship of input and output signals. The transfer function, which is defined as the ratio of the output signal to the input signal, is placed within the block representing each element.

First, let us learn how the mathematics of block diagrams can be used.

4.4 Mathematics of Block Diagrams

There are two universally accepted standards for the mathematical notation used in process control presentations: time domain and frequency domain. Most of the time, process control is described in the time domain. This means that the system is described in terms of changes as a function of time. However, transfer functions are better expressed in the frequency domain. In this book, the following conventions will be used. For the generalized mathematical representation of an element or a system's transfer function, we will use the frequency domain, *G(s)*, where *s* is the Laplace transform operator that indicates taking the derivative with respect to time, *d/dt*. *1/s* implies integration $\int \ldots dt$. In its simplest form, *G(s)* may imply simple algebraic multiplication, and in more complicated (time-dependent) situations, it represents a differential equation. Appendix C provides an explanation of this mathematical transformation and its advantages for those who are interested in the mathematical derivations.

The following is a transfer function of a first-order differential equation (first-order lag), where K is the steady-state gain, τ is the time constant, and s is the Laplace operator.

$$G(s) = \frac{K}{(1+\tau s)} \tag{4-4}$$

In block diagram notation, this may be represented as:

$$\longrightarrow \boxed{\frac{K}{(1+\tau s)}} \longrightarrow$$

By convention, all time domain functions are in lowercase and Laplace functions are in capitals.

4.5 Gain and Sensitivity

Sensitivity is always the ratio of a change in the output variable to a change in the input variable. For example, a process gain or sensitivity of 5.0°F/(lb/min) means that an increase of 1.0 lb/min in the steam flow will increase the hot water temperature by 5°F.

Gain is also defined as the change in output divided by the change in input. The difference is that, strictly speaking, gain has the same units for input and output, meaning that the gain ratio is dimensionless. *Gain* and *sensitivity* are often used interchangeably, and the term's meaning is clear within context. The term *gain* is often used when different input and output units are present, particularly when speaking of process gain.

The transfer function can be as simple as a constant. In such case, we refer to the constant as a gain. The sensitivity of a measuring device is defined by the ratio of the output signal change to the change in the measured variable. Sensitivity has the engineering units of the output signal divided by the engineering units of the input signal. Clearly, the greater the output signal changes for a given input change, the greater the sensitivity of the measuring element. Sensitivity is a steady-state ratio, and, in effect, it is a steady-state characteristic of the element.

Example 1

A pressure transmitter is calibrated with an input range from 0 to 100 psi and transmits a 0 to 5 V signal.

- Draw a block diagram of the transmitter.
- What is the transfer function of the transmitter?
- When the transmitter transmits 2.5 V, what pressure is the transmitter reading?

Solution

The transmitter has a single input and a single output; it can be represented using a single block.

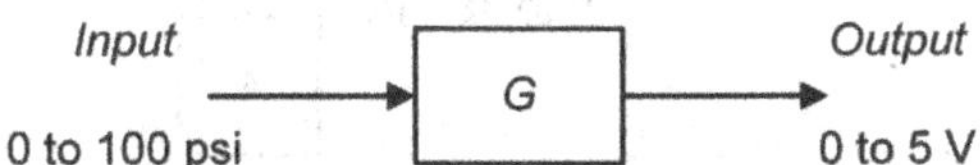

The transfer function can be calculated by the ratio of output over input.

$$G = \frac{output}{input} = \frac{(5-0)\text{ V}}{(100-0)\text{ psi}}$$

$$= \frac{5\text{ V}}{100\text{ psi}} = 0.05\text{ V/psi}$$

In this case, the transfer function also represents the sensitivity of the transmitter.

$$sensitivity = G = 0.05\text{ V/psi}$$

We can now use the transfer function to calculate the transmitter's output at 2.5 V.

$$TF = \frac{output}{input}$$

$$input = \frac{output}{TF} = \frac{output}{\frac{5\text{ V}}{100\text{ psi}}} = output \times \frac{100\text{ psi}}{5\text{ V}}$$

$$= 2.5\text{ V} \times \frac{100\text{ psi}}{5\text{ V}} = 50\text{ psi}$$

Example 2

A pressure transmitter is calibrated with a range from 0 to 100 psi and transmits a 4 to 20 mA signal.

1. Draw a block diagram of the transmitter.
2. What is the transfer function of the transmitter?
3. When the transmitter transmits 12 mA, what pressure is the transmitter reading?

Note that this is similar to Example 1; however, the output has an offset of 4 mA.

Solution

Because the transmitter has an output that has a bias (offset) of 4 mA, we must indicate this in the block diagram. The output has a span of (4 – 20) mA = 16 mA.

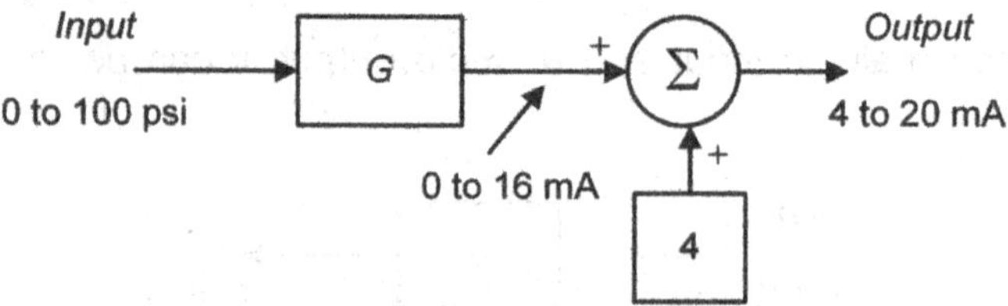

The transfer function can be calculated by the ratio of output over input.

$$G = \frac{(20-4)\text{ mA}}{(100-0)\text{ psi}} = \frac{16\text{ mA}}{100\text{ psi}} = 0.16\text{ mA/psi}$$

$$output = G + 4 = 0.16\text{ mA/psi} + 4$$

Note that we cannot algebraically calculate the ratio of $\frac{output}{input}$ due to the 4 mA offset.

It is often convenient and conventional that when working with transfer functions, we ignore any offsets or bias so that we can derive the required ratio. When this is done, we are implicitly only calculating the output change.

In this case, the sensitivity of the transmitter is:

$$sensitivity = G = 0.16\text{ mA / psi}$$

We can now use the transfer function and block diagram to calculate the transmitter's output at 12 mA, by first accounting for the 4 mA bias: (12 – 4) mA = 8 mA.

$$TF = \frac{output}{input}$$

$$input = \frac{output}{TF} = \frac{output}{\frac{16\text{ mA}}{100\text{ psi}}} = output \times \frac{100\text{ psi}}{16\text{ mA}}$$

$$input = 8\text{ mA} \times 0.16\,\frac{\text{psi}}{\text{mA}} = 50\text{ psi}$$

Example 3

A simple closed-loop system that has a gain of *G* is shown by the block diagram below.

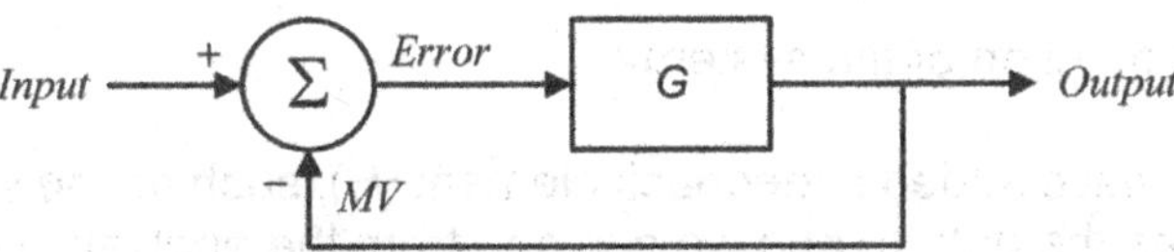

Derive the transfer function of the system.

Solution

$$error = input - MV$$

$$MV = output$$

$$\begin{aligned} output &= G \times error \\ &= G \times (input - MV) \\ &= G \times input - G \times output \end{aligned}$$

$$output + G \times output = G \times input$$
$$output(1+G) = G \times input$$
$$\frac{output}{input} = \frac{G}{1+G}$$

Negative feedback as shown in the preceding example is the foundation on which all automatic control is built. Negative feedback can provide for stable system operation with the following four characteristics:

1. Gain is reduced.
2. Stability is increased.
3. Response time is slowed down.
4. Bandwidth is increased.

These characteristics will become clear in our future study of automated process control.

Example 4

Analyze the following example.

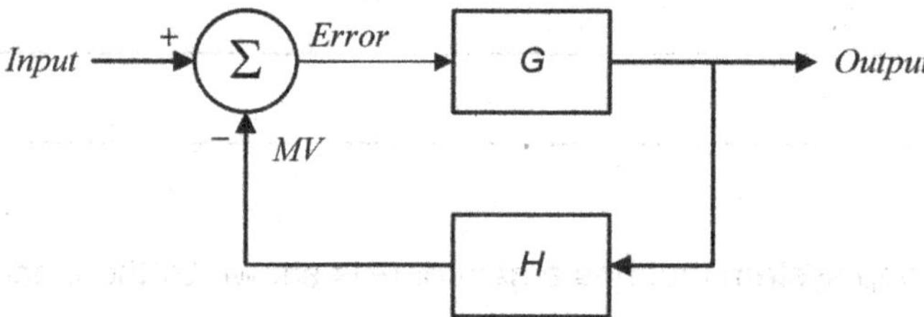

Derive the transfer function of the system.

In this example, we have added a feedback element (*H*), such as the sensor and/or transmitter that measures the output, and we can see from the analysis the effect this has on the *input/output* relationship of the system.

Solution

$$error = input - MV$$
$$MV = H \times output$$
$$output = G \times error$$
$$= G \times (input - MV)$$
$$= G \times input - G \times H \times output$$

$$output + G \times H \times output = G \times input$$
$$output(1 + G \times H) = G \times input$$
$$\frac{output}{input} = \frac{G}{1 + GH}$$

The transfer function may be considered an operator because it operates on the *input* to give an *output*. Using the rules given (summarized in Table 4-1 and Table 4-2), blocks may be combined into a single block and a pick-off point or a summation junction may be shifted ahead of or behind a block, all for the purpose of simplifying system analysis and diagram reduction.

There are no universally accepted standards for the mathematical notation used in process control presentations, and as a result, practice varies. For those who are mathematically impaired, do not feel you must enjoy all these details to understand the fundamental principles that unfold throughout the subsequent chapters of this book.

Example 5

Now we will take a detailed look at an actual control loop, taken from industry. Figure 4-3 shows a constant volume of water (represented by h) stored in a tank to provide a reserve for process equipment that withdraws water at fluctuating rates, called the *demand.* Automatic level control is used to keep the tank level high and to avoid overflowing the tank. A transmitter (LT1) measures the water level and sends a signal that corresponds to the level to a controller (LC1). A transmitter is an instrument that passes on the information it receives from a sensor. The controller directs the motor-driven control valve, which, in turn, manipulates the water supply flow to the tank as required to keep the level constant. The controlled variable is the water level; the manipulated variable is the flow of water to the storage tank.

Provide a detailed analysis of the system.

Solution

The same system can be represented in block diagram form as shown in Figure 4-4. Here we also added the actual signal levels of each element.

Note the similarity between this block diagram and the previous one for the cruise control. Most control systems are similar.

In this diagram, we have combined the comparator and the control, giving us the controller. This is done because the control and comparator are housed in a single unit in

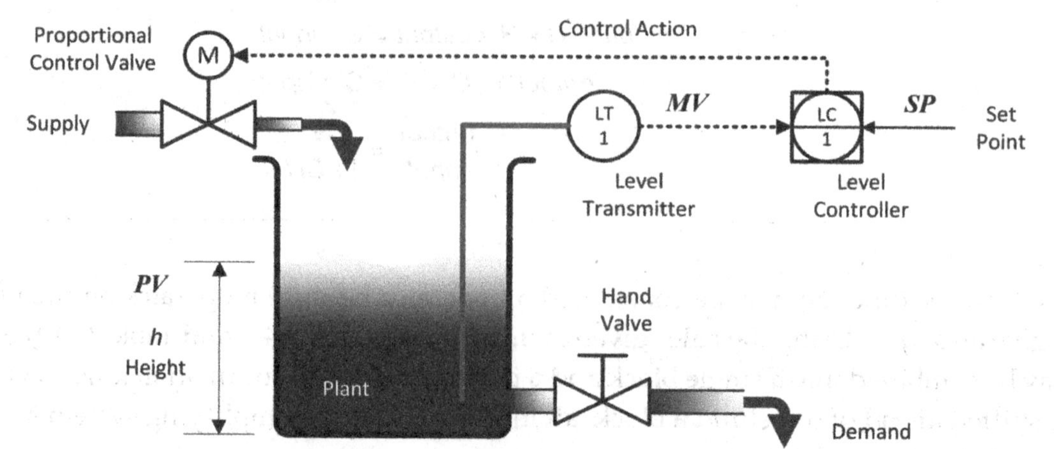

Figure 4-3. Water level control in a tank.

commercial controllers. Note also that the comparator is expanded to clearly show that the error signal is mathematically related to the *SP* and the *MV* as follows:

$$error = SP - MV$$

The control action that opens or closes the proportional valve is controlled by the error multiplied by the control mode (in this example multiplied by 1 for simplicity). The electrical signals used are in the range of 4 to 20 mA. Because the control valve is also controlled using a signal in the range of 4 to 20 mA, 4 mA must be added to the control signal.

The control action (*CA*) can then be calculated as:

$$CA = error \times control\ mode + 4$$

We will use the block diagram for our detailed analysis.

Consider that the tank in Figure 4-3 is 15 m high and initially empty, and the hand valve is closed. The tank is to be filled to a height of 10 m. The tank has no liquid leaving and has a proportional valve at the source to facilitate the filling process. The level transmitter

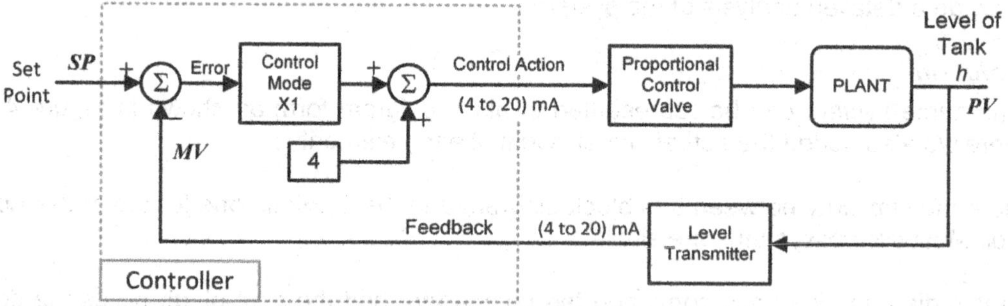

Figure 4-4. Block diagram: controlling the water level in a tank.

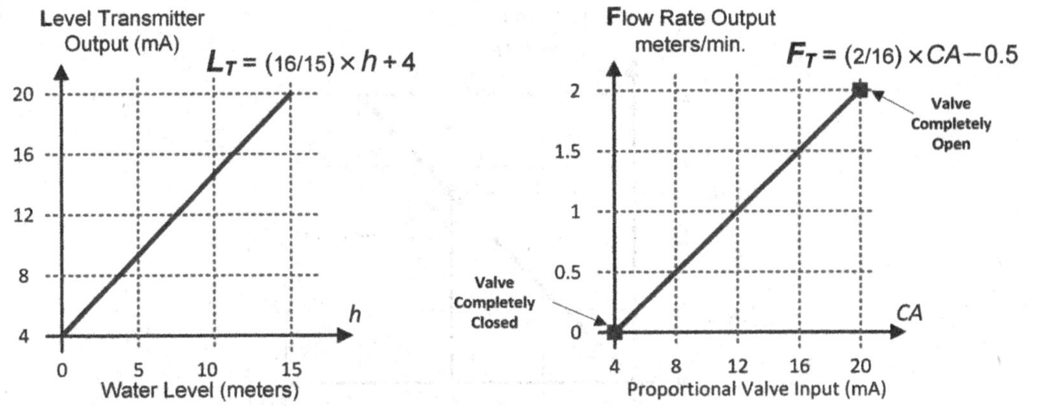

Figure 4-5. Relationship between the level transmitter and valve response.

produces a 4 to 20 mA output for a 15 m change in liquid in the tank. This means that the level transmitter has a sensitivity or transfer function of:

$$L_T = \frac{output}{input} = \frac{(20-4)\text{mA}}{15\text{ m}} = 1.067\,\text{mA/m}$$

The characteristic of the transmitter is shown in Figure 4-5a. Note that the transmitter produces a signal in the range from 4 to 20 mA, meaning it has an offset of 4 mA. The 4 to 20 mA signals are typical for transmitters used in industry.

The transmitter function (including the 4 mA offset) can be expressed mathematically as:

$$L_T = 1.067\,\text{mA/m} \times h + 4\text{ mA}$$

Figure 4-5b shows the relationship of the plant (height of water level) to the proportional control valve. The input to the control valve is again a signal of 4 to 20 mA. The output flow rate is specified in level (meters per minute, reflecting the process output, namely, level rather than in cubic meters per minute). The transfer function or sensitivity of the valve is then:

$$F_T = \frac{output}{input} = \frac{2\text{ m/min}}{(20-4)\text{ mA}} = 0.125\frac{\text{m/min}}{\text{mA}}$$

The graph can be expressed mathematically as:

$$flow\ rate = 0.125\frac{\text{m/min}}{\text{mA}} \times control\ action - 0.5$$

It is evident that our plant is composed of the tank and a proportional valve. The transmitter in this case converts liquid level to a corresponding current output. This output provides the required feedback, and the controller summing junction adds the *SP* to the

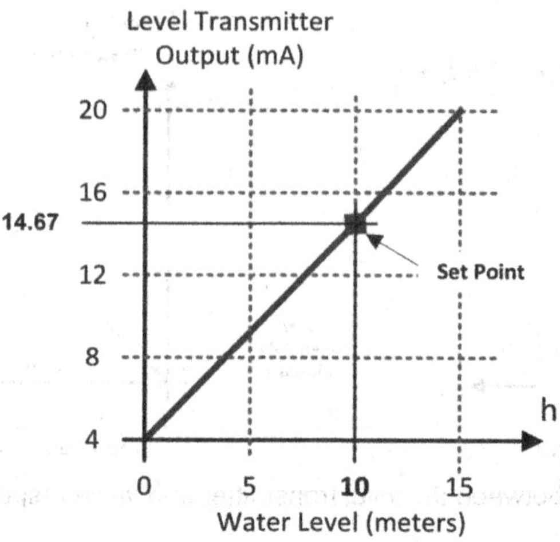

Figure 4-6. Relationship between level and the transmitter with the required SP shown.

inverted feedback (negative feedback) to create the error. The error is multiplied by 1 to become the control action (control current) for the proportional valve. This signal will open the valve and begin to fill the tank with liquid. Note that in this control system, the *controlled variable* is liquid level, yet the *control medium* is current. This fact illustrates the importance of transmitters in the field of process control. To obtain 10 m of liquid in the tank, we require an *SP* of 14.67 mA. This is illustrated by examining the transmitter characteristic. See Figure 4-6.

At time $t = 0$ minutes, the following conditions exist:

$$
\begin{aligned}
SP &= 14.67 \text{ mA } (10 \text{ m}) \\
h &= 0 \text{ (tank is empty)} \\
MV &= \frac{16}{15} \times h + 4 \text{ mA} = 1.067 \times 0 + 4 \text{ mA} \\
MV &= 4 \text{ mA} \\
error &= SP - MV = 14.67 - 4 = 10.67 \text{ mA} \\
control\ action &= error \times CA = 10.67 \times 1 + 4 = 14.67 \text{ mA} \\
flow\ rate &= 0.125 \times control\ action - 0.5 \\
&= 0.125 \times 14.67 - 0.5 = 1.33 \text{ m/min}
\end{aligned}
$$

This means that after 1 minute, the fluid in the tank will have reached a height of 1.33 *m* and the error now has changed. The following summarizes the new conditions:

The SP stays the same, but the error decreases, meaning the tank is filling up.

$$
\begin{aligned}
SP &= 14.67 \text{ mA} \\
MV &= 6.67 \text{ mA } (1.33 \text{ m}) \\
error &= 14.67 - 6.67 = 8 \text{ mA}
\end{aligned}
$$

$$\begin{aligned} \textit{control action} &= 8\times1+4 = 12\,\text{mA} \\ \textit{flow rate} &= 0.125\times \textit{control action} - 0.5 \\ &= 0.125\times12-0.5 = 1\,\text{m/min} \end{aligned}$$

Note that the flow rate is now slightly reduced from 1.33 to 1 m/min.

Repeating this calculation at 1-minute intervals, we see that the flow rate continuously decreases while the tank fills up, the error gets smaller until the tank is full, and the error is zero. At that point, the valve will be completely closed. See Table 4-1.

From Table 4-1, it is evident that this control system is trying to attain the level of 10 m indicated by our *SP*. The system is accomplishing this by diminishing the error to zero and in doing so will eventually close the proportional valve. Our control action in this case is equal to the error of the system, and it should be clear that if we increased our gain in our control network, our system would diminish the error more quickly. This would require that our control system provide a means for the designer to change the control system's speed of response.

If we plot the *SP* and plant output as a function of time and plot the control action (error, in this case) as a function of time, we obtain the graphs presented in Figure 4-7.

It is interesting to see the input as a step input and the plant output as an exponential response. We may recall that this type of response is also exhibited when we charge a capacitor in a RC circuit with a step input. In a frequency sense, our plant is allowing a band of frequencies to pass and is attenuating higher frequencies. As mentioned earlier,

Table 4-1. Parameter changes.

Time (min)	Set Point (SP)		Measured Value MV (mA)	Error SP – MV (mA)	Control Action (mA)
	(m)	(mA)			
0	10	14.67	4.00	10.67	14.67
1	10	14.67	6.67	8.00	12.00
2	10	14.67	8.67	6.00	10.00
3	10	14.67	10.17	4.50	8.50
4	10	14.67	11.29	3.38	7.38
5	10	14.67	12.14	2.53	6.53
6	10	14.67	12.77	1.90	5.90
7	10	14.67	13.24	1.42	5.42
8	10	14.67	13.60	1.07	5.07
9	10	14.67	13.87	0.80	4.80
10	10	14.67	14.07	0.60	4.60

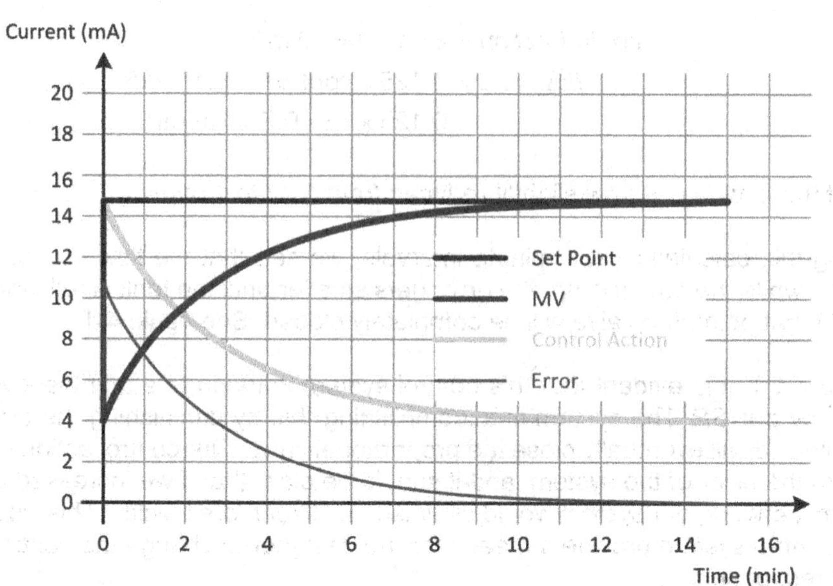

Figure 4-7. Relationship between the MV and control action.

an increase in gain in the control network will promote a faster response in our control system. It appears that our bandwidth is increasing because the response is faster, where the bandwidth represents the frequencies passed by our plant. It appears that we could design or analyze our control system in the frequency domain, which would be independent of time. This domain is used extensively in advanced motion control to study system stability.

In Figure 4-7, the control action (or error in this case) is decreasing with time. Note that the plant output is always continuous because physical components are involved. The control action was discretized for ease of calculation and to generate a finite set of data points for ease of illustration. This form of control represents the method of computerized control, and human beings attack a problem in the same fashion as a computer with rather limited speed.

4.6 Process Gain

A definition is often a good place to start, and that is particularly true when trying to understand the role of process gain. *Process gain* (G_p) is defined as how far the measured process variable (PV) moves to a change in controller output (CO). However, let us break down the definition of process gain a bit further.

Process gain is one of the model parameters that describes how a process behaves in response to dynamic changes. As mentioned before, the process gain quantifies how much the PV changes when the CO changes.

4.7 Summary

Each functional component in a control system can be described mathematically by a transfer function (TF), where TF = output/input. Transfer functions of individual elements or components in a system can be represented using block diagrams, and can

Table 4-2. Fundamental block diagram identities.

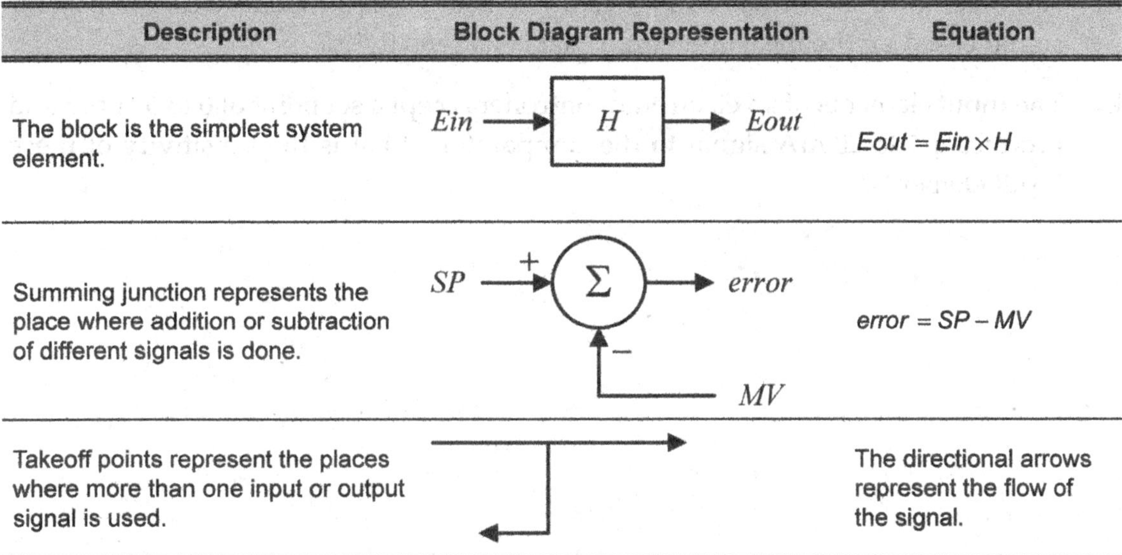

Description	Block Diagram Representation	Equation
The block is the simplest system element.	Ein, H, Eout	$Eout = Ein \times H$
Summing junction represents the place where addition or subtraction of different signals is done.	SP, +, Σ, error, −, MV	$error = SP - MV$
Takeoff points represent the places where more than one input or output signal is used.		The directional arrows represent the flow of the signal.

Table 4-3. Basic block diagram manipulations.

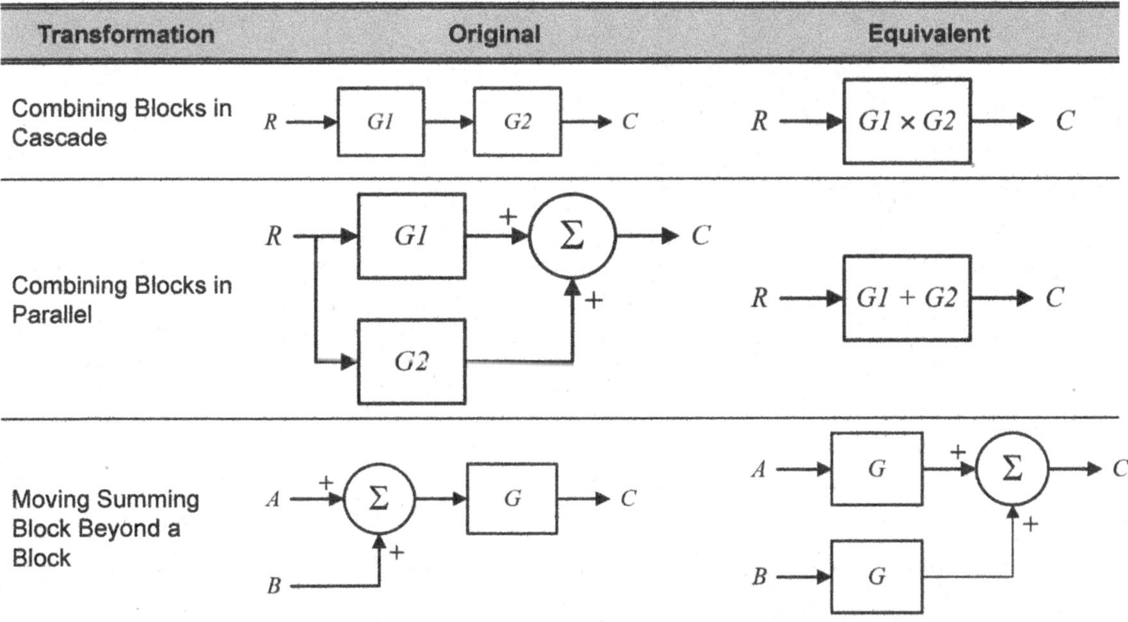

Transformation	Original	Equivalent
Combining Blocks in Cascade	R, G1, G2, C	R, G1 × G2, C
Combining Blocks in Parallel	R, G1, G2, +, Σ, +, C	R, G1 + G2, C
Moving Summing Block Beyond a Block	A, B, +, Σ, +, G, C	A, B, G, G, +, Σ, +, C

be mathematically combined to calculate overall system performance. A true transfer function includes time-dependent and steady-state characteristics, whereas a useful simplification considers only steady-state conditions. Table 4-2 summarizes basic block diagram symbols, and Table 4-3 summarizes some basic block diagram manipulations.

Exercises

4.1 A temperature sensor has a range of 0°F to 200°F and transmits a 3 to 15 psig air signal based on the measurement. What is its sensitivity?

4.2 The input elements of an electronic controller accept a set point of 0 to 60 gpm and produce a 4 to 20 mA signal to the comparator. What is the sensitivity of these input elements?

5 Dynamics of Systems

The term *dynamics* refers to how a system's output changes when the input changes or how a system reacts to a change in set point, load, or disturbances. Dynamics is often referred to as *transient*, meaning the temporary change in how a process reaches a new steady state. It is important for the automation professional to develop an overall understanding and appreciation of dynamic behavior of the process, its element, and the process control loop. This chapter describes many of the general concepts and principles of dynamics and its associated terminology.

Learning Objectives

- Know the general response characteristics of a first-order lag component that has been subjected to a step input
- Determine graphically the time constant for a first-order lag system that has been driven by a step input
- Distinguish the difference between real dead time and apparent dead time
- Identify and quantify the dead time on a process response curve
- Understand the effects of process lags and dead times on loop process dynamics
- Understand and determine process gain and sensitivity

5.1 Dynamic Components

Dynamics describes how a component or system responds to change over time. The various elements represented by blocks of a feedback control loop have different types of dynamic behavior, and it is important that you gain insight into these process dynamics.

Every change in input will produce a corresponding change in output.

Many of the individual components of the process control loop have no significant time-dependent behavior, that is, there is neither lag nor lead in their operation. When the input to a component changes, the output of the component changes instantaneously, for all practical purposes.

This type of component, illustrated in Figure 5-1, is referred to as a *nondynamic* component. In effect, it can be described as a pure gain. For all practical purposes, the output changes instantaneously when the input changes. The output is always directly proportional to the input, and this proportionality constant is referred to as the *gain K* of the component. The term *gain* does not necessarily imply that the output is greater than the input. *Gain* is also used when the output is smaller than the input. When this occurs, it is called *attenuation*. In that case, K will be less than one ($K < 1$).

Components that do not exhibit dynamic behavior are referred to as *zero-order systems*.

Many components do exhibit dynamic behavior. Typically, their output lags or leads any input or alters the input significantly; lag tends to be far more common than lead. When such time-dependent behavior is encountered, it can appear as shown in Figure 5-2. Systems and components that exhibit such dynamic behavior are referred to as *first-order* or *second-order* systems, or *higher-order* systems. The order refers to the number of elements capable of storing material or energy within the system.

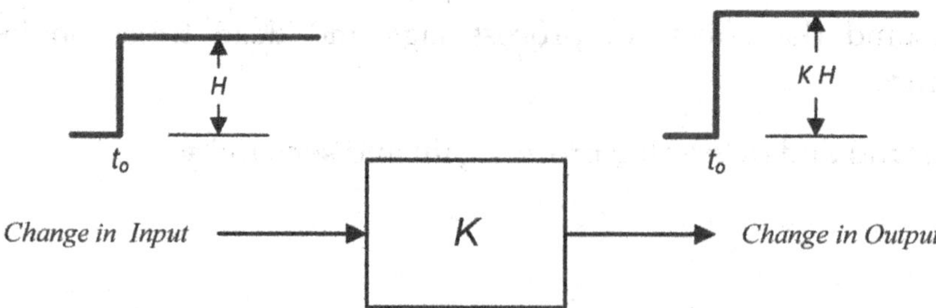

Figure 5-1. Nondynamic behavior of a component.

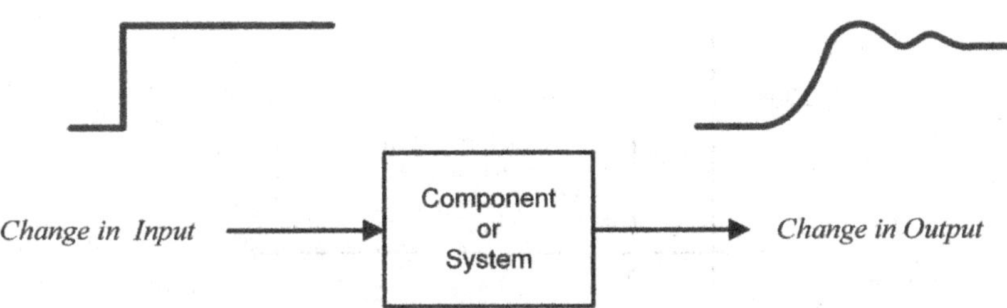

Figure 5-2. Dynamic behavior of a component.

Block diagrams are often represented as nondynamic components, that is, when the input changes, the signal at the output tends to change relatively instantaneously. Components are designed so they have insignificant time lags associated with them. However, in many cases, this is not possible—specifically, in valves and pneumatic transmission systems, and the process itself.

To a lesser extent, both the sensor and the actuator have their own dynamics, but in the typical case, the dynamics of these individual components is much less than that of the process itself. All process loops are functionally the same, and in general, they follow the layouts presented in Chapter 4. Process dynamics vary significantly from one individual loop to another, and the practitioner must gain some appreciation and insight into the dynamics of an individual loop in order to install, maintain, and tune the loop to provide optimum control.

Quite often, students get distressed about the need to analyze dynamic performance mathematically or quantitatively. This frustration is understandable, but it does not eliminate the necessity for dealing with process dynamics. Today, much of the mathematics is done by computer, and the practitioner only needs to understand what the computer is doing.

Process control is obviously needed only in situations that are changing; that is, if nothing is changing, you do not need any control. Things that are changing are doing so with respect to time, and it is important to understand their dynamic behavior. As a result, to understand process control, one must study and appreciate process dynamic behavior.

Most, if not all, processes have dynamic lag, and when disturbances enter the process, they will produce an effect on the controlled variable that is dynamic in nature. The same thing can be said of the manipulated flow as it enters the process.

5.2 First-Order Lag

The first-order lag is the most common type of dynamic component or system encountered in process control, or in nature for that matter. To understand it, it is helpful to

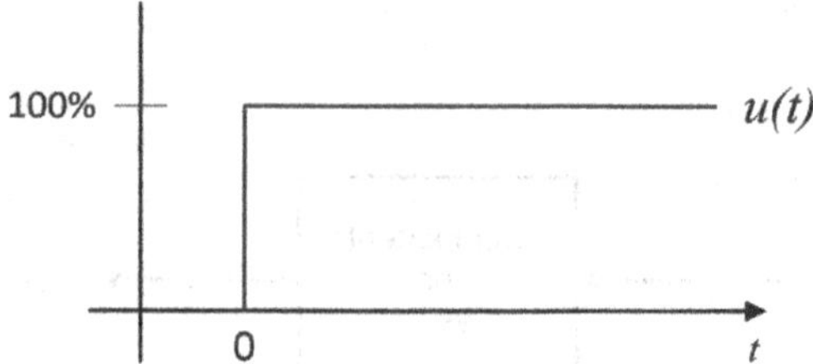

Figure 5-3. Unit step function.

look at response curves when the component under study is subjected to a step input, such as shown in Figure 5-3. This type of input is representative of a change in input, or a change in demand.

The advantage of using such a step input designated $u(t)$ as a forcing function or driving force is that the input is at steady state for the time previous to the change, and then the input is *instantaneously* switched to a new steady-state value. This is similar to turning a light from off to on. When the resulting output or response curve is inspected, you observe the transition of the component as it passes from one steady state to a new steady state.

For a first-order lag component, the response to a step input is as illustrated in Figure 5-4.

The component whose response is shown in Figure 5-4 is called a *first-order lag* because the output lags behind the input and the differential equation that describes the underlying mathematics is a linear first-order differential equation. This is the most common type of dynamic component encountered in practice. In addition to being referred to as *first-order lag,* these components are often referred to as *linear lags* or *exponential*

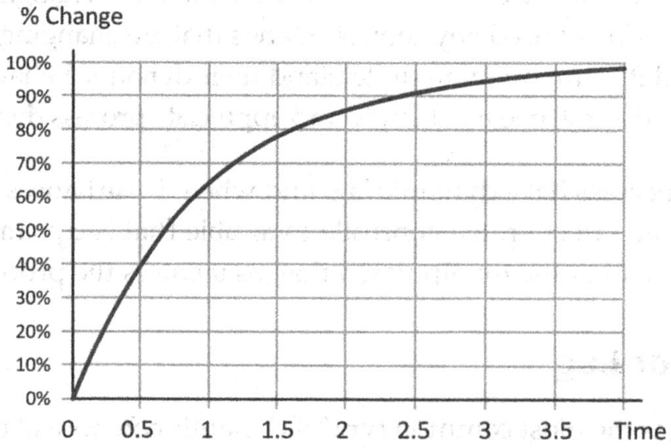

Figure 5-4. The response of a first-order lag component to a step input.

transfer lags. They are characterized by their ability to store material or energy, and the dynamic shape of the response curves is described by a time constant.

5.3 Time Constants

The time constant, designated by the Greek letter τ (tau), is meaningful both in a physical sense and in a mathematical sense. In a physical sense, if you look closely at the first-order response curve in Figure 5-4, you can see that the response is always falling off; that is, the rate of response is a maximum at time 0 in the very beginning of the curve and is continuously decreasing from that time onward. If the system were to continue to change at its maximum response rate (which occurs at the origin), it would reach its final value (or 100%) in one time constant as is shown in Figure 5-5. In a mathematical sense, this represents the time it took to reach 63.2% of the change.

At any point on the curve, a tangent would also reach its final value at one time constant as is shown in Figure 5-5.

The step response of a first-order lag is measured in terms of its time constant. The time constant is the duration of time in seconds (or minutes) required for the output to reach 63.2% of the final steady-state output. During each additional time constant, the output will again reach 63.2% of the remaining change.

During the first time-constant interval, the output changed by an amount equal to $0.632 \times 100\% = 63.2\%$. This means that a remaining change of $100\% - 63.2\% = 36.8\%$ is required to reach the steady-state or final value. During the second time interval, the

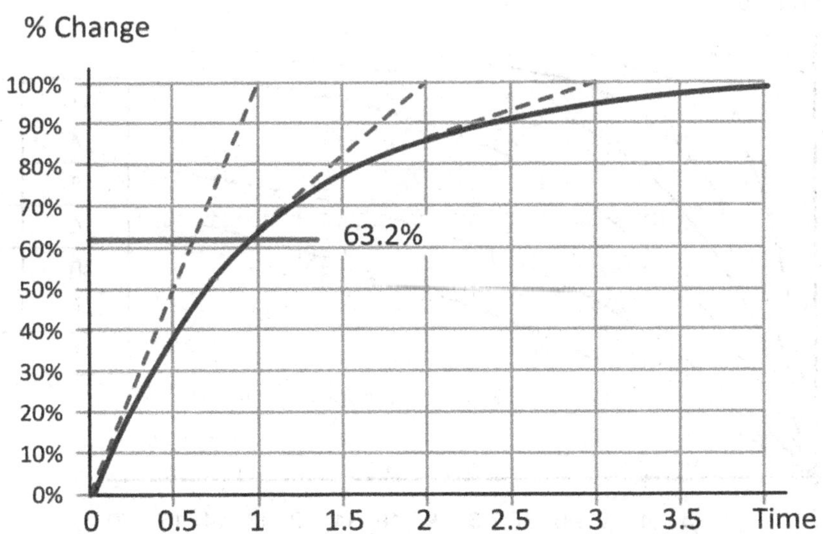

Figure 5-5. Time constant.

Table 5-1. Response of a first-order lag to a step change.

Elapsed Time	% of Total Response	% Response Remaining	63.2% of Response Remaining
1 τ	63.2	36.8	23.2
2 τ	86.4	13.6	8.6
3 τ	95.0	5.0	3.16
4 τ	98.16	1.84	1.16
5 τ	99.32	0.68	0.429

output again increases by 63.2% or $0.632 \times 0.368 = 0.232$ or 23.2%. Thus, after two time constants, the output will reach $0.632 + 0.232 = 0.864$, or 86.4% of the final value. After five time constants, the output will have reached 99.3% of the final value.

Table 5-1 provides the numerical value of the changes that take place in a first-order lag response to a step input. In the first time constant of elapsed time, 63.2% of the total response will occur. Within the next time constant, 63.2% of the remaining 36.8% of the change will take place, and so on. Theoretically, of course, the response never reaches 100%, but it does approach it asymptotically.

The time constant provides specific dynamic insight into the rapidity with which the system or component responds. It is an unambiguous measure of the speed of response. Figure 5-6 shows six different response curves, all for first-order lags, but each of the six has a different time constant. As the figure illustrates, as the time constant gets larger, the response gets slower.

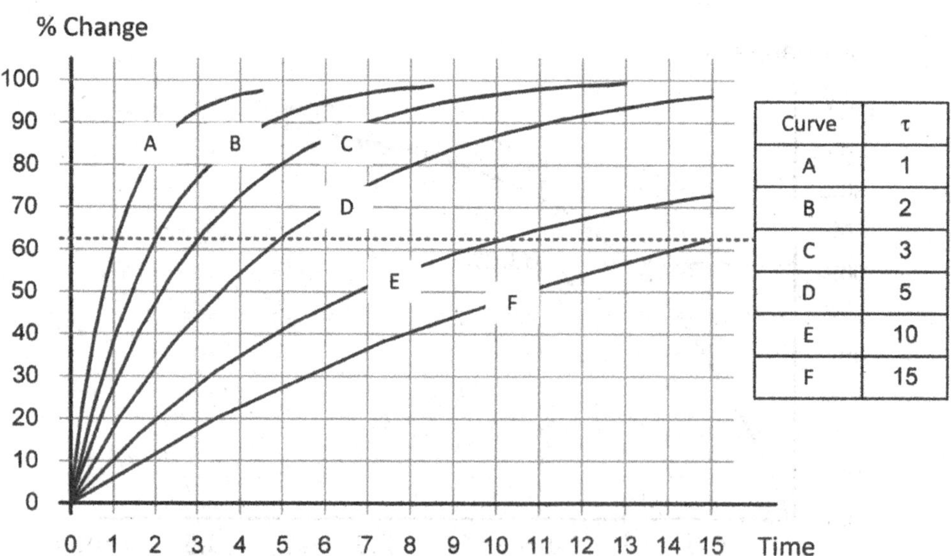

Figure 5-6. First-order lag responses to a step input.

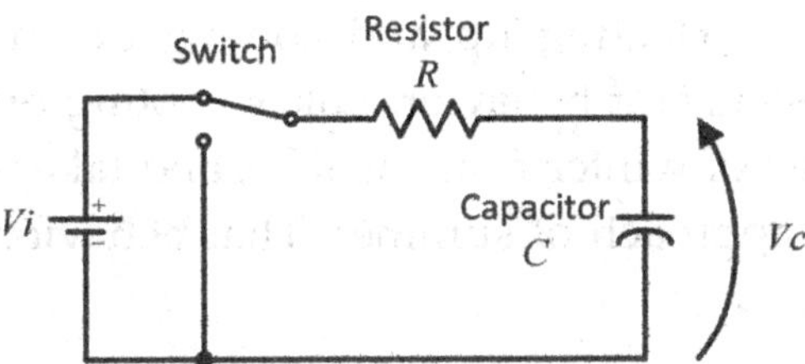

Figure 5-7. Electrical RC circuit.

Frequent use is made of the time constant concept. It is a concept that should not be entirely new to the student, and it forms the basis for understanding simple electrical circuits. Figure 5-7 illustrates a simple electrical resistor-capacitor (RC) network.

The basic problem is to derive the equation that describes the charging of the capacitor when a voltage (V_i) is suddenly applied to the circuit by closing the switch, that is, applying a step input (V_i). The voltage across the capacitor (V_C) builds up and is given by a linear first-order lag equation. The product of the resistance (R) and the capacitance (C) is the time constant for this simple network. If one checks the units involved, the product of (RC). has the units of time. In this example, the capacitor is the energy-storing component.

This physical analogy can also be used to gain insight into time constants of other elements of systems. To do so, it is advantageous to use the reciprocal of resistance, which is conductance. Mathematically:

$$conductance = \frac{1}{resistance} \tag{5-1}$$

For many physical systems, time constants may be viewed more easily as capacitance divided by conductance:

$$\begin{aligned} time\ constant\ (\tau) &= resistance \times capacitance \\ &= \frac{capacitance}{conductance} \end{aligned} \tag{5-2}$$

Time constants are everywhere. They are found in many different kinds of systems, including the following:

- **Electrical systems** – All resistor-capacitor and resistor-inductor circuits exhibit time-constant behavior.
- **Mechanical systems** – Many motors exhibit time-constant behavior. For example, when a motor speeds up, the measurements of speed reveal time-constant behavior.

- **Thermal systems** – Heating up and cooling down in simple thermal systems shows time-constant behavior. Lakes cooling off after the summer slow down the approach of winter near the lake, and lakes slowly heating up in the spring retard the approach of summer. That behavior can be explained using time-constants.
- **Physiological systems** – The inner ear has dynamics that can be explained using two time constants.
- **Psychological systems** – Psychologists have measured the time constant that determines how much learned material a person retains as a function of time after learning.

For example, observe the first-order lag response of a bare bulb thermometer immersed in an agitated bath (a step input). In this case, the capacitance for the bulb is the mass of bulb material (which must change in temperature) times its weighted average heat capacity. The conductance is the heat transfer coefficient between the bath liquid and the bulb, times the area through which heat transfer takes place. By dividing this capacitance by this conductance, we can find the natural time constant of the bulb expressed in units of time. For most control loops, we can introduce a step input by using a *bump test*. This can be done by changing the set point by 5% to 10% from its required value.

It is important to get a general idea of the physical meaning of time constants. With practice, you can inspect an individual component or process and achieve some appreciation of its capacity to store material or energy, as well as of the conductance of material or energy into or through the component or process. In this way, you achieve specific insight into the time constants involved and, therefore, better understand the process dynamics of an individual component or process. Such understanding will evolve and assist you in process control applications.

You can quantitatively determine the time constant by measuring how much time the system or process needed to change by 63.2% (approximately two-thirds) of its final value in response to a step input change.

The physical significance of a first-order lag is that it represents the action at a place of mass or energy storage. The general form of the transfer function (*H*). for a first-order lag is as follows:

$$H = \frac{output(s)}{U(s)} = \frac{K}{\tau s + 1} \tag{5-3}$$

The parameter K is called the *steady-state gain*. The parameter τ is the time constant, and s is the Laplace operator. The function $U(s)$. is the step input.

We can solve this equation with a step input using the inverse Laplace transforms. The derivation is included in Appendix C and results in the following exponential equation that was shown in Figures 5-4 and 5-5:

$$outout(t) = 1 - e^{-t/\tau} \tag{5-4}$$

The time constant parameter (τ). is an indication of how quickly the output comes to an equilibrium following a step change in the input. (Note that we said, "an indication of," not "the time it takes to come to equilibrium.") The time constant can be stated in any convenient time units, such as seconds (s), minutes (min), or even hours (h). For industrial process control work, it is usually most convenient to express τ. in minutes.

Control systems consist of a variety of different types of components. In a temperature control loop, for example, the process and the primary sensor are thermal systems, the temperature transmitter and controller are electrical systems, and the final control element consists of fluid flow (pneumatic), a mechanical system, or an electrical system. This variety of components within a control system complicates the analysis, tuning, and design of the control loop. Fortunately, the behavior of these diverse systems is determined by the same four common elements: resistance, capacitance, inertia (or inductance in electrical systems), and dead time. The effects of these basic elements in one system are analogous to the effects of the same elements in any other system. Thus, thermal resistance and capacitance have the same effect in a thermal system as electrical resistance and capacitance have in an electrical system.

5.4 Elements of Systems

Most control system components fit into one of the following five types:

1. Electrical
2. Mechanical
3. Liquid flow
4. Gas flow
5. Thermal

A control system may include two, three, or even all five types of components. The behavior of the components of one type is analogous to the behavior of components of any other type. This analogous behavior is determined by the four *basic elements* that are common to the five types of components. These four elements are

1. resistance,
2. capacitance,
3. inertia (or inductance), and
4. dead-time.

Some knowledge of these basic elements for the five component types will clarify the behavior of control system components and control loops.

For each type of component, the four elements are defined in terms of three variables.

1. Quantity of material, energy, or distance
2. A driving force or potential that tends to move or change the quantity variable
3. Time

For example, in a liquid flow component, the quantity variable is the volume of liquid that is moved, and the potential variable is the pressure drop that tends to cause the liquid to flow. Table 5-2 names the three variables for each of the five types of components.

The following sections describe each of the four elements in their context.

Table 5-2. The three variables used to define the four elements for each type of component.

	Variable		
Type of Component	**Quantity**	**Potential**	**Time**
Electrical	Charge	Voltage	Second
Liquid Flow	Volume	Pressure	Second
Gas Flow	Mass	Pressure	Second
Thermal	Thermal (heat) energy	Temperature	Second
Mechanical	Distance	Force	Second

Resistance

Resistance is an opposition to the movement or flow of material or energy. All materials permit heat to pass through them, but not equally well. Some materials, such as building bricks or a wool coat, conduct heat poorly; they have high resistance and are called *thermal insulators.* Other materials, such as a metal spoon or a copper wire, conduct heat well; they have low resistance and are good *thermal conductors.* The thermal resistance depends on the type and the amount of material in the path of the energy flow. Similarly, for electricity, there are insulators and conductors. The flow of materials also may be inhibited by a resistance (or friction in mechanical systems). Just as a narrow street impedes the flow of traffic, so is a small-diameter pipe more of a hindrance to liquid flow than a large-diameter pipe. Resistance dissipates energy in the form of heat and is an irreversible process.

The *greater the resistance* of the process, the *more difficult it is to control* because resistance is the characteristic that impedes the flow of material or energy.

Resistance is measured by the amount of potential required to make a unit change in the quantity moved each second.

$$resistance = \frac{potential}{flow} \tag{5-5}$$

Summary

Electrical resistance is the increase in the voltage across two terminals of a component required to move one more coulomb of charge through the component each second (or to increase the current by 1 A). Resistance is a measure of the opposition to current flow in an electrical circuit.

Liquid flow resistance is the increase in the pressure drop between two points along a pipe required to increase the flow rate through the pipe by 1 m^3/s.

Gas flow resistance is the increase in the pressure drop between two points along a pipe required to increase the flow rate through the pipe by 1 kg/s.

Thermal resistance is the increase in temperature difference across a wall section required to increase the heat flow through the wall section by 1 J/s.

Mechanical resistance is the change in the force applied to an object required to increase the velocity of the object by 1 m/s.

To summarize again, resistance is the amount of potential required to produce one unit of

- electric current,
- liquid flow rate,
- gas flow rate,
- heat flow rate, or
- mechanical velocity.

Capacitance

Capacitance is the ability to store energy in the form of material, energy, or distance required to make a unit change in potential.

Capacitance does not dissipate energy in the form of heat; rather, it stores and releases energy from and to the rest of the system. Capacitance is the property that resists a change in potential.

For example, capacitance is the number of liters of water required to *change* a tank level by 1 meter. In other words, to make a change in a controlled variable, some quantity of a manipulated variable must be supplied or removed; that quantity divided by the change is the capacitance. Capacitance expresses the relationship between a change in quantity and its corresponding change in potential.

$$capacitance = \frac{quantity}{potential} \tag{5-6}$$

Capacitance, unlike resistance, does not dissipate energy in the form of heat; rather, it *stores and releases energy* from and to the rest of the system, providing the system with a damping effect.

Capacitance should not be confused with capacity, which is the total material or energy holding ability of a device. If we say a tank holds 100 L, we are stating its capacity. If we say that 100 m^3 of liquid must be added to a tank to increase the pressure at the bottom of the tank by 1 N/m^2, we are stating the capacitance of the tank.

Figure 5-8 shows two tanks with the same capacity but different capacitances. As far as level control is concerned, the process is more sensitive in the tall tank than it is in the short tank because the capacitance of the tall tank is smaller; for the same change

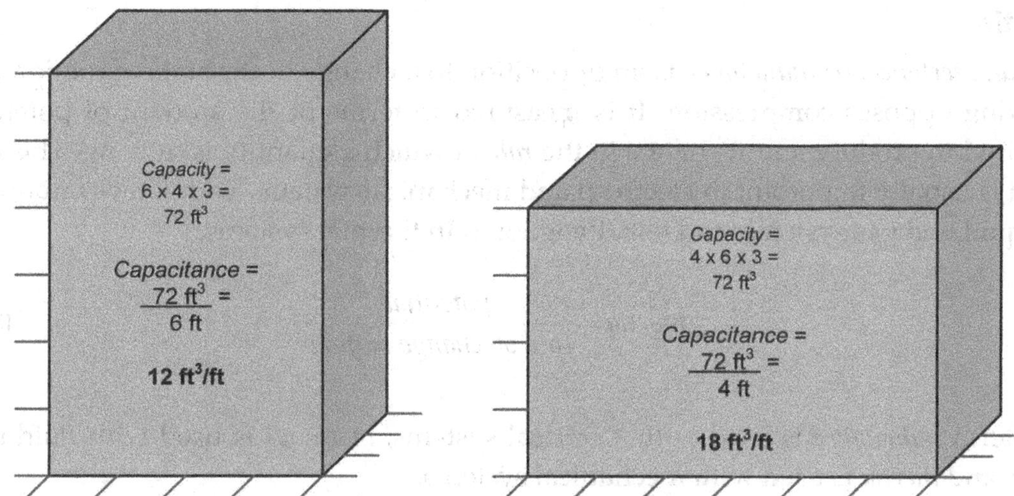

Figure 5-8. Tank capacity and capitance.

of volume in both tanks, the level in the tall tank will have a larger change. It has a faster response but is more difficult to control within a given range. Increasing the capacitance for a given control range improves the stability of the controlled variable in response to an upset.

For the same volume change of liquid in both tanks, the level in the short tank is more stable.

Summary

Electrical capacitance is the coulombs of charge that must be stored in a capacitor to increase its voltage by 1 V.

Liquid capacitance is the cubic meters of liquid that must be added to a tank to increase the pressure at the bottom of the tank by 1 Pa.

Gas capacitance is the kilograms of gas that must be added to a tank to increase the pressure of the tank by 1 Pa.

Thermal capacitance is the amount of thermal (heat) energy that must be added to an object to increase its temperature by 1°C.

Mechanical capacitance is the amount of compression of a spring required to increase the spring force by 1 N.

Capacitance, then, is the amount of material, energy, or distance required to make a unit change in potential.

Inertia

Inertia, inertance, or *inductance* is an opposition to a change in the state of motion like a spring opposes compression. It is measured in terms of the amount of potential required to produce a unit change in the *rate* at which a quantity is moving. This element is usually important in electrical and mechanical systems, sometimes important in liquid and gas systems, and usually ignored in thermal systems.

$$inertia = \frac{potential}{rate\ of\ change\ of\ flow} \tag{5-7}$$

The term *inductance* is used with electrical systems, *inertance* is used with fluid systems, and *inertia* is used with mechanical systems.

Summary

Electrical inductance is the increase in voltage across an inductor required to increase the current by 1 A each second.

Liquid flow inertance is the increase in the pressure drop between two points along a pipe required to accelerate the flow rate by 1 m^3/s.

Mechanical inertia is the increase in force required to produce an acceleration of 1 m/s.

Inertia, inertance, or inductance is the amount of potential required to increase

- electric current by one unit per second,
- liquid flow rate by one unit per second,
- gas flow rate by one unit per second, or
- velocity by one unit per second.

Table 5-3 provides a broader view of the time constant analogy between the various physical systems.

5.5 Dead Time

Another form of response often encountered in process control applications is a time delay, often called *dead time*.

Table 5-3. Time constant analogy between basic physical systems.

Variable	Electrical	Liquid Level	Thermal	Pressure
Quantity	coulomb	cubic foot	Btu	cubic foot
Potential or Force	volt	foot	°F	$\frac{\text{pound}}{\text{square inch}}$
Flow Rate	$\frac{\text{coulombs}}{\text{seconds}}$ = amperes	$\frac{\text{cubic feet}}{\text{minutes}}$	$\frac{\text{Btu}}{\text{minutes}}$	$\frac{\text{cubic feet}}{\text{minutes}}$
Resistance (R)	$\frac{\text{volts}}{\text{coulombs/second}}$ = ohms	$\frac{\text{feet}}{\text{cubit feet/min}}$	$\frac{\text{°F}}{\text{Btu/min}}$	$\frac{\text{pound/square inch}}{\text{cubic feet/min}}$
Capacitance (C)	$\frac{\text{coulombs}}{\text{volts}}$ = farads	$\frac{\text{cubic feet}}{\text{foot}}$	$\frac{\text{Btu}}{\text{°F}}$	$\frac{\text{cubic feet}}{\text{pound/square inch}}$
Time Constant	RC (second)	RC (minute)	RC (minute)	RC (minute)

Dead time is the time interval between the time a signal appears at the input of a component and the time the corresponding response appears at the output. A pure dead-time element does not change the magnitude of the signal; only the timing of the signal is changed. Dead time occurs whenever mass or energy is transported from one point to another. It is the time required for the mass or energy to travel from the input location to the output location. Dead time is also called *transport time, pure delay,* or *distance-velocity lag.* Control becomes increasingly more difficult as dead time increases.

If v is the velocity of the mass or energy and D is the distance traveled, the dead-time delay is equal to the distance divided by the velocity:

$$dead\ time = \frac{distance}{velocity}$$

or

$$t_d = \frac{D}{v} \tag{5-8}$$

The effect of dead time is to delay the input signal by the dead-time delay (t_d).

Dead time can also be represented (without providing formal justification) with the following transfer function (DT) using the s operator.

$$DT = e^{-\theta s} \tag{5-9}$$

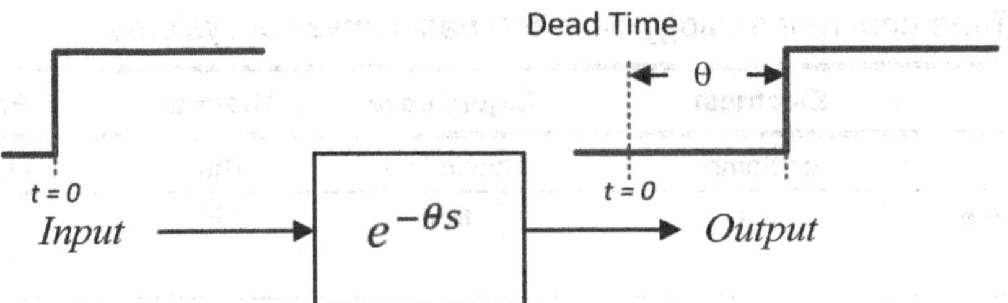

Figure 5-9. Dead time θ for a nondynamic element.

This function describes a system or process that has the input–output response shown in Figure 5-9. The parameter θ represents the amount of dead time. For industrial process control, it is usually expressed in minutes, but it can be expressed in any convenient time units.

Such dead times are among the most difficult situations to control and will be covered in Chapter 19. The difficulty stems from the fact that during this period (the dead time), there is absolutely no response and therefore there is no information available to initiate any corrective action.

For specific insights into how a dead time might occur in practice, Figure 5-10 illustrates a simple feedback control system in which steam is injected into a water tank to produce hot water.

There is a thermo-bulb in the outlet stream to measure the temperature of the exit hot water. The control system increases or decreases the manipulated variable, the steam flow, to produce the outlet water at the desired temperature.

The difficult issue arises when one begins to debate where this temperature bulb should be located, in other words, where should the feedback sensor be installed? It is tempting to say that it should be installed farther down the outlet pipe to be closer to the location where the water will be used. This rationale sounds effective because it is at the point of usage that the temperature of the water is important. The water temperature in the tank itself or at the point of tank exit is less important. This type of reasoning, while appealing, can be unsuccessful. As the temperature bulb is moved farther and farther down the outlet pipe, a dead time is created in the feedback loop before the sensor. In this case, dead time is equal to the distance down the pipe to the point at which the bulb is installed, divided by the velocity of the water inside the pipe. As a result of such a dead time encountered in the feedback loop, overall loop control deteriorates, and it may reach a point where it is impossible to achieve any feedback control.

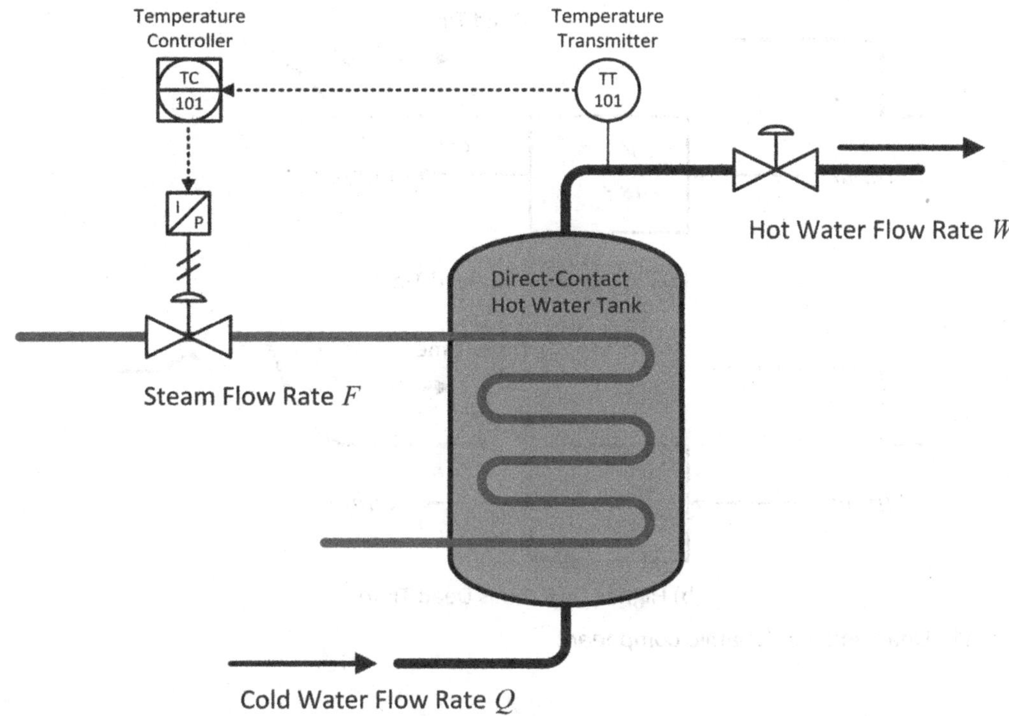

Figure 5-10. Sample system showing dead time.

Often dead times are referred to by other names, such as *distance-velocity lags, transportation lags,* or *time delays.* These descriptive names can be appreciated in terms of the example shown in Figure 5-10.

Often, dead times are unavoidable in actual practice, for example, in a rotating kiln, in a plug flow reactor, or in similar situations. However, there are other cases where the designer's lack of insight and lack of forethought causes the sensor to be installed in an inappropriate location, and, for no good reason, too often a dead time is inadvertently placed into the feedback control loop. Such mistakes can cause serious operating problems, and unstable and difficult control situations. Sometimes such problems occur as a result of the type of reasoning discussed for the hot water tank; sometimes such problems are a result of a desire to place a sensor in a spot where maintenance might be achieved more easily. Whatever the reason that dead times occur, they should be avoided at all costs wherever possible. They are without doubt one of the most difficult dynamic elements to control.

5.6 First-Order plus Dead Time

Dead times often occur in combination with nondynamic components. They may occur with first-order lag as shown in Figure 5-11, or they may be found in combination with

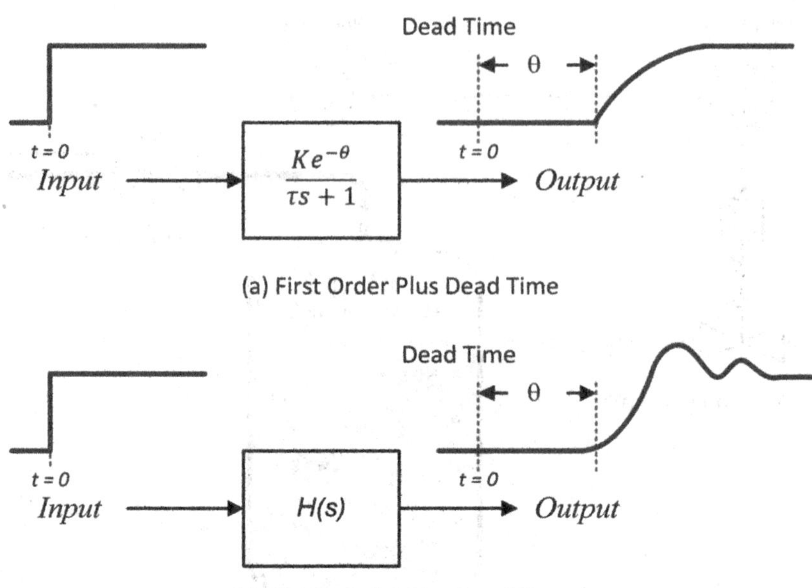

Figure 5-11. Dead time for dynamic component.

higher-ordered lags, also shown in Figure 5-11. The transfer function of a first-order plus dead time (FOPDT) is:

$$H = \frac{K}{\tau s + 1} \times e^{-\theta} = \frac{Ke^{-\theta}}{\tau s + 1} \tag{5-10}$$

Where K is the gain, τ is the time constant, θ is the dead time, and s is the Laplace operator.

Whenever dead time is encountered, you should be able to identify it from the process response curve by a simple measurement of the amount of time that transpires before any output response occurs after an input change to the system.

5.7 Higher Order Systems

Of course, not all process dynamics can be neatly characterized by first-order lags. Often a situation is encountered in which a step input will produce a response curve, such as shown in Figure 5-11(b).

Note that the maximum rate of change for the output response curve does not occur at the origin but at some later time. This implies that a higher order system is involved, and a first-order lag is inadequate to characterize the system.

Higher order systems can be the result of several different situations.

- Several first-order lag processes may be encountered in series.
- The installed feedback controller may introduce a characteristic differential equation which, when considered in series with the other system components, makes the overall description of the system higher-ordered.
- Mechanical or fluid components of the system may be subject to accelerations or to inertial effects (usually this is a minor possibility).
- The process may be a distributed process that gives a response curve that can be described only by higher-ordered differential equations or by partial differential equations.

Note that not all higher order systems behave in an oscillatory manner. This is discussed in Section 5-9, "Sensor Dynamics," and illustrated in Figure 5-16.

When first-order lags are encountered in series, the way in which they are connected in series has a lot to do with the resulting process dynamics. They may be connected so as to produce independent stages or interacting stages; these are illustrated in Figure 5-12.

For example, both of the tanks in Figure 5-12 taken separately would give first-order lag responses to a change in inlet flow rate. However, the manner in which they are connected causes their combined response to differ between case (a) and case (b). In case (a), the height of liquid in the second tank does not influence the flow of liquid out of the first tank, but in case (b), it is clear that the height of liquid in the second tank

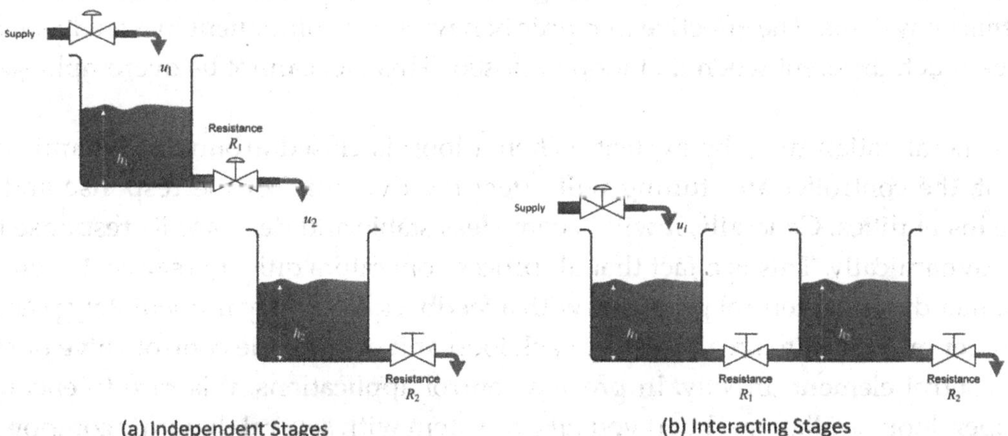

Figure 5-12. Independent and interacting stages.

will influence the rate at which liquid flows out of the first tank. When stages are interacting with one another, the resulting process response will always be more sluggish than expected from simple first-order lags connected as independent stages. This case is of special importance in thermal systems because they are usually sluggish to begin with. Section 5-9, "Sensor Dynamics," will discuss this in more detail.

Typically, when multiple first-order systems are in the process loop, they will exhibit an apparent dead time as shown in Figure 5-11.

It is sometimes difficult to formulate a precise process model of higher order systems, but we can often approximate the response with a FOPDT form. This requires three parameters:

1. Process gain
2. Apparent dead time
3. Estimated time constant

We will discuss obtaining numerical values of these parameters in a later section.

When stages are interacting and have two storage elements like in Figure 5-12(b), these systems are referred to as *second-order.*

5.8 Closed-Loop versus Open-Loop Response

In the examples previously presented, attention was devoted to the open-loop response curve of an individual component or system. The implication is that the feedback loop was not closed and that you were looking at the open-loop behavior of individual components or systems. The effective dynamic behavior of a component in a feedback loop will be much different when that loop is closed. This fact cannot be overemphasized.

One generalization must be explicit. When a loop is closed around a dynamic component, the controller and tuning will affect the dynamics of the response and can cause instabilities. Generally, it will become less stable and decrease its response time more dynamically. This is a fact that all process operators quickly realize. When they encounter dynamic control problems with a feedback loop, they immediately place the controller on *manual* to break the feedback loop and operate the control valve or other final control element directly. In process control applications, it is rare to encounter any open-loop oscillations, but if you take a system with several dynamic components, place these components together in a loop, and then close the loop, it is quite possible for the system to have an output response curve that exhibits oscillations. Figure 5-13 illustrates some of the descriptive terms used in characterizing systems that exhibit oscillations.

To illustrate the effects of closing a process control loop, study the two response curves shown in Figure 5-14. Assume there are two first-order lags with an overall gain of 4 and that these first-order lags have equal time constants. Comparing the two curves, we see that the response to the same step input, closing the loop, has an impressive change. This illustrates the need to make significant differentiations between open-loop and closed-loop response considerations.

The point has been made that a loop is less stable when it is closed. This generalization can be carried a step further. If you increase the gain on a loop that has been closed, the loop progressively becomes less stable; it will exhibit more and more oscillatory behavior. In Figure 5-14, if the gain on the loop had been higher, when the loop was closed the oscillation shown in the closed-loop response curve would have been larger. As you gradually increase the gain, you increase the response of the loop. This is desirable because it decreases the rise time, speeding up the controlled variable's transition due to set point changes, and provides a more rapid response to disturbances. But as the gain is increased higher and higher, the loop becomes less and less stable and, therefore, there must be a trade-off between the two phenomena. In effect,

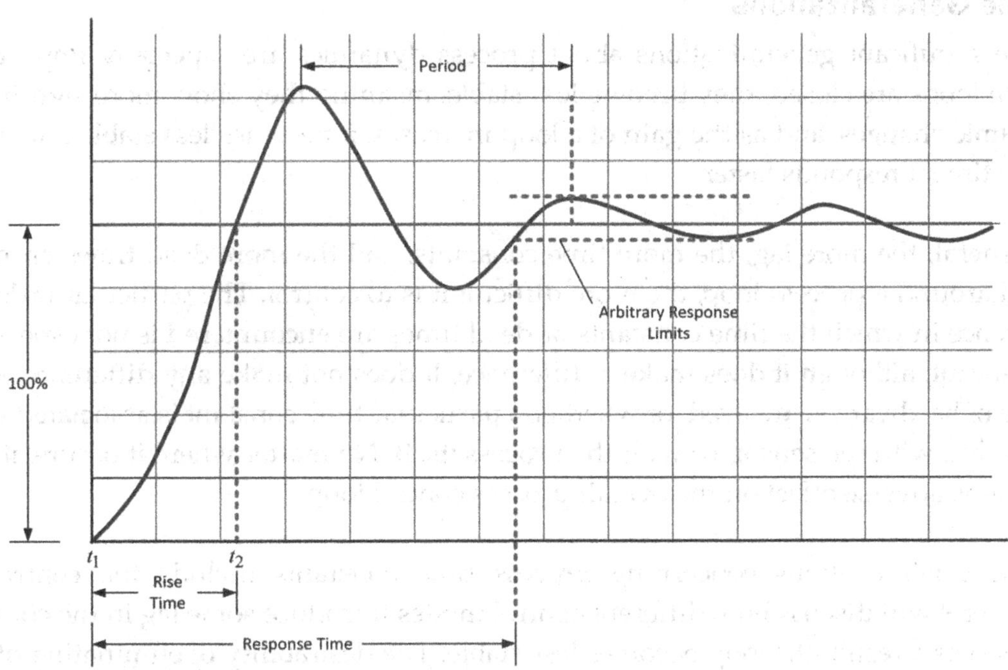

Figure 5-13. Descriptive terms for underdamped response curves.

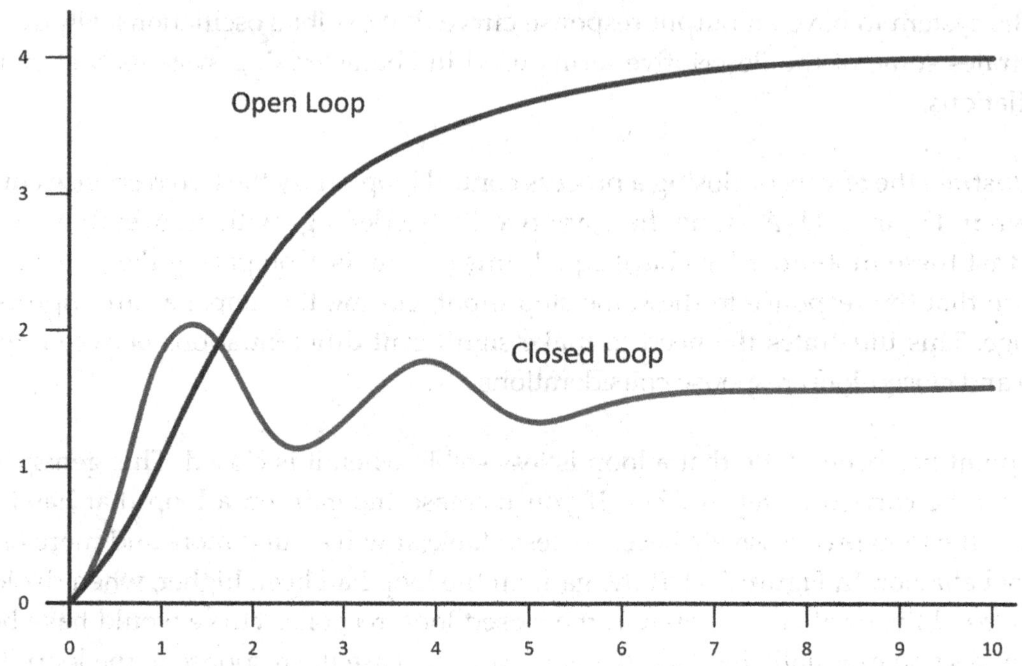

Figure 5-14. Open-loop versus closed-loop response.

the optimum gain is the one that gives the proper trade-off between speed of response and stability for the loop.

Some Generalizations

Some significant generalizations about process dynamics are especially important. When loops are closed, they become less stable, meaning they show more significant dynamic changes, and as the gain of a loop increases, it becomes less stable but at the same time it responds faster.

In general, the more lag, the more time constants, and the more dead times encountered around a process loop, the more difficult it is to control. The particular order or sequence in which the time constants or dead times are encountered is not especially significant, although it does make a difference. It does not make any difference what piece of hardware is involved, or whether a particular time constant is associated with the valve, with the sensor, or with the process itself. No matter where it occurs, it has the same adverse effect on the overall process control loop.

These generalizations concerning process time constants include the controller. Chapter 9 will discuss how different control modes introduce some lag in the controller, and as a result, the loop becomes less stable. The desirability of eliminating offset is often a justification, but we must appreciate the penalties involved in its usage.

There are cases in which the process may be so rapid that it, in effect, must be filtered or slowed down to increase controllability. An important case is liquid flow control loops. In these, the process itself has almost instantaneous response and, as a result, a lot of high-frequency noise is transmitted around the loop. This is undesirable, and to counteract this, filtering is added to the measured value of the process value in the flow controller to provide some damping in the form of filtering in the loop. In effect, lag is introduced via the controller to smooth out some of the noise being transmitted around the loop.

The point is that the more process lag, specifically, the more time constants and dead times around a loop, the worse the situation gets (except as noted in the previous paragraph). In addition, the more this process lag is distributed around the loop, the worse the control case. Put another way, the more the process lag is concentrated in a specific component, the better the control situation. For example, a first-order lag with a time constant of 10 min would not be too difficult to control. However, if this same process lag were broken into two time constants of 5 min each, the control would be somewhat more difficult. If it were broken into 10 time constants of 1 min each, it would be even more difficult.

Little can be done about process dynamics in many situations. The dynamic behavior of the process usually is established by the function it serves and the process characteristics, in other words, the transfer function of the process itself. It is expected that the practitioner will design and install a transmission system that operates sufficiently fast so that it does not contribute significant lag to the overall process control loop. This leaves the final element, often a control valve, and the sensor where dynamic choices are often made—sometimes without thinking. The dynamic behavior of these two components is especially important, and many control loops are made unnecessarily difficult to operate simply because the installed sensor exhibits sluggish behavior. Or, the installed final control element, such as a control valve, is such that its lag is a significant contributing factor to the poor performance of the loop. The goal should be to have a good appreciation of the dynamic behavior of all the hardware installed.

5.9 Sensor Dynamics

As pointed out previously, it is important to gain an understanding of sensor dynamics. Quite often the primary measuring element's speed of response is one of the most important factors affecting the operation of a feedback control loop. The rate at which the controller can detect changes in the controlled variable depends directly on the response of the primary measuring element and is critical to the overall stability and controllability of the process loop.

To increase our understanding of sensor dynamics, refer to Figure 5-15, which shows a thermocouple connected to thermometer. For analysis purposes, suppose that we immerse this thermocouple into an agitated constant temperature bath as shown.

The thermocouples will make the transition from ambient temperature to the temperature of the bath, and the thermometer needle will rise as is illustrated. The temperature response will be a first-order lag response similar to the curve shown, that is, it is exponential and approaches the bath temperature gradually. This type of response is typical for dynamic sensor elements.

The study of the dynamic behavior begins by defining the time constant that characterizes the dynamic behavior of the response curve. It is the time necessary for the response curve to reach 63.2% of its final value. In Figure 5-15, the time constant for the thermocouple is approximately 12 min. We can use the term to make some quantitative comments about sensor dynamics.

The signal shown in Figure 5-15 is typically available for transmission, that is, there will be some lag introduced into the feedback loop by the sensor. Such lag should be minimized wherever possible. Fast sensors make it possible for the controller to function in a timely manner. Sensors with large time constants are slow and degrade the overall operation of the feedback loop. The dynamic characteristics of sensors should be considered in their selection and installation.

Many sensors will exhibit higher-ordered responses, such that the maximum change in process response rate does not occur at the origin like in a first-order system. Higher order sensors are quite common. Figure 5-16 shows some typical effects of thermowells on thermometer bulb response.

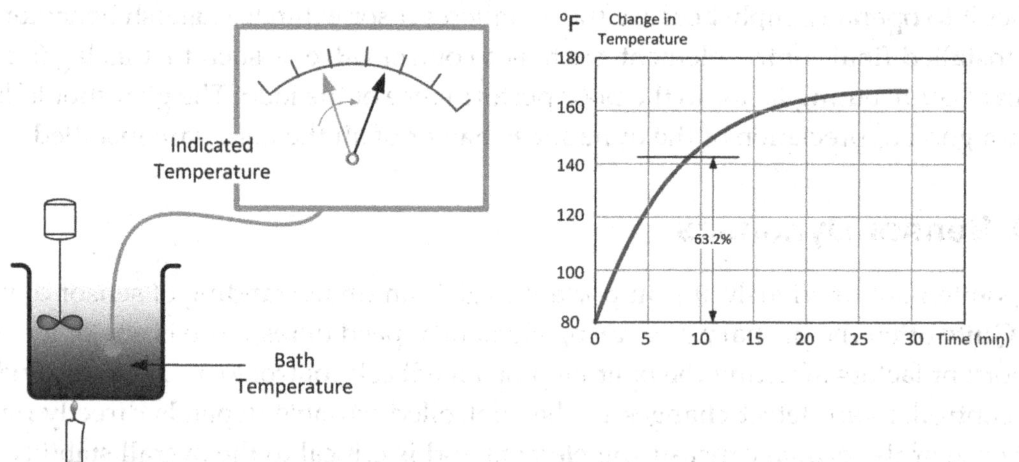

Figure 5-15. Thermometer experiment and response curve.

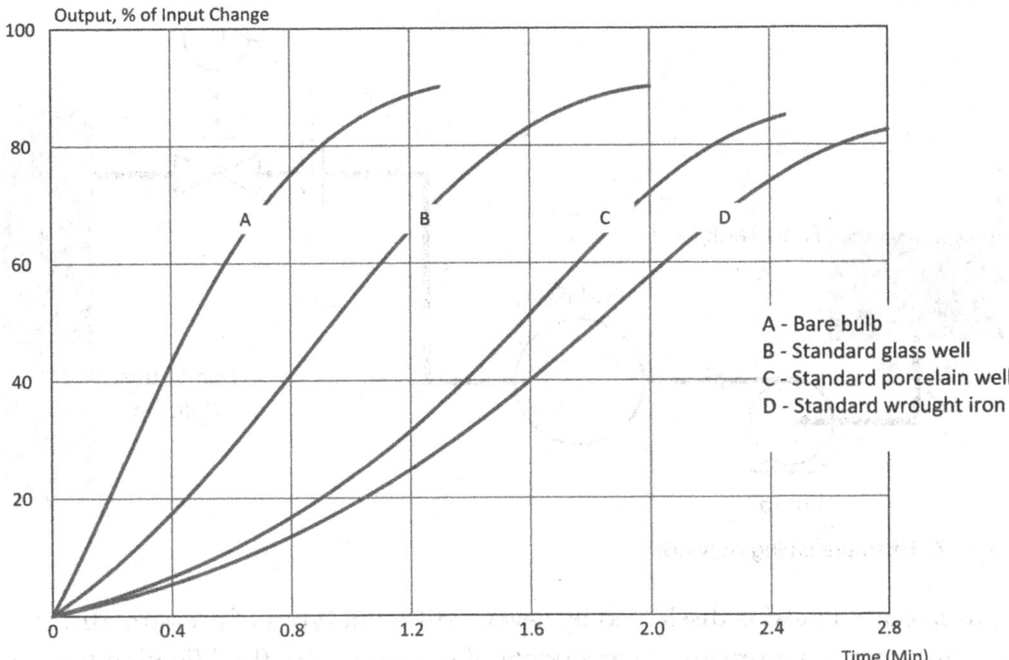

Figure 5-16. Effects of thermowells.

Curve A is a plain bare bulb with first-order lag dynamics. As the bulb is provided with more protection, the response gradually becomes more sluggish and starts to exhibit the S-shaped response of higher-ordered lags. In these cases, of course, much more lag is involved, and the process becomes more difficult to control.

5.10 Actuator Dynamics

All actuators, and specifically valve and valve positioners, being mechanical in nature, are a second important contributor to lag and dead time. Again, an understanding of the dynamic characteristics of a control valve installed as a part of a total piping network is important. The dynamic concepts involved significantly affect the valves' operating characteristics and the total performance of the feedback control system. A piping network is shown in Figure 5-17. In this case, a centrifugal pump applies suction on a feed tank and pumps a liquid stream through a preheater into a distillation column. There is a liquid-flow control loop that serves to control the actual liquid flow rate into the distillation column.

If you inspect the total distribution of head loss throughout this piping network, you will get some insight into the dynamic characteristics of the valve. The centrifugal pump might have a total characteristic head curve like that shown in Figure 5-18.

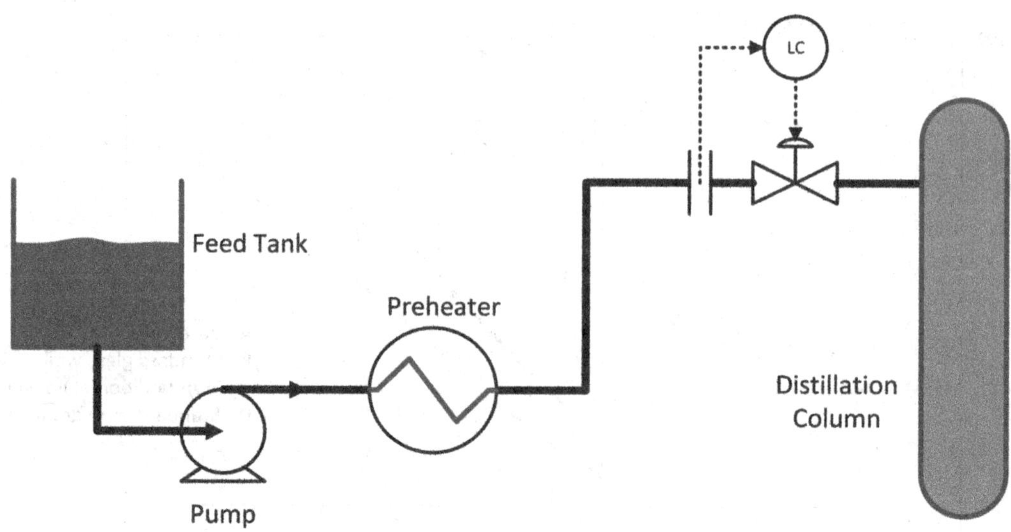

Figure 5-17. Example piping network.

This pump output head is dissipated by several different types of pressure drops. Many of these are static in nature and independent of the amount of fluid flowing through the piping network. One example of such static loss is the potential energy that is required to raise the liquid to the feed point on the column. A second static loss is illustrated by the operating pressure of the column itself, which must be overcome by the inlet liquid feed.

In addition to these static losses, there are many *dynamic losses* involved in the piping network besides the losses associated with the control valve. These additional

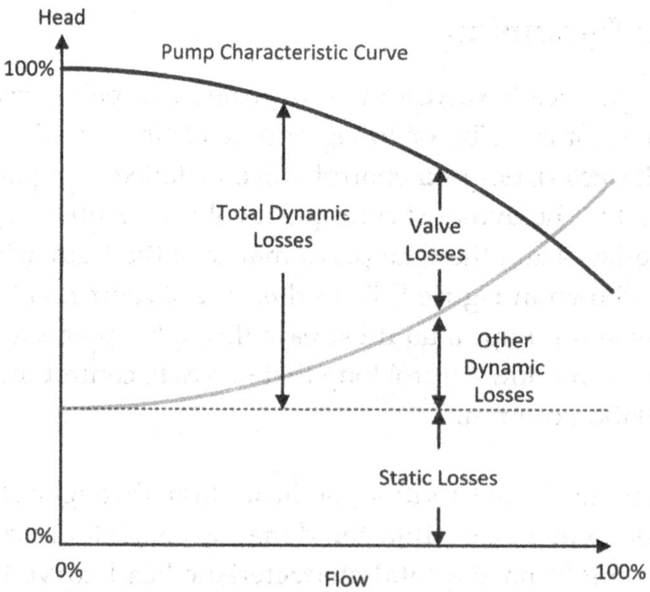

Figure 5-18. Head distribution in the piping network.

dynamic losses tend to increase as the square of the flow rate through the piping network. These losses are the friction losses in the piping, flow through the pre-heater, and flow through the orifice plate. After the static and dynamic losses, all the other head losses of the flowing fluid must be dissipated across the control valve.

In a real sense, the control valve is a bottleneck but one that is designed and installed on purpose. If there is no bottleneck, there is no control. As flow increases and a smaller percentage of the total dynamic drop occurs across the valve, the piping network—not the valve—tends to become the bottleneck.

When most valves are sold, the customer obtains a characteristic curve for the valve, which illustrates the valve flow characteristics.

5.11 Process Gain and Time Constant

The steady-state gain, or simply the process gain (G_p), is one of the most important parameters of a process. It is a measure of the sensitivity of the process output to a change in its input. The gain is defined as the steady-state change in output divided by the change in input that caused it:

$$G_P = \frac{\textit{change in output}}{\textit{change in input}} \tag{5-11}$$

The change in output is measured after the process reaches a new steady state (see Figure 5-19).

The units of process gain are transmitter output (TO) divided by controller output (CO). For a given process, the numerical value of the gain is the same whether it is expressed in mA/mA (electronic controller), psi/psi (pneumatic controller), or percent TO per percent CO (%TO/%CO). The units most commonly used with modern digital controllers are %TO/%CO. Because the controller gain is dimensionless, the process gain to be used in the tuning formulas must also be dimensionless.

The gain defined by the Equation 5-11 includes the gains of the transmitter, the process, and the control actuator. This is because, as illustrated in Figure 5-20, these three blocks are essentially combined into one. It is common practice, however, to express the transmitter signal in the engineering units of the measured variable, in which case it is necessary to convert the value of the gain to dimensionless units. This is illustrated in the next example.

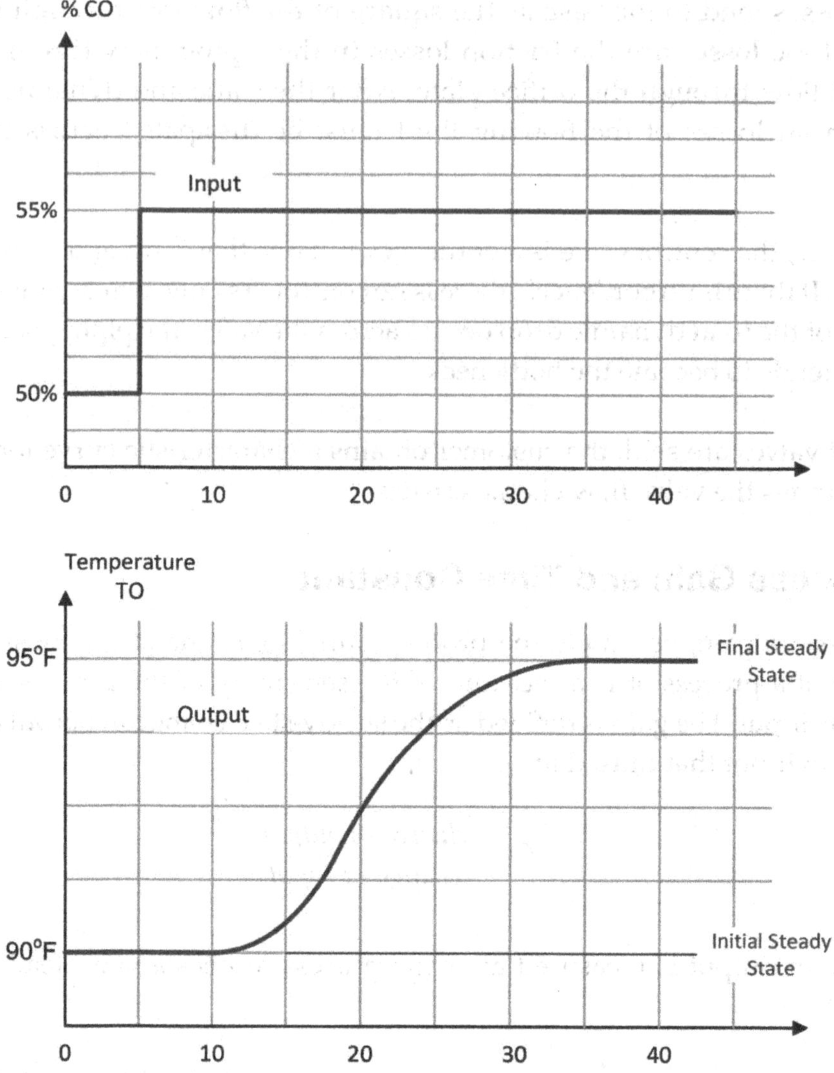

Figure 5-19. Process reaction curve.

Estimation of the Gain from the Step Response

Figure 5-20 shows a typical setup for performing a step test.

The step test in Figure 5-19 shows that a 5% change in CO causes a steady-state change in temperature from 90°C to 95°C. First, the change in temperature must be converted to a percentage of the TO range. Assume the transmitter range for the steam heater is 50°C to 150°C. Thus, the change in the TO signal is as follows:

$$(95-90)°\text{C}\frac{(100-0)\%\ \text{TO}}{(150-50)°\text{C}} = 5\%\ \text{TO}$$

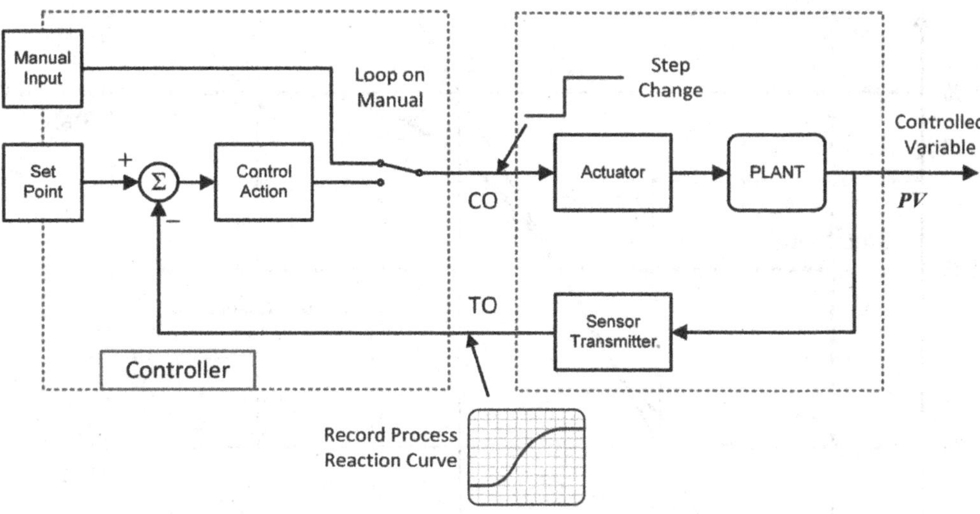

Figure 5-20. Typical setup for determining process reaction curve.

Then, the dimensionless process gain is as follows:

$$G_p = \frac{5\%\ \text{TO}}{5\%\ \text{CO}}$$

By using percent of range as the units of the signals, the value of the gain is equally valid for electronic, pneumatic, and computer-based controllers.

This example illustrates that it is important to keep track of the units of the gain when calculating process gain. This becomes especially important when tuning control systems.

Estimating Time Constant and Dead Time

Just as the gain is a measure of the steady-state sensitivity of the controlled process, the time constant and the dead time are measures of its dynamic response. The time constant is an indication of how long it takes the process to reach a new steady state after the initial change in output is detected. The dead time is a measure of how long it takes for the initial change to be detected after its occurrence. The ratio of the process dead time to its time constant is often used as an indication of the controllability of a feedback control loop.

In a first-order system, the time constant can be easily obtained. However, when dealing with a higher order system and dead time, determining the time constant is more demanding. Often a second-order or higher order system with dead time is modeled after a first-order plus dead time signal as shown in Figure 5-21. This figure shows a typical higher order system and its first-order plus dead time approximation.

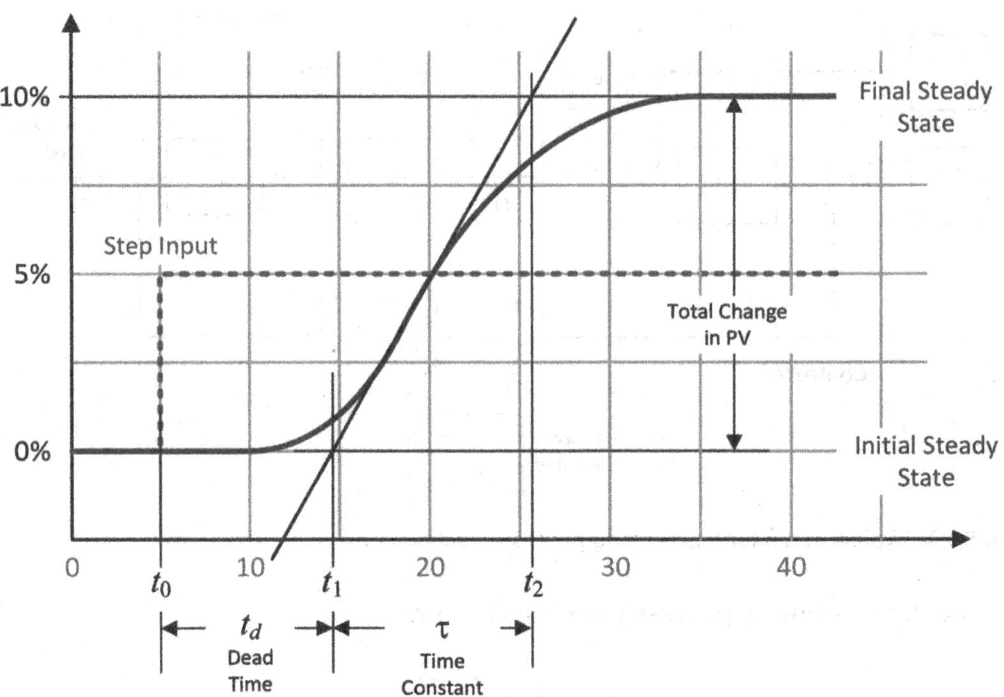

Figure 5-21. Tangent method.

There are several methods for estimating the process time constant and dead time from the step response. The first of these methods was originally proposed by Ziegler and Nichols[1] and is often referred to as the *tangent method*. Other methods are the *tangent-and-point method* and the *two-point method*. Both give more reproducible results than the tangent method. The constructions that are required to estimate the time constant and the dead time are shown in Figure 5-21, which is basically a reproduction of the step response in Figure 5-19, but it shows the constructions needed to analyze it. As mentioned earlier, for a first-order system the rate of response is at a maximum at the very beginning of the curve, and if the system were to continue to change at its maximum rate of change, it would reach its final value in one time constant.

Tangent Method

The tangent method is based on the maximum rate of change of a first-order system and requires you to draw a tangent to the process response curve line at the point of

1. John G. Ziegler (1909–1997) was an American control engineer who made significant contributions to the field of control theory. He is well known for his research paper on the Ziegler-Nichols method, co-authored with Nathaniel B. Nichols. Nathaniel B. Nichols (1914–1997) was an American control engineer and is well known for his book *Theory of Servomechanisms*.

maximum rate of change or its inflection point, as shown in Figure 5-21. The dead time (t_d). is then defined as the distance in time from the start of the input (t_0). to where the tangent line crosses the initial steady-state value (t_1).

The time constant τ is defined as the distance in the time axis between the point where the tangent crosses the initial steady state of the process variable (t_1). and the point where the tangent crosses the final steady-state value (t_2).

Mathematically: $t_d = t_1 - t_0 \quad \tau = t_2 - t_1$

where τ is the process time constant and t_d is the process dead time.

The basic drawback of the tangent method is that drawing the tangent is difficult and not accurately reproducible, which creates significant variance in the estimates of the process time constant and dead time. If the data is available in discrete points, a computer algorithm could be used to identify the point of maximum rate of change or inflection using differentiation. Another problem with the tangent method is that its estimate of the process time constant tends to be rather large compared to the definition of a standard first-order time constant, and thus it results in tighter controller tuning than the tangent-and-point and two-point methods.

Tangent-and-Point Method

The tangent-and-point method differs from the tangent method in the estimate it provides of the time constant, but it estimates the dead time in the same way. In this method, it is necessary to determine the point at which the step response reaches the conventional 63.2% of its total steady-state change. This point is marked as t_3. in Figure 5-22. The time constant is then the time between the point where the tangent line crosses the initial steady state and the point where the response reaches 63.2% of the total change. Thus, the time constant is calculated as:

$$\tau = t_3 - t_1$$

The tangent-and-point method results in a shorter estimate of the time constant and thus is used in more conservative controller tuning than the tangent method. However, notice that both estimates of the dead time and the time constant are dependent on how the tangent line was drawn. This is because the 63.2% point fixes only the sum of the dead time and the time constant, making each estimate dependent on the location of the tangent line, which is the least reproducible step of the procedure. Because of

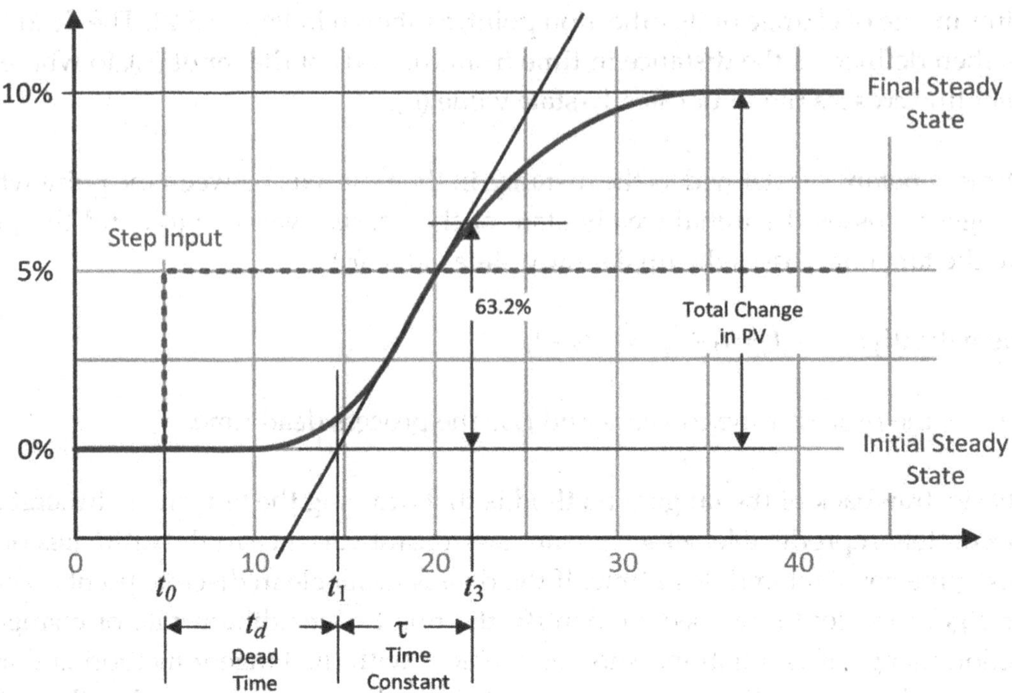

Figure 5-22. Tangent-and-point method.

this, Cecil Smith[2] proposed the two-point method, which does not require the tangent line to be drawn.

Two-Point Method

The two-point method makes use of the conventional first-order time constant of 63.2% point defined in the tangent-and-point method as well as one other point where the step response reaches 20% of its total steady-state change. These points (t_0, t_3, t_4). are marked in Figure 5-23. Actually, any two points in the region of maximum rate of change of the response will do, but the two points Smith preferred resulted in the following simple estimation formulas for the time constant and the dead time:

$$\tau = 1.29 \times (t_3 - t_4) \quad \text{and} \quad t_d = t_3 - \tau$$

Expressing this in terms of the two points, we get:

$$\tau = 1.29 \times (t_{63\,\%} - t_{20\,\%}) \quad \text{and} \quad t_d = t_{63\,\%} - \tau$$

2. Cecil L. Smith, PhD, is president of Cecil L. Smith, Inc., a firm specializing in automation technology that was founded after he left the faculty of Louisiana State University. Smith has been listed among *InTech* magazine's 50 most influential industry innovators.

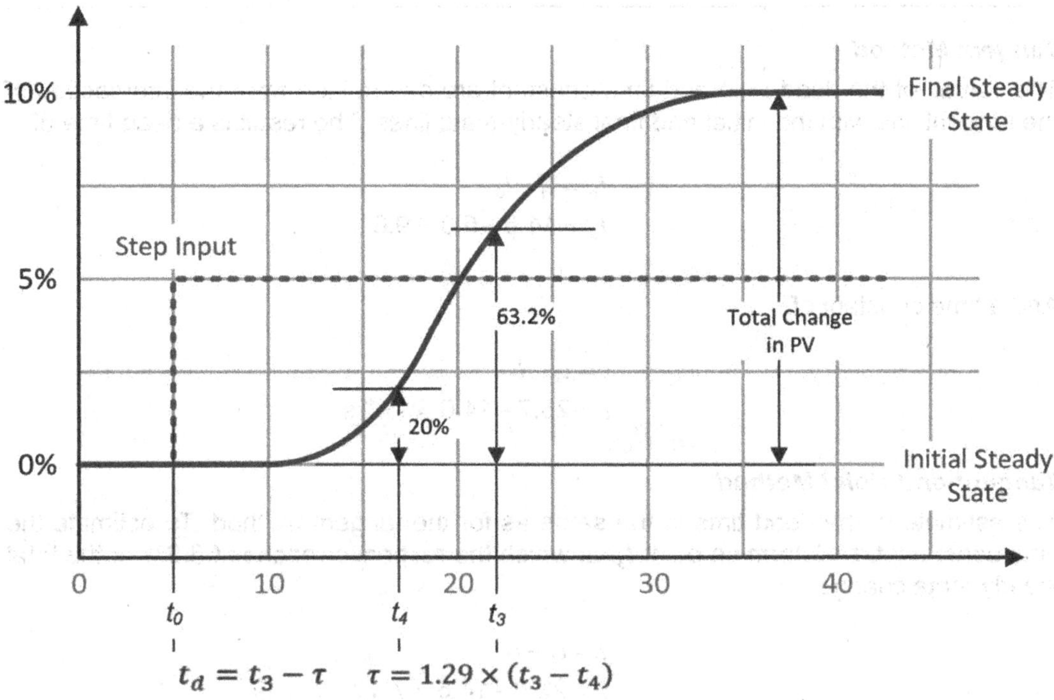

Figure 5-23. Two-point method.

The reason the two points should be in the region of the maximum rate of change is that otherwise small errors in the ordinate would cause large errors in the estimates of t_3. and t_4.

Of the three methods, the two-point method is the easiest to use.

Example

This example illustrates the three methods for determining the dynamic parameters of the process from the step response based on the following values taken from Figures 5-21, 5-22, and 5-23:

$t_0 = 5.0$	$t_1 = 14.6$	$t_2 = 25.7$	$t_3 = 22.0$	$t_4 = 16.9$

Using these values, we will calculate the dead time and time constant using each method for comparison.

Tangent Method

The values of the dead time and time constant are determined from the intersection of the tangent line with the initial and final steady-state lines. The result is a dead time of:

$$t_d = t_1 - t_0$$
$$t_d = 14.6 - 5.0 = 9.6 \text{ s}$$

And a time constant of:

$$\tau = t_2 - t_1$$
$$\tau = 25.7 - 14.6 = 11.1 \text{ s}$$

Tangent-and-Point Method

The estimate of the dead time is the same as for the tangent method. To estimate the time constant, first determine point t_3. at which the response reaches 63.2% of the total steady-state change:

$$\tau = t_3 - t_1$$
$$\tau = 22.0 - 14.6 = 7.4 \text{ s}$$

Two-Point Method

In addition to the 63.2% point, which was determined in the previous method, we also need the 20% point, t_4:

$$\tau = 1.29 \times (t_3 - t_4)$$
$$\tau = 1.29 \times (22.0 - 16.9)$$
$$\tau = 1.29 \times 5.1 = 6.6 \text{ s}$$

and

$$t_d = t_3 - \tau$$
$$t_d = 22.0 - 6.58 = 15.4 \text{ s}$$

The following table compares the three methods for the above example:

Method	Dead Time t_d	Time Constant τ
Tangent	9.6 s	11.1 s
Tangent-and-Point	9.6 s	7.4 s
Two-Point	15.4 s	6.6 s

Compared to the tangent-and-point method, the two-point method results in longer estimates of the dead time and shorter estimates of the time constant, but it is more reproducible because it does not require the tangent line to be drawn. This feature is particularly useful when the response takes the form of sampled values stored in a computer. In this case, the values can be directly determined by a computer, programmed to provide estimates of the time constant and the dead time from the recorded step response data.

To see how accurately the FOPDT approximates the actual response curve, we can superimpose the first-order curves generated from the FOPDT calculations to see how well they fit. This is illustrated in Figure 5-24.

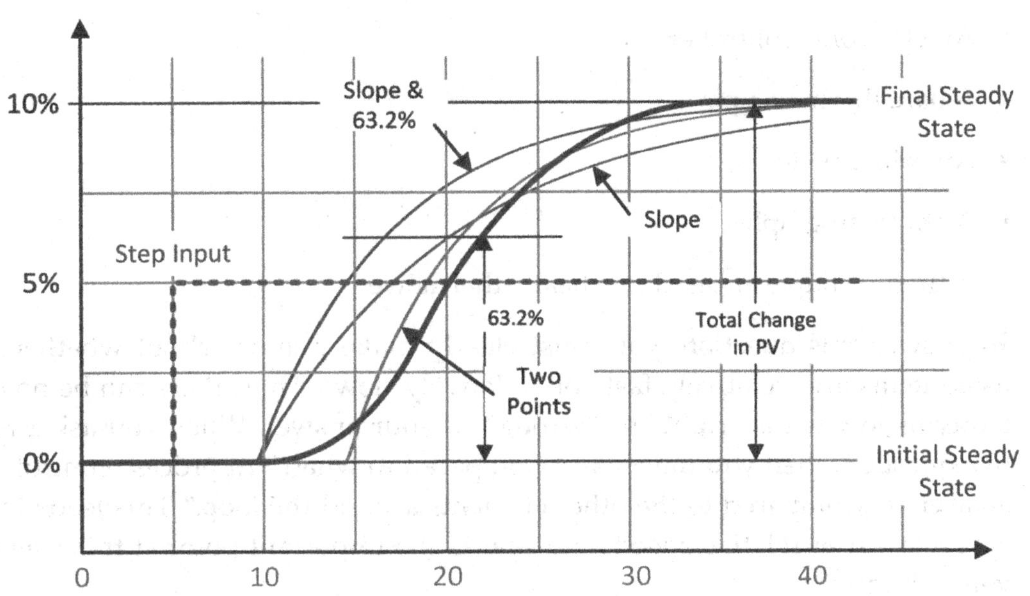

Figure 5-24. Comparison of different FOPD approximations.

5.12 Summary

- First-order dynamic behavior was studied in terms of gain, time constant, and dead time.
- A first-order system with time delay is a common empirical description of many stable elements of process dynamics.
- Time constant is defined in terms of resistance and capacitance, independent of the type of element.

Exercises

5.1 Estimate whether the dynamic characteristics of the following items of hardware would be significant in the total dynamics of a feedback control loop:

- The input elements in a controller
- An electronic transmission system
- A large, pneumatically operated control valve
- A 2000 ft pneumatic transmission system
- A pneumatic controller
- An electronic controller
- A bare thermocouple
- An orifice meter
- A chromatograph
- A thermocouple encased in a heavy thermowell

To answer this question, you must clearly make guesses about whether the listed items are "relatively fast" or "relatively slow." Thus, there can be no certainty as to what is "right" or "wrong" in your answer. When you ask a relative question, then you must ask, "Compared to what?" In process control, the answer is "Compared to the other elements around the loop." This leads to an appreciation of relative speed, and that is the important point at this stage of your education.

5.2 In general terms as well as for typical industrial applications, rank the following processes as (a) responding very rapidly to input changes, (b) responding at a moderate rate, or (c) responding very slowly:

- Liquid flow in a line
- Gaseous flow in a line
- Liquid level in a small tank
- Liquid pressure in an enclosed tank
- Gaseous pressure in a large tank
- Composition in a large distillation column

- Temperature in a liquid-filled tank
- Your coworkers at 5:00 p.m.

The comment at the end of Exercise 5.1 about making guesses applies here as well.

5.3 Refer to the immersion of the bare bulb in the bath in Figure 5-15. How would the process response curve be different if the thermometer bulb were not bare but instead encased in a protective thermowell?

5.4 A process chromatograph is used as a sensor to measure the composition of a liquid stream. It operates on a discrete basis, that is, it takes a sample and analyzes it and establishes a signal for feedback purposes. It then takes a new sample and starts its cycle over again. What effect does this mode of operation have on loop operation as compared to a continuous sensor, such as a thermocouple or an orifice meter?

6 Process Characteristics

The characteristics of a process are an expression of the relationship between the manipulated variable (process input) and the controlled variable (process output).

Understanding the characteristics of a process enables proper analysis, modeling, and tuning of the process.

Learning Objectives

- Understand the general response characteristics of typical processes
- Determine the characteristics of the following types of processes:
  - Integral
  - Self-regulating (first-order lag)
  - Second-order lag (resonance and damping ratio)
- Derive the time-domain and process transfer function equation

6.1 Introduction

One of the most important aspects of controlling a process is to know and understand the process. Adequately understanding the characteristics of a process will provide valuable information in controlling it. We will next investigate different process characteristics and how to identify each.

To design, analyze, or simulate a process, we must understand how the process works and be able to clearly model the process mathematically. By developing mathematical models of processes, we can simulate their responses and clearly determine their characteristic response. Process and system modeling will be covered in detail in Chapter 18. This chapter will outline fundamental characteristics of some of the basic processes.

There are three major classifications of process response:

1. Integrating
2. Self-regulating
3. Runaway

Each of these process types can be defined by its response to a step change in input or the manipulated variable.

An integrating process responds by ramping either up or down at a rate proportional to the magnitude of the final control element's step change.

A self-regulating process responds to a step change in the final control element's status by settling to a new stable value.

A runaway process responds by ramping either up or down at a rate that increases over time, headed toward complete instability without some form of corrective action from the controller. Runaway processes will be discussed in Chapter 18.

The characteristics of a process are an expression of the relationship between the manipulated variable (input) and the controlled variable (output). In other words, the process or plant can be expressed as a transfer function and is referred to as the *process* or *plant transfer function*. System modeling is primarily based on determining mathematical relationships of the process.

Most control components and systems can be classified by means of determining their characteristic response to a step input. The following five classifications are prominent:

1. Integral
2. First-order lag

3. Second-order lag
4. Dead time
5. First-order lag plus dead time

Process characteristics can be expressed in several ways. In this chapter, we will use the time domain as in previous chapters, but we will use the Laplace transform to express the process transfer function for convenience.

The objective of a control system is to maintain a desired value of a controlled variable in a process. To do this, the controller adjusts the manipulated variable.

We will also provide and compare thermal, liquid flow, gas flow, mechanical, and electrical examples of the five types of processes.

A process or component is characterized by the relationship between the input signal and the output signal. This input/output (I/O) relationship determines the requirements of the controller. If the I/O relationship of the process is completely defined, optimum controller parameters can be determined. If the I/O relationship is poorly defined, many adjustments of the controller parameters may be required; therefore, the optimum settings must be determined during start-up and system tuning.

The step response graph is the time graph of the output signal following a step change in the input signal from one value to another.

The time-domain equation defines the size versus time relationship between the input signal and the output signal. Time-domain equations are expressed in terms of the basic elements of a process or plant.

The transfer function of a process is obtained by transforming the time-domain equation into a frequency-domain algebraic equation and then solving for the ratio of output over input. The transfer function defines the gain and time relationship difference versus the input and output signals.

The characteristics of a process depend on its basic elements (resistance, capacitance, inertia, and dead time) as discussed in Chapter 5, not on the type of system (thermal, liquid flow, gas flow, mechanical, or electrical). Two different systems may have the same process characteristics, that is, they may have the same equations. This means we can extend our knowledge of one system to other systems that have the same process characteristics. This becomes useful for process modeling and simulation.

6.2 The Integral or Ramp Process

An *integral* (also called *ramp*) process is characterized as having a single capacitive element configured such that the outflow of material or energy is independent of the amount of material or energy stored in the capacitive element.

The simplest integral process is that of filling a tank with liquid. Consider the example in Figure 6-1.

For a constant inflow rate, q_{in} (m^3/s), the liquid level will rise linearly as shown. The term *ramp process* is derived from the ramp-like shape of the output response graph. The faster the inflow rate, the steeper the slope of the ramp.

During a certain time interval (Δt), the amount of liquid volume in the tank will change by an amount (ΔV) equal to inflow rate multiplied by Δt.

Mathematically: $\Delta V = (q_{in})\Delta t$ in cubic meters (m^3)

The change in the liquid level in the tank (Δh) is equal to the change in volume (ΔV) divided by the cross-sectional area of the tank (A).

Mathematically:

$$\Delta h = \frac{\Delta V}{A}$$

then

$$\Delta h = \frac{(q_{in})\Delta t}{A} \tag{6-1}$$

The rate of change of the level in the tank is equal to the change in level (Δh) divided by the time interval (Δt).

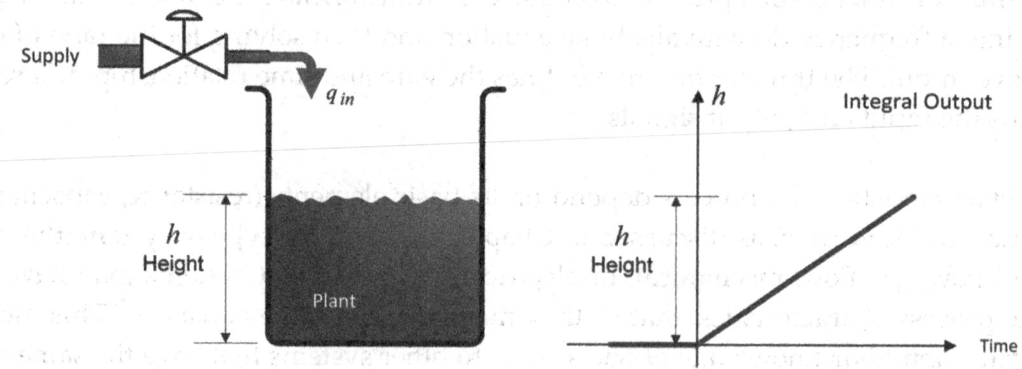

Figure 6-1. Single capacitive integrating process.

$$rate\ of\ change = \frac{\Delta h}{\Delta t}$$

Example 1

If the level changed 0.6 m during a time interval of 100 s, the average rate of change of level would be:

$$\frac{0.6\ m}{100\ s} = 0.006\ \text{m/s}$$

The rate of change of level can then be expressed as follows.

From Equation 6-1 we get:

$$\frac{\Delta h}{\Delta t} = \frac{q_{in}}{A}$$

Different time intervals can be used to determine the rate of change. The average of 0.006 m/s could have resulted from a change of 0.06 m during a time interval of 10 s or a change of 0.006 m during a 1 s interval. When the time interval diminishes and approaches zero, we call this *an instant of time*, and the relationship can be expressed in what is referred to as its *differential form*.

We could add up or accumulate all the instances of time of Equation 6-1 to calculate the change in height (Δh) over a specific period of time from t_1 to t_2 mathematically:

$$\Delta h(t) = \frac{1}{A}\sum_{t_1}^{t_2} q_{in}(t)\Delta t \qquad (6\text{-}2)$$

As the time interval (Δt) diminishes to an instant of time, the *average rate* of change becomes the instantaneous rate of change. In mathematics, the instantaneous rate of change is called the *derivative* and denoted with the letter *d*.

If the time interval (Δt) in Equation 6-2 diminishes to an instant, the average rate of change of level becomes the instantaneous rate of change denoted as dh/dt.

In other words:

As $\Delta t \rightarrow 0$ then $\frac{\Delta h}{\Delta t} \rightarrow \frac{dh}{dt}$

When we sum instances of time, we refer to this as integration from time interval t_1 to t_2. It is usually denoted as:

$$dh(t) = \frac{1}{A}\int_{t_1}^{t_2} q_{in}(t)dt \qquad (6\text{-}3)$$

> This equation indicates the behavior of an integral process. The symbol ∫ denotes integration. *Integration* basically means add up slices of width Δt of a specific variable. See Section 1-8 in Chapter 1.

Liquid Level

However, no consideration has been given in the above description as to how the liquid is removed from the tank. Removing liquid from the tank can be done either by installing a positive displacement pump or by installing a valve at the bottom of the tank. Installing a positive displacement pump will be analyzed next, followed by installing a valve at the bottom of the tank.

A positive displacement pump provides a constant output flow rate of q_{out} (see Figure 6-2).

The quantity of stored material or energy remains constant if, and only if, the inflow rate is equal to the outflow rate. If the inflow rate is greater than the outflow rate, the quantity stored will increase at a rate proportional to the difference between inflow (q_{in}) and outflow (q_{out}). If the inflow rate is less than the outflow rate, the quantity stored will decrease at a rate proportional to the difference. The input flow rate is the input signal to the integral process, but the output flow rate is *not* the output signal. Rather, the output is the amount of energy stored in the tank, in other words, its capacitance. The output of this process is the variable that represents the amount of energy stored. In this example, the liquid level (h) represents the amount of material or energy stored

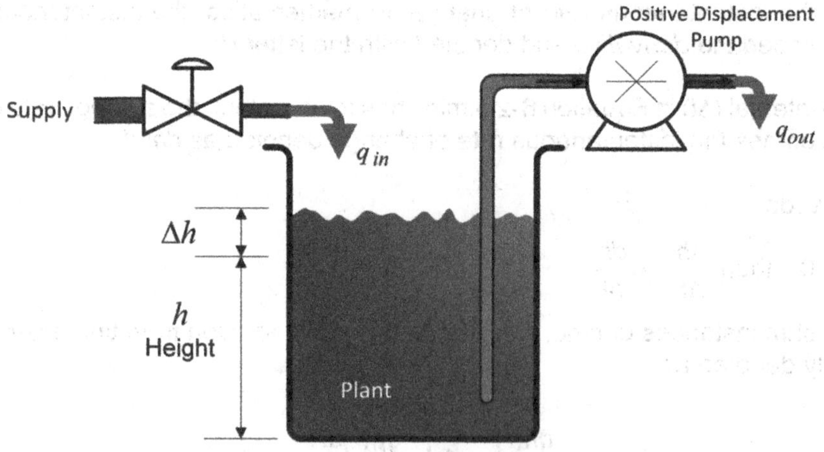

Figure 6-2. Liquid-level integral process with positive displacement pump.

in the tank (its capacitive element). Applying a change in input flow can be analyzed by using a step input. Figure 6-3 shows the response of an integral process subjected to a step input.

Before the step change, the input flow rate is equal to the output flow rate and the level is constant. After the step change, the input flow rate is greater than the output flow rate, and the level increases at a constant rate. The step response of an integral process is measured by the integral action time constant (*Ti*), which is the number of seconds (or minutes) required for the output to reach the same percentage change as the input.

Equation 6-1 can now be restated as:

$$\Delta h = \frac{\Delta V}{A} = \frac{(q_{in} - q_{out})_{avg}}{A} \Delta t \tag{6-4}$$

For convenience, we will define *q*(*t*) as follows:

$$q(t) = q_{in} - q_{out}$$

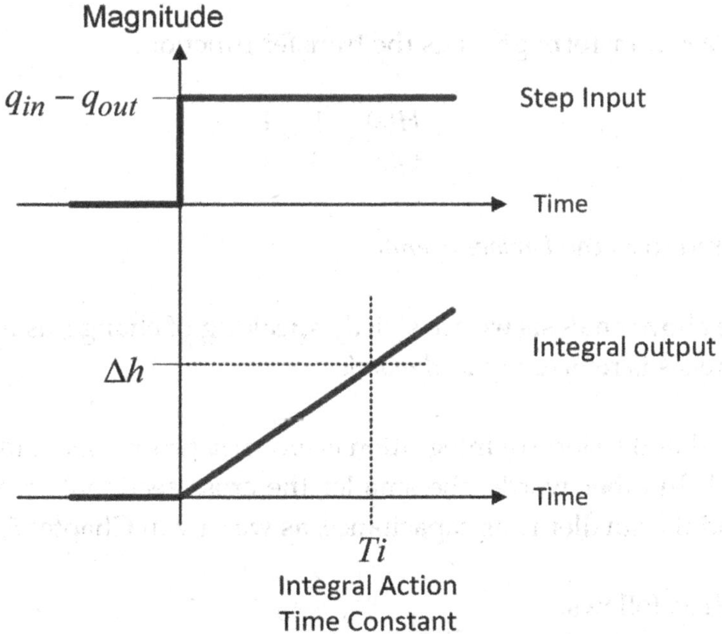

Figure 6-3. Response of an integral process to a step change in the input signal.

In other words, $q(t)$ is the difference between the input flow rate and the output flow rate. If time interval Δt begins at time t_0 and ends at time t_1, then

$$\Delta t = t_1 - t_0$$

and

$$\Delta h = h(t_0) - h(t_1)$$

We can also define

$$q_{avg} = q_{in} - q_{out}$$

The term q_{avg} represents the average accumulation of liquid in the tank over the interval Δt. The accumulation of liquid as an integral from t_0 to t_1 is as follows:

$$\Delta h = \frac{1}{A}(q_{in} - q_{out})(t_1 - t_0)$$

or

$$\Delta h = \frac{1}{A} q_{avg} \Delta t \tag{6-5}$$

again as $\Delta t \rightarrow 0$

$$\Delta t \rightarrow dt$$

and

$$\Delta h \rightarrow dh$$

then

$$dh(t) = \frac{1}{A}\int_{t_0}^{t_1} q(t)dt \tag{6-6}$$

Taking the Laplace transform gives us the transfer function.

We get

$$\frac{H(s)}{Q(s)} = \frac{1}{A} \times \frac{1}{s} \tag{6-7}$$

where s is referred to as the *Laplace operator.*

Note that in the above analysis we are strictly speaking of change; as indicated earlier, this type of analysis is referred to as *dynamic.*

It is noteworthy that the slope of integration is *inversely* proportional to the tank cross-sectional area A. In other words, the smaller the cross-sectional area, the faster the tank will fill and the smaller is its capacitance, as we saw in Chapter 5, Figure 5-8.

We can define Ti as follows:

$$Ti = A\left(\frac{FS_{out}}{FS_{in}}\right) \tag{6-8}$$

where

A = cross-sectional area of the tank in square meters (m²)
FS_{in} = full-scale input range in cubic meters per second (m³/s)
FS_{out} = full-scale output range in meters (m)

If we specify both q_{in} and q_{out} in percent of full scale, then Equation 6-5 becomes:

$$\Delta h = \frac{1}{Ti}(q_{in} - q_{out})(t_1 - t_0) \tag{6-9}$$

The equations for our liquid level integral process are summarized in Table 6-1.

Example

A liquid level integral process has the following parameters:

Tank height: 4 m
Tank diameter: 1.5 m
Initial level: 0.9 m
$q_{out} = 0.006$ m³/s
$q_{in} = 0.008$ m³/s
$FS_{in} = 0.01$ m³/s

Determine the time-domain equation, the integral action time, and the level of the tank after 10 s.

Table 6-1. Liquid level integral process.

Time-domain equation	$\Delta h = \frac{1}{Ti}(q_{in} - q_{out})(t_1 - t_2)$
Transfer function	$\frac{H(s)}{Q(s)} = \frac{1}{Tis}$
where	$\Delta h(t)$ = change in height during time t in meters (m) t = time in seconds (s) q_{in} = normalized input flow rate in percent of FS_{in} q_{out} = normalized output flow rate in percent of FS_{out}
and	$Ti = A\left(\frac{FS_{out}}{FS_{in}}\right)$
where	A = cross-sectional area of the tank in square meters (m²) FS_{in} = full-scale input range in cubic meters per second (m³/s) FS_{out} = full-scale output range in meters (m)

Solution

Area

$$A = \frac{\pi d^2}{4} = \frac{\pi(1.5)^2}{4}$$
$$= 1.767\ \text{m}^2$$

The tank has a height of 4 m. This is also FS_{out}.

Integral time

$$Ti = A\frac{FS_{out}}{FS_{in}} = 1.767\frac{4}{0.01}$$
$$= \mathbf{706.8}\ \text{s}$$

Or

$$Ti = \frac{706.8\ \text{s}}{60\ \text{s/min}}$$
$$= \mathbf{11.78}\ \text{min}$$

$$\Delta h = \frac{1}{A}(q_{in} - q_{out})(t_1 - t_2)$$
$$= \frac{1}{1.767}(0.008 - 0.006)(10 - 0)$$
$$= 0.01132\ \text{m}$$

This value represents the increase in level after 10 s.

The final level in the tank will then be the initial level plus the change in level:

$$0.9 + 0.01132 = \mathbf{0.91132}\ \text{m}$$

We started the chapter by stating that the characteristics of a process depend on its basic elements (resistance, capacitance, inertance, and dead time). They do not depend on the type of system (liquid, thermal, electrical, or mechanical), and two different systems may have the same process characteristics. This is illustrated by analyzing a mechanical sheet loop system.

Sheet Loop

A second integral process is illustrated in Figure 6-4. This system consists of a sheet loop (integral) process. The input signal is the input sheet velocity, and the output signal is the distance from the belt to the bottom of the sheet loop.

In both integral processes, the change in the output signal during the time interval from t_0 to t_1 is equal to the integral of the difference between the input flow rate and the output flow rate divided by the integral action time constant (*Ti*).

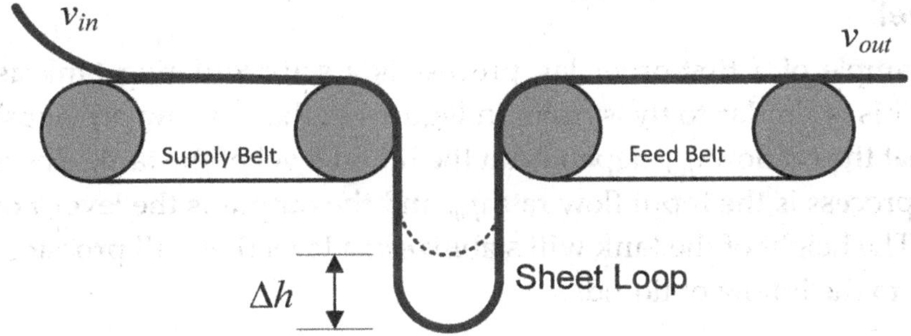

Figure 6-4. Sheet loop integral process.

Table 6-2. Sheet integral process.

Time-domain equation	$\Delta h = \frac{1}{Ti}(v_{in} - v_{out})(t_1 - t_2)$
Transfer function	$\frac{H(s)}{V(s)} = \frac{1}{Ti} \times s$
where	$\Delta h(t)$ = change in height during time t in meters (m) v_{in} = input sheet velocity in percent of FS_{in} v_{out} = output sheet velocity in percent of FS_{out} t = time in seconds (s) Ti = integral action time constant in seconds (s)
and	$Ti = 2\left(\frac{FS_{out}}{FS_{in}}\right)$
	FS_{in} = full-scale input range in meters (m) FS_{out} = full-scale output range in meters (m)

The equation for the sheet loop integral process is obtained from Equations 6-1 and 6-2 by replacing q_{in} and q_{out} with v_{in} and v_{out}.

The equations for our sheet integral process are summarized in Table 6-2 and again specified in percentage change.

6.3 First-Order Lag Process

We studied the dynamic response of the first-order lag process in Chapter 5. In this section, we will look at some typical processes that exhibit first-order lag response.

A first-order lag process is characterized as having a single capacitive element configured such that the outflow of material or energy is proportional to the amount of material or energy stored in the capacitive element. For each input flow rate, there is a corresponding amount of stored material or energy that will produce an output flow rate equal to the input. First-order lag processes are also referred to as *self-regulating processes*.

Liquid Level

An ideal example of a first-order lag process is a self-regulating tank as shown in Figure 6-5. This is similar to the system in Figure 6-2 that we saw previously, with the exception that the outflow q_{out} depends on the liquid level of the tank. The input of the liquid level process is the input flow rate q_{in}, and the output is the level h of the liquid in the tank. The height of the tank will stabilize at a level that will produce an outflow that is equal to the inflow of liquid.

This process produces an output flow rate q_{out} that depends on the liquid level in the tank. The level in the tank will depend on the input flow rate q_{in}. This process is a self-regulating process, meaning that the outflow will eventually self-regulate to be equal to the input flow, and this equilibrium will depend on the liquid height in the tank.

All first-order lag processes have the same characteristics and have a first-order lag response to a step input, as shown in Figure 6-6.

Before the step change, the input flow rate is equal to the output flow rate and the level is maintained at a constant level. The step change consists of increasing the input flow rate. If q_m represents the percentage increase in the input flow rate, then the output

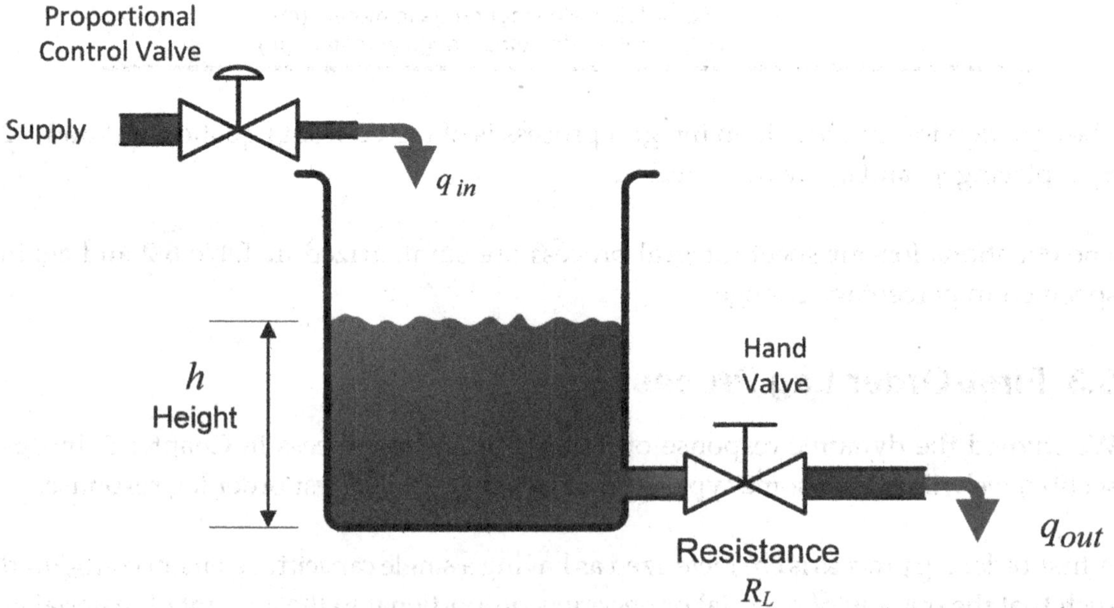

Figure 6-5. Liquid level first-order lag process.

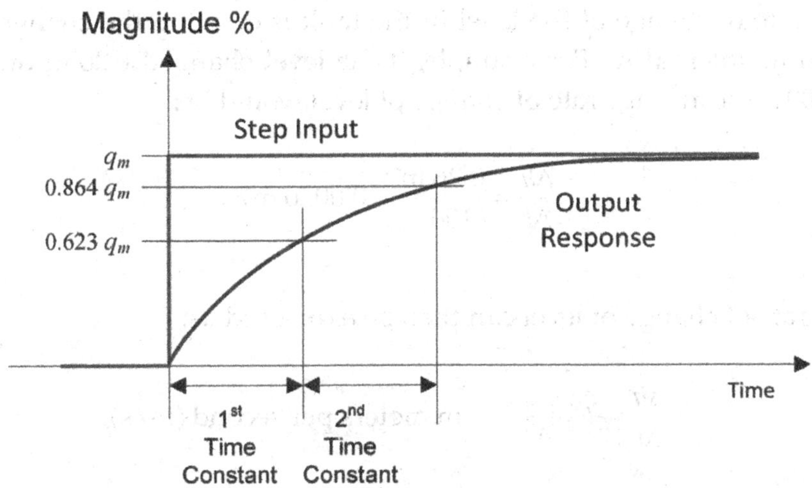

Figure 6-6. Step response of a first-order lag process.

flow rate is proportional to the level, which does not change immediately. The input is greater than the output, so the level will increase at a rate proportional to the difference. As the level increases, the difference between the input and the output decreases. This in turn reduces the rate at which the level increases. The result is the output response curve of Figure 6-6.

The response is a first-order lag and is measured in terms of the system's time constant (τ). The system's time constant is the duration of time in seconds (or minutes) required for the output to reach 63.2% of its final output after the system no longer changes. This value is called the system's *steady-state* value.

If the inflow rate (q_{in}) is greater than the outflow rate (q_{out}), the liquid level will rise. If the inflow rate is less than the outflow rate, the level will fall. During a certain time interval (Δt), the amount of liquid in the tank will change by an amount (ΔV) equal to the average difference between the inflow rate and the outflow rate multiplied by Δt. Mathematically:

$$\Delta V = avg(q_{in} - q_{out})\Delta t \quad \text{in cubic meters (m}^3\text{)}$$

The change in the liquid level in the tank (Δh) is equal to the change in volume (ΔV) divided by the cross-sectional area of the tank (A). Mathematically:

$$\Delta h = \frac{\Delta V}{A} = \frac{avg(q_{in} - q_{out})\Delta t}{A} \quad \text{in meters (m)}$$

The average rate of change of the level in the tank is equal to the change in level Δh divided by time interval Δt. For example, if the level changed 0.26 m during a time interval of 100 s, the average rate of change of level would be:

$$\frac{\Delta h}{\Delta t} = \frac{0.26 \text{ m}}{100 \text{ s}} = 0.0026 \text{ m/s}$$

The average rate of change of level can then be expressed as:

$$\frac{\Delta h}{\Delta t} = \frac{q_{in} - q_{out}}{A} \quad \text{in meters per second (m/s)}$$

Different time intervals may be used to determine the average rate of change. The average of 0.0026 m^3/s could have resulted from a change of 0.026 m during a time interval of 10 s or a change of 0.0026 m during a 1 s interval. Again, when the time interval diminishes and approaches zero, the relationship can be expressed as a differential equation.

As time interval Δt diminishes to an instant of time, the average rate of change becomes the instantaneous rate of change. In mathematics, the instantaneous rate of change is called the *derivative.*

If the time interval (Δt) in the preceding equation diminishes to an instant of time, the average rate of change of level becomes the instantaneous rate of change and is denoted as dh/dt. This follows the same reasoning as that for the integral flow process.

In other words:

As $\Delta t \rightarrow 0$

$$\text{then} \quad \frac{\Delta h}{\Delta t} \rightarrow \frac{dh}{dt}$$

$$\text{and} \quad \frac{dh}{dt} = \frac{q_{in} - q_{out}}{A} \tag{6-10}$$

If the flow out of the tank is laminar,[1] the output flow rate (q_{out}) can be calculated from Equation 5-5 (*resistance = potential/flow*). The potential is the pressure exerted by the

1. Laminar flow was discussed in Chapter 1.

liquid in the tank at the point where the valve is connected. This pressure, when expressed as a height of a liquid column, is called *head*. Head pressure can be calculated as follows:

$$P_H = \rho \times g \times h \quad \text{in pascals (Pa)}$$

The valve provides resistance to the flow. The flow rate can be calculated by the following equation:

$$q_{out} = \frac{P_H}{R_L} = \frac{\rho \times g \times h}{R_L} \tag{6-11}$$

where

q_{out} = liquid flow rate	in cubic meters per second (m^3/s)
ρ = liquid density	in kilograms per cubic meter (kg/m^3)
g = gravitational acceleration	in meters per square second (m/s^2)
h = liquid level	in meters (m)
R_L = laminar flow resistance	in pascals second per cubic meter ($Pa \cdot s/m^3$)

Substituting Equation 6-10 in Equation 6-11 gives us:

$$\frac{dh}{dt} = \frac{q_{in} - \dfrac{\rho g h}{R_L}}{A}$$

$$A\frac{dh}{dt} = q_{in} - \frac{\rho g h}{R_L}$$

Separating h we get

$$\frac{R_L}{\rho g} A \frac{dh}{dt} = \frac{R_L}{\rho g} q_{in} - h$$

or

$$R_L\left(\frac{A}{\rho g}\right)\frac{dh}{dt} = \left(\frac{R_L}{\rho g}\right) q_{in} - h$$

The term $A/\rho g$ is the capacitance (C_L) of the liquid tank. Multiplying the tank's capacitance by its resistance (R_L) will give us the time constant (τ) of the self-regulating liquid tank.

$$\tau_L = R_L C_L = R_L \frac{A}{\rho g}$$

The term $R_L/\rho g$ is called the *steady-state gain* (G) of the system.

Substituting these terms into the preceding equation gives us the final form of the differential equation for the liquid tank.

$$\tau\frac{dh}{dt} + h(t) = G \times q_{in} \tag{6-12}$$

where

$\tau = \dfrac{R_L A}{\rho g}$ in seconds (s)

$G = \dfrac{R_L}{\rho g}$ in seconds per square meter (s/m^2)

Taking the Laplace transform, we can derive the transfer function as follows.

$$\begin{aligned}\tau\frac{dh}{dt} + h(t) &= G \times q_{in}\\ \tau \times sH(s) + H(s) &= G \times Q_{in}(s)\\ H(s)(\tau s + 1) &= G \times Q_{in}(s)\\ H(s) &= \frac{G \times Q_{in}(s)}{\tau s + 1}\end{aligned}$$

$$\frac{H(s)}{Q_{in}(s)} = \frac{G}{\tau s + 1} \tag{6-13}$$

Example 5 in Chapter 4 is a detailed example of a first-order, closed-loop, level control system.

The equations are summarized in Table 6-3.

Thermal

A second representative first-order lag process is shown in Figure 6-7.

In this process, a jacketed kettle is used to heat a liquid. The liquid temperature is maintained uniformly throughout the kettle using a mixer. The input to the process is the jacket temperature ($\Delta\theta_j$) provided by the steam entering the jacket, and the output of the system is the liquid temperature (θ_L) in the kettle. As steam passes through the

Table 6-3. Liquid flow first-order lag process.

Time-domain equation	$\tau\frac{dh}{dt} + h(t) = G \times q_{in}$
Transfer function	$\frac{H(s)}{Q_{in}(s)} = \frac{G}{\tau s + 1}$
where	G = steady-state gain of the process $h(t)$ = output height at time t in meters (m) q_{in} = input flow rate in cubic meters per second (m^3/s) τ = time constant in seconds (s)
and	$\tau = R_L C_L$
	$G = \left(\frac{R_L}{\rho g}\right)$
	R_L = liquid resistance in pascal second per cubic meters ($Pa \cdot s/m^3$)
	C_L = liquid capacitance in cubic meters per pascal (m^3/Pa)

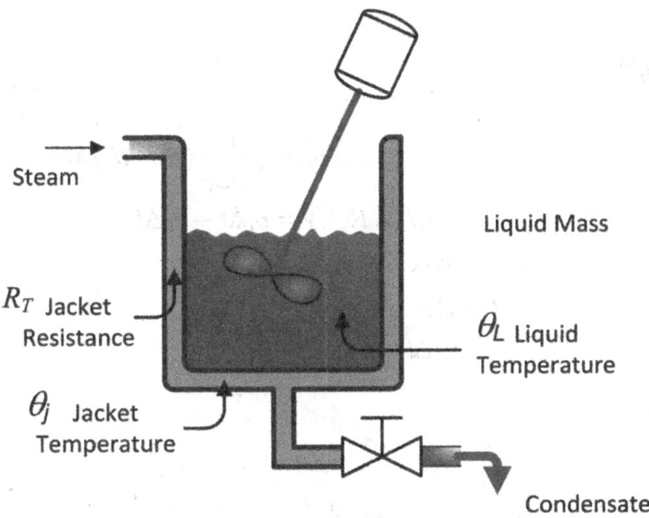

Figure 6-7. Thermal first-order lag process.

jacket, an amount of heat (Δq) is transferred to the liquid. The liquid temperature will then increase by an amount ($\Delta\theta_j$). This increase in temperature represents the amount of heat transferred from the jacket to the liquid divided by the thermal capacitance (C_T) of the liquid in the tank.

$$\Delta\theta_L = \frac{\Delta q}{C_T}$$

The amount of heat (Δq_L) transferred to the liquid depends on the thermal resistance (R_T) of the wall between the steam and the liquid, on the temperature difference ($\theta_j - \theta_L$) between the jacket and the liquid, and the time interval (Δt). Mathematically:

$$\Delta q = \frac{(\theta_j - \theta_L)}{R_T}\Delta t$$

Combining this with the previous equation, we get:

$$\begin{aligned} \Delta\theta_L &= \frac{\Delta q}{C_T} \\ &= \frac{\frac{(\theta_j - \theta_L)}{R_T}\Delta t}{C_T} \\ &= \frac{1}{R_T C_T}(\theta_j - \theta_L)\Delta t \end{aligned}$$

Solving for θ_j we get:

$$\begin{aligned} \Delta\theta_L &= \frac{1}{R_T C_T}(\theta_j - \theta_L)\Delta t \\ \Delta\theta_L \times R_T C_T &= \theta_j \Delta t - \theta_L \Delta t \\ \Delta\theta_L \times R_T C_T + \theta_L \Delta t &= \theta_j \Delta t \\ \frac{\Delta\theta_L \times R_T C_T + \theta_L \Delta t}{\Delta t} &= \theta_j \end{aligned}$$

then

$$R_T C_T \frac{\Delta\theta_L}{\Delta t} + \theta_L = \theta_j$$

and as $\Delta t \to 0$, we get the differential equation:

$$R_T C_T \frac{d\theta_L}{dt} + \theta_L = \theta_j$$

Again, the time constant τ is given by $\tau = R_T C_T$ giving:

Table 6-4. Thermal first-order lag process.

Time-domain equation	$\tau \frac{d\theta_L}{dt} + \theta_L = \theta_j$
Transfer function	$\frac{\theta_L(s)}{\theta_j(s)} = \frac{1}{\tau s + 1}$ Note that the steady-state gain equals 1.
where	θ_L = output temperature (liquid) in degrees Celsius (°C) θ_j = input temperature (jacket) in degrees Celsius (°C) τ = time constant in seconds (s)
and	$\tau = R_T C_T$ R_T = thermal resistance in kelvins per watt (K/W) C_L = thermal capacitance in joules per kelvin (J/K)

$$\tau \frac{d\theta_L}{dt} + \theta_L = \theta_j$$

and a transfer function of

$$\frac{\theta_L(s)}{\theta_j(s)} = \frac{1}{\tau s + 1}$$

This equation is again a first-order lag process with a time constant τ equal to $R_T C_T$, and a steady-state gain equal to 1.

The equations are summarized in Table 6-4 and specified in percentage change.

Gas Pressure

A gas pressure first-order lag process is shown in Figure 6-8.

In this process, the input is the mass flow rate of gas entering the tank (w_{in}) and the output is the pressure (p_{out}) of the gas in the tank. The gas process is also described by a first-order differential equation. Let R_R be the gas flow resistance of the outlet and C_R be the capacitance of the tank. Then:

$$R_R C_R \frac{dP}{dt} + P = R_R w_{in}$$

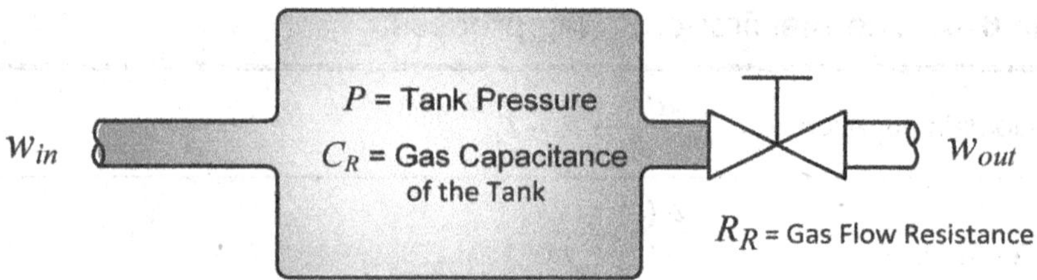

Figure 6-8. Gas pressure first-order lag process.

Table 6-5. Gas pressure first-order lag process.

Time-domain equation	$\tau\frac{dP}{dt}+P=R_R w_{in}$
Transfer function	$\frac{P_{out}(s)}{W_{in}(s)}=\frac{1}{\tau s+1}$
where	P_{out} = output temperature (liquid) in pascals (Pa) W_j = input temperature (jacket) in Pa τ = time constant in seconds (s)
and	$\tau = R_T C_T$; note that the steady-state gain equals 1. R_R = inflow resistance in pascal second kilograms (Pa·s·kg) C_R = tank capacitance in kilograms per pascal (kg/Pa)

This is the equation of a first-order lag process with a time constant equal to $\tau = R_R C_R$ and a steady-state gain equal to 1.

The equations are summarized in Table 6-5.

Blending

A blending, first-order lag process is illustrated in Figure 6-9.

In this system, a constant flow rate of q in cubic meters per second passes through the tank. Note that in this process, the input flow rate and output flow rate are the same. This can be achieved either by restricting the output flow as in the first-order lag liquid flow example or by maintaining the input flow and output flow using pumps. The input of the process is the concentration (c_{in}) of the incoming fluid. The output of the process is the concentration (c_{out}) exiting the blending tank. The purpose of the system is to ensure that the outgoing fluid of the tank has a uniform concentration (c_{out}).

In analyzing this process, a material balance is used to determine the time-domain equation. In a time interval (Δt), an amount of liquid equal to $q \times \Delta t$ is added at the

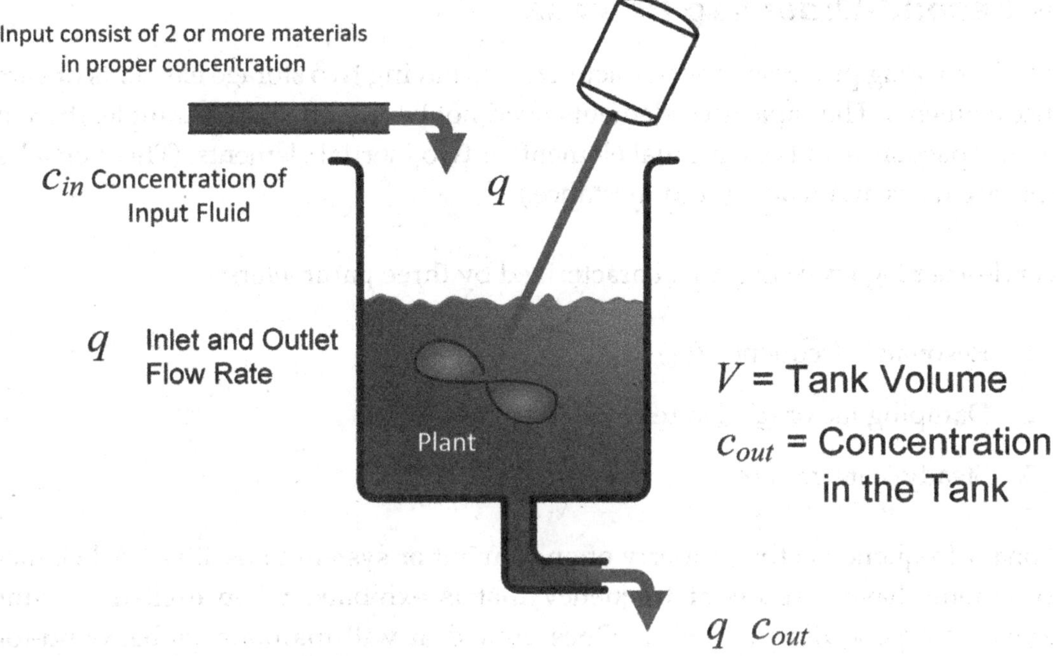

Figure 6-9. Blending, first-order lag process.

inlet, and an equal amount is removed at the outlet. The amount of component A added to the tank is equal to $q \times \Delta t \times c_{in}$. The amount of component A removed at the same time is $q \times \Delta t \times c_{out}$. The difference is the increase of component A in the tank, which is equal to $q \times V \times \Delta c_{out}$ where V is the volume in the tank.

$$\textit{amount of buildup} = \textit{amount input} - \textit{amount removed}$$

$$V \times \Delta c_{out} = q \times \Delta t \times c_{in} - q \times \Delta t \times c_{out}$$

Rearranging, we get

$$\frac{V}{q} \times \frac{\Delta c_{out}}{\Delta t} + c_{out} = c_{in}$$

and as $\Delta t \to 0$, we get

$$\frac{V}{q}\frac{dc_{out}}{dt} + c_{out} = c_{in}$$

This is again the equation of a first-order lag process with a time constant ($\tau = V/q$) and a steady-state gain of 1.

6.4 Second-Order Lag Process

Second-order lag processes are characterized as having two storage elements or capacitance elements. The capacitive elements need not be the same; for example, there can be one capacitance and one inertial element, or two inertial elements. (The inertial elements are mass, inductance, and inertance.)

Second-order lag processes are characterized by three parameters:

1. Resonance frequency (ω_o)
2. Damping factor (ζ, also referred to as ratio)
3. Steady-state gain (K)

Resonant frequency is the property of an element or system to oscillate. All elements and systems have a resonant frequency that is exhibited when excited. A simple example is a pendulum or swing. Once excited, it will maintain its back-and-forth movement at its resonant frequency. As it swings back and forth, its amplitude will diminish slowly, depending on its damping factor.

All second-order lag processes have similar characteristics; however, depending on the value of their frequency, damping factor and steady-state gain will react to a step input differently. Specifically, the damping factor determines the response curve as shown in Figure 6-10. A second-order lag process's dynamic response to a unit step input depends on whether it is overdamped, critically damped, or underdamped.

$\zeta > 1$ The system is overdamped.

$\zeta = 1$ The system is critically damped.

$0 \leq \zeta < 1$ The system is underdamped.

Second-order systems, when underdamped, also exhibit a ringing characteristic as shown in Figure 6-11.

Also note that-second-order systems, unlike first-order systems, have an apparent dead time.

Second-order systems can be described by the following differential equation:

$$\frac{d^2y(t)}{dt^2} + 2\zeta\omega_o \frac{dy}{dt} + \omega_o^2 \tag{6-14}$$

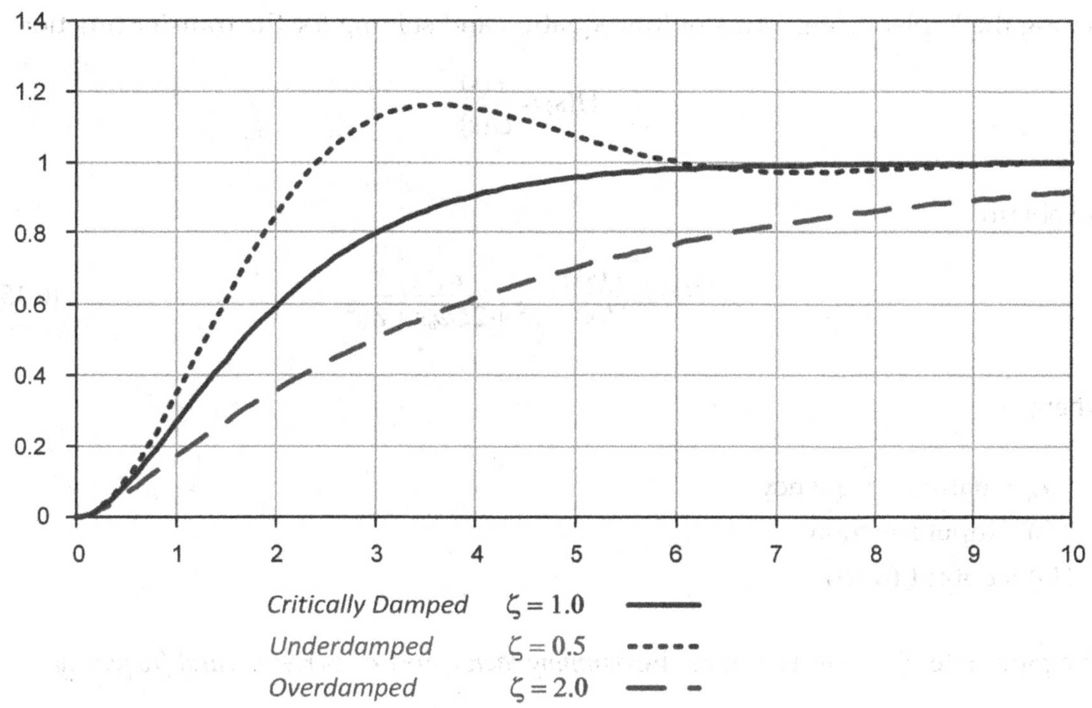

Figure 6-10. Effect of damping ratio (ζ) of a second-order system.

where

ω_o = natural frequency
$y(t)$ = input function
$u(t)$ = output function

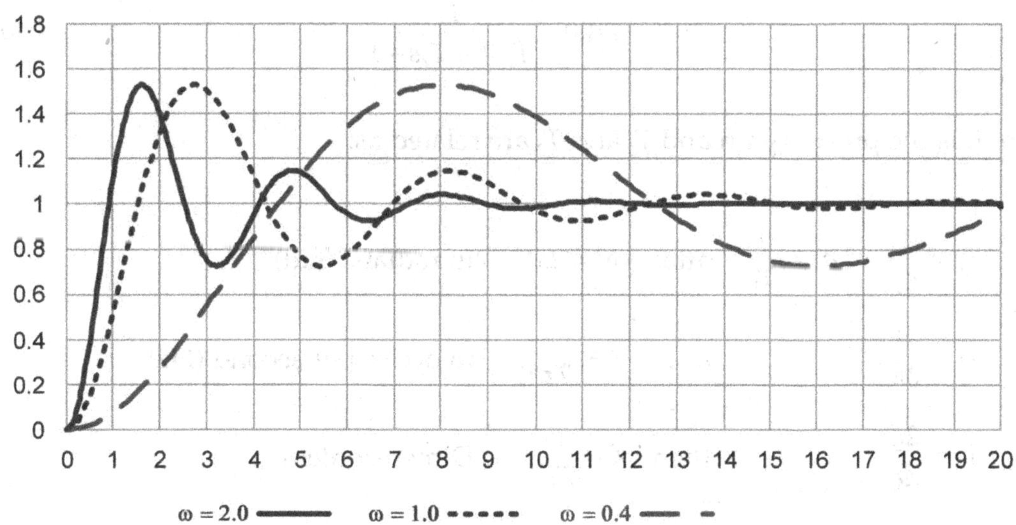

Figure 6-11. Effect of frequency (ω) of a second-order system.

Taking the Laplace transforms of this equation and solving for the transfer function:

$$H(s) = \frac{Y(s)}{U(s)}$$

we obtain

$$H(s) = \frac{Y(s)}{U(s)} = \frac{\omega_o^2}{s^2 + 2\zeta\omega_o s + \omega_o^2} \tag{6-15}$$

where

ω_o = natural frequency
$Y(s)$ = input function
$U(s)$ = output function

The parameter ζ is referred to as the *damping factor,* and ω_o is the *natural frequency.*

In process modeling, the equation is better written and understood using the following:

$$H(s) = \frac{Y(s)}{U(s)} = \frac{1}{\dfrac{s^2}{\omega_o^2} + \dfrac{2\zeta s}{\omega_o} + 1}$$

Then the transfer function of a process with gain is:

$$H(s) = \frac{K}{T_1^2 s^2 + T_2 s + 1} \tag{6-16}$$

where K is the process gain and T_1 and T_2 are related as:

$T_1 = \dfrac{1}{\omega_o}$ $\quad \omega_o = \dfrac{1}{T_1}$ and $\omega_o = 2\pi f$ in radians (rad)

then $T_1 = \dfrac{1}{2\pi f}$ and $f = \dfrac{1}{2\pi T_1}$ in cycles per second (Hz)

and $T_2 = \dfrac{2\zeta}{\omega_o}$ then $\zeta = \dfrac{T_2}{2T_1}$ Dimensionless

The second-order process time constant is the speed at which the output response reaches a new steady-state condition. An overdamped second-order system is typically the combination of two first-order systems.

A typical second-order process is two interacting tanks as shown in the next example and Figure 6-12.

Example: Two-Tank Level Control

In the chemical processing industry, often a two-level liquid level control system is used. The setup is shown in Figure 6-12.

The setup in Figure 6-12 consists of two tanks. Tank 1 receives liquid via a control valve and passes the liquid to Tank 2 through a connecting pipe at the bottom of the tank, and the liquid is drawn out of Tank 2 through a similar pipe. A level transmitter is located in Tank 2 that provides the required feedback to control the level in this tank. This setup allows drastic changes in Tank 1 while minimizing disturbances to Tank 2.

Variables:

$C_1 = 1$	capacitance of Tank 1
$C_2 = 1$	capacitance of Tank 2
$h_1(t)$	height of liquid in Tank 1
$h_2(t)$	height of liquid in Tank 2
$u_{in}(t)$	input flow rate
$R_1 = 0.5$	resistance out of Tank 1
$R_2 = 0.25$	resistance out of Tank 1

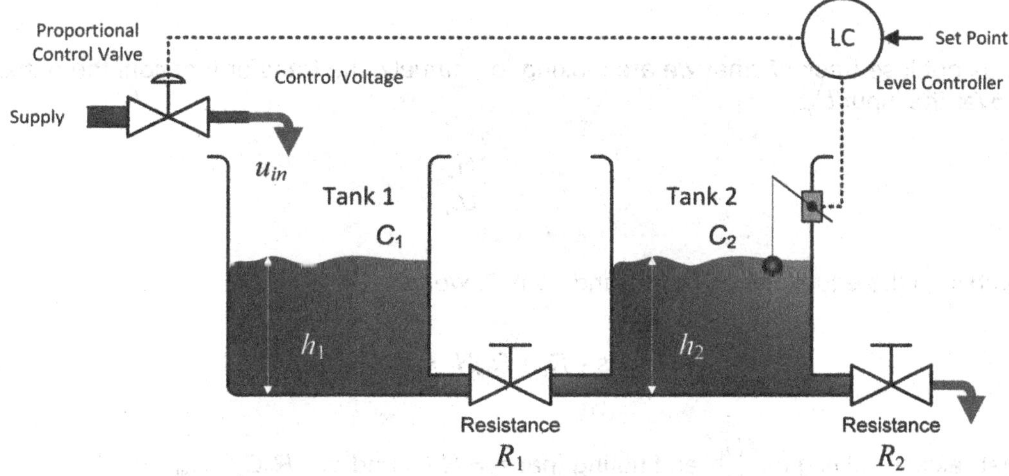

Figure 6-12. Two-tank system.

Tank 1 and Tank 2 are of equal size, $1m^3$

We can write the following equations for each tank:

Tank 1

$$C_1\frac{dh_1}{dt} = u_{in} - q_1$$

$$q_1(t) = \frac{h_1 - h_2}{R_1}$$

$$C_1\frac{dh_1}{dt} = u_{in} - \frac{h_1 - h_2}{R_1}$$

$$R_1C_1\frac{dh_1}{dt} = R_1u_{in} - h_1 + h_2$$

Note that the time constant for Tank 1 is $\tau_1 = R_1C_1$

Taking the Laplace transforms, we get:

$$\tau_1H_1s = R_1U_{in} - H_1 + H_2$$

Solving for H_1, the tank's transfer function, we get:

$$\tau_1H_1s + H_1(s) = R_1U_{in} + H_2$$

$$H_1(\tau_1s + 1) = R_1U_{in} + H_2$$

$$H_1 = \frac{R_1U_{in} + H_2}{(\tau_1s + 1)}$$

Tank 2

$$C_2\frac{dh_2}{dt} = q_1 - q_2$$

$$q_2 = \frac{h_2}{R_2}$$

$$C_2\frac{dh_2}{dt} = \frac{h_1 - h_2}{R_1} - \frac{h_2}{R_2}$$

$$R_1R_2C_2\frac{dh_2}{dt} = R_2(h_1 - h_2) - R_1h_2$$

Note that the time constant for Tank 2 is $\tau_2 = R_2C_2$

$$\tau_2R_1H_2s = R_2(H_1 - H_2) - R_1H_2$$

$$\tau_2R_1H_2s = R_2H_1 - R_2H_2 - R_1H_2$$

$$\tau_2R_1H_2s + R_2H_2 + R_1H_2 = R_2H_1$$

$$(\tau_2R_1s + R_2 + R_1)H_2 = R_2H_1$$

$$H_2 = \frac{R_2H_1}{(\tau_2R_1s + R_2 + R_1)}$$

Let us not lose track of what we are looking for, namely, the transfer function: the output H_2 over the input U_{in}:

$$\frac{H_2}{U_{in}}$$

Combining the equations of Tank 1 and Tank 2, we get:

$$(\tau_2R_1s + R_2 + R_1)H_2 = R_2\frac{R_1U_{in} + H_2}{(\tau_1s + 1)}$$

Algebraically solving for $\frac{H_2}{U_{in}}$ and noting that $\tau_1 = R_1C_1$ and $\tau_2 = R_2C_2$, we get:

$$\frac{H_2}{U_{in}} = \frac{R_2}{\tau_1\tau_2 s^2 + s(C_1R_2 + \tau_1 + \tau_2) + 1}$$

The detailed algebraic derivation can be found in Chapter 18.

Substituting values: $\tau_1 = R_1C_1 = 0.5 \times 1 = 0.5$ and $\tau_2 = R_2C_2 = 0.25 \times 1 = 0.25$

$$\frac{H_2}{U_{in}} = \frac{0.25}{0.5 \times 0.25\, s^2 + (1 \times 0.25 + 0.5 + 0.25)s + 1}$$

$$\frac{H_2}{U_{in}} = \frac{0.25}{0.125\, s^2 + 1s + 1}$$

Substituting these into:

$$H(s) = \frac{K}{T_1^2 s^2 + T_2 s + 1}$$

then $T_1^2 = 0.125$ and $T_2 = 1$

$$T_1 = \sqrt{0.125} = 0.353$$

and

$$\zeta = \frac{T_2}{2T_1} = \frac{1}{2 \times 0.353} = 1.42 \qquad \omega = \frac{1}{T_1} = \frac{1}{0.353} = 2.83 \text{ rad} \qquad \text{or} \qquad f = \frac{1}{2\pi T_2} = \frac{1}{2\pi \times 0.353} = 0.45 \text{ Hz}$$

Because the damping factor is greater than 1, we know that the system will not exhibit any ringing and it can be approximated as we did in Chapter 5 as a first-order plus dead-time system.

Why the emphasis on second-order systems? It is evident that most physical systems or plants exhibit a differential equation of order greater than 2, and it would seem a futile exercise to study this system at any depth. The reason is quite simple: mathematically, we can disassemble any higher order system into a series of first- and second-order systems. The first-order system provides a decaying contribution, while the second-order system can provide the oscillations and overshoot that may occur. This implies that if we fully understand first- and second-order systems, then we understand higher order systems as well. Often for practical applications, it is also possible to approximate higher order systems with second-order systems.

Summary

- An integral process is characterized as having a single capacitive element configured such that the outflow of material or energy is independent of the amount of material or energy stored in the capacitive element.
- Integrating processes are characterized by a ramping of the process variable in response to a step change in the set point or load change. This integration occurs as a result of either mass flow imbalance or energy flow imbalance in and out of the process.
- A first-order lag process is characterized as having a single capacitive element configured such that the outflow of material or energy is proportional to the amount of material or energy stored in the capacitive element. These types of processes are self-regulating and exhibit a first-order exponential response to a step input.
- Second-order processes are characterized as having a two-capacitive or energy-storing element. Second-order systems can exhibit an oscillating response to a step input.

7

Sensors, Transducers, and Transmitters

For process control to function, you must be able to measure the process variable. The performance quality of a feedback control system is directly dependent on the validity of the process variable measurement. This measured value must be transmitted to the controller in a timely fashion so that corrective action may be calculated and transmitted to the final control element. This chapter provides a brief study of some of the more common devices used in measuring process variables and the operation of these measuring and transmission elements.

Learning Objectives

- Understand the difference between a sensor, a transducer, and a transmitter
- Understand more fully the role and functioning of the sensors in a feedback control loop
- Define *accuracy*, *precision*, and *sensitivity*
- Understand the 4 to 20 mA current loop

7.1 The Sensor, Transducer, and Transmitter

One critical problem in designing and installing a feedback control loop is specifying the measuring system that will obtain a measurement of the controlled variable. The measuring element not only provides a measurement of the process variable, but it also provides the feedback that causes the change of the process variable by means of the feedback control loop. The change of variable takes place because the process variable itself is not the actual signal that is transmitted back to the comparator. The transmitter produces an output signal whose steady-state value has a predetermined relationship to the controlled variable. A transmitter is not required from the standpoint of either measurement or control. The transmitter is an operating device that forwards the measured value of the controlled variable to a more convenient and/or a more centralized location, for example, in a remote control room. The transmitter output signal is generally in a standardized form, such as 3 to 15 psig air pressure or 4 to 20 mA DC. From a hardware standpoint, the measurement function and the transmitter function are often both incorporated into a single device.

Some of the most common process variables are temperature, pressure, flow, composition, liquid level, speed, and position. Some of these variables, such as pressure, can be measured relatively directly, while others, such as temperature, can only be measured indirectly.

There is no consensus as to what distinguishes a sensor from a transducer. For this book, we will make the following distinction: an *electrical sensor* is a device that converts an energy form directly to an electrical output signal. Typical examples are thermocouples, resistors, and capacitors. An *electrical transducer* is a device composed of a sensor, whose output is modified or conditioned to suit a specific application. The transducer receives information in one physical form, modifies the information or its form, or both, and sends an electrical output signal. Depending on the application involved, a transducer in a functional sense can be a primary measuring element (a sensor), a transmitter, a relay, a converter, or some other device. Many of these terms—*sensor, transducer, converter,* and *transmitter*—have broad and sometimes overlapping meanings, and quite often the installation of a piece of hardware will dictate the appropriate descriptive term. Confusion can be minimized if the practitioner focuses on the function involved rather than the term. To summarize:

- *Sensors* are devices that detect variables, such as pressure, temperature, movement, electrical signals, radiant energy, and thermal or magnetic energy, and provide mechanical or electrical outputs that are directly related to the magnitude of the variables.

- *Transducers* are devices that can change one form of energy to another. For example, a resistance thermometer converts temperature into electrical resistance. The term *transducer* applies most often in control loops to such instruments that convert an electrical signal to a pneumatic signal or a pneumatic signal to an electric signal.

- *Transmitters* are used to convert a signal's format for the purpose of transmitting that signal, for example, an electronic circuit that converts a voltage signal into a standard 4 to 20 mA current signal, or an electromechanical circuit that converts the voltage to a 3 to 15 psi signal. Often the term *transmitter* is used to describe the entire measuring device, such as *pressure transmitter* or *temperature transmitter.*

  Transmitters are usually used when the measurement is to be transmitted a distance away from the controller.

Most measuring systems also provide for the following signal processing functions:

- Provide filtering to remove noise
- Change the level or value of the signal
- Change the signal from one form to another (e.g., voltage to current or current to pneumatic)
- Change the operating characteristic with respect to time
- Convert analog and digital signals from one to the other

7.2 Selection of Measuring Devices

Many factors must be carefully considered before selecting a sensor or measuring device for a specific process control loop. There are no hard-and-fast rules for making such decisions, but there are several factors that should be considered:

- What are the normal and maximum/minimum ranges of the process variable?
- What accuracy, precision, and sensitivity are required? (These terms are defined in detail in the next section.)
- What sensor dynamics can be tolerated?
- What is the required reliability?
- What costs are involved, such as purchasing, installation, and operating costs?
- Are there special installation problems, for example, corrosive fluids, explosive mixtures, size and shape constraints, and remote transmission questions?

With an extensive list of important factors to keep in mind, selecting a particular sensor for a specific installation can be complex. As a result, whole books have been devoted to topics such as to how to measure temperature, how to measure flow rate, how to measure composition, and how to measure other common process variables. To study these subjects, you may want to refer to these books. ISA publishes many of them. For now, we will not explore in further detail the specific methods of sensor selection. Instead, we will focus on them only to the extent that they affect the basic performance and dynamics of the control loop. In other words, we will discuss understanding the validity of measurements and clearly understanding the manufacturers' specifications for sensors, transducers, and transmitters.

7.3 Accuracy and Precision

Central to the study of instrumentation is to properly identify measurements and ensure that the measuring device accurately quantifies the process variable. That means when taking measurements, we must have a certain degree of confidence that the measurement is valid. Of course, no measurement is perfect, and there will always be some deviation between the actual value of what is measured (the measurand) and the actual measurement. This deviation must simply be small enough so that the measured value can be used for its intended purpose.

Several standard terms are used to specify the validity of a measurement.

Error

The error of a measurement is defined as the difference between the measured value and the true value of the measurand:

$$error = measured\ value - true\ value \tag{7-1}$$

Error in this technical usage does not imply that there is any mistake in the measurement process, although mistakes can cause errors. Normally, we never really know the error of a measurement because we do not know the true value. If the true value of the measurand were known, there would be no need to make the measurement. There are two notable exceptions.

1. In the process of calibration, where measurements are made of measurands whose values are independently known
2. In the process of repair, troubleshooting, and design verification, where measurements are made and compared to expected and/or theoretical values

What we can do is estimate the uncertainty of the measurement. The uncertainty is an estimate, with some level of confidence of the limits of error in the measurement. For example, it might be stated that with 95% confidence, the uncertainty of a temperature measurement is ±1°C. This means that in less than 5% of the cases where we have made such a measurement, the error will be greater than 1°C. Narrow uncertainty intervals are usually achieved by using calibrated, high-quality measuring instruments or systems.

Errors in measurements generally fall into two categories: bias errors (fixed or systematic errors) and precision errors (random errors). Although both types of error degrade the validity of the data, their causes are different, and different actions must be taken to minimize them.

Bias Error

Bias errors are consistent and repeatable. For example, a measuring system might give a consistent 10% high reading. In general, if the same measuring system is used in the same manner more than once to measure the same value of the measurand, the bias error will be the same each time.

The first major source of a bias error is the result from calibration of the measurement system. If the calibration process has some error, that error will be carried into the measurement as a bias error. Even the most exact calibration will result in some residual bias error. This is known as a *calibration error.* One common source of this type of calibration error is nonlinearity. Many modern systems are assumed to have a linear relationship between the input and the output, although in reality they are nonlinear, and the nonlinearity of the system will cause errors.

The second major source of bias error results from the use of a measuring system in a particular application where the insertion of the measuring device alters the measurand. For example, connection of a temperature-measuring device to a surface may change the local temperature of the surface. Such errors are known as *loading errors.* Consider placing a mercury-in-glass thermometer into a beaker of water. If the beaker and the thermometer are initially at different temperatures, energy will be exchanged between them and the measured temperature will be neither the initial water temperature nor the initial thermometer temperature (but usually closer to the water temperature). The thermometer is an intrusive measurement device and produces a loading error. Some measuring devices with negligible loading errors are called *nonintrusive.* For example, devices are available that measure temperature by sensing the infrared radiation emitted. Such devices have a negligible effect on the measured temperature and are said to be *nonintrusive.*

A third major bias error results because the measuring system is affected by variables other than the measurand. For example, a thermometer used to measure the air temperature in a room may read too low, due to thermal radiation effects, if the walls are cooler than the air. A related error is the spatial error. If the measurand varies in a spatial region and yet a single measurement or limited numbers of measurements are used to determine the average value for the region, a spatial error will exist.

Bias errors are often not obvious because the measuring system will show clear and consistent changes in output following changes in the measurand. Yet, the measurement can still have significant bias error. In setting up a measurement system, considerable time may be required to detect and minimize bias errors. Bias errors in the measuring system may be detected and reduced by the process of calibration. Some bias errors caused by use of the measuring system in a specific application may be reduced by analytical correction of the data for unwanted effects.

Precision Errors

Precision errors are those caused by a lack of repeatability in the output of the measuring system. The distinction between bias and precision errors is shown graphically in Figure 7-1.

The scatter in the data represents the precision error, and the deviation between the average of the readings and true value demonstrates the bias error. The precision error in a single measurement can be estimated as the difference between the single reading and the average of all readings of the same value of measurand, which distinguishes the precision error from the bias error as follows:

$$\textit{precision error} = \textit{reading} - \textit{average of readings} \tag{7-2}$$

$$\textit{bias error} = \textit{average of readings} - \textit{true reading} \tag{7-3}$$

For these estimates of bias and precision errors to be reasonable, the number of readings forming the average must be sufficiently large to eliminate the effects of precision error in individual measurements on the average.

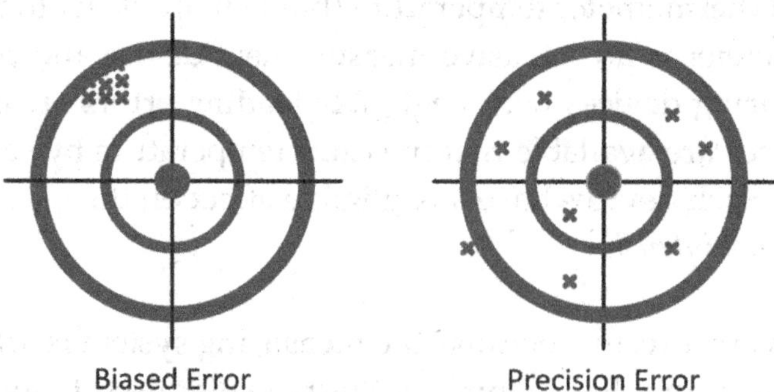

Figure 7-1. Target showing the distinction between bias and precision error.

Precision errors can originate from the measuring system itself, from the system being measured, or from the environment. Precision errors are usually caused by uncontrolled variables in the measurement process. For example, the performance of a pressure transmitter may be slightly sensitive to its temperature. If we do not measure or control the temperature (as we would do with a temperature transmitter), performance measurements may show certain variability or spread. One important environmental cause of precision error is electrical noise. Measuring systems usually operate in an environment of electrical and magnetic fields caused by sources such as building wiring, motors, and local radio and TV stations, and wireless communication such as Wi-Fi and cell phones. These electrical and magnetic background fields can significantly affect readings by randomly altering voltages in measuring systems and interconnecting wiring. This is referred to as *electromagnetic interference* (EMI).

Precision errors can often be minimized by eliminating uncontrolled variables and properly shielding or grounding the measuring system. Remaining precision errors may be amenable to statistical analysis; for example, multiple readings can be averaged. The following example demonstrates how to estimate the bias and maximum precision errors for a calibration test of a temperature transmitter.

Example

In a calibration test, a digital thermometer has been used to take 10 measurements of the temperature that is known to have a true value of 61.1°F. The readings are:

59.8	60.5	61.0	60.6	59.9	59.6	60.2	60.9	60.3	59.9

Estimate the bias and maximum precision errors caused by the thermometer.

Solution

First, determine the average of the 10 readings:

$$\textit{average reading} = \frac{\Sigma \textit{ all reading}}{\textit{number of reading}}$$

The symbol Σ indicates summation.

$$\begin{aligned}\Sigma &= 59.8 + 60.5 + 61.0 + 60.6 + 59.9 + 59.6 + 60.2 + 60.9 + 60.3 + 59.9 \\ &= 602.7\end{aligned}$$

The average reading can then be calculated as follows.

$$\textit{average reading} = \frac{602.7}{10} = 60.3°\text{F}$$

rounded to one decimal place. Then the estimate of the bias error is

$$bias\ error = 60.3 - 61.1 = -0.8°F$$

To estimate the maximum precision error, we must determine the reading that deviates the most from the average reading. This is the reading of 59.6°F. The maximum precision error is then:

$$max\ precision\ error = 59.6 - 60.3 = -0.7°F$$

Note that this simple statement of maximum precision error might not adequately describe precision errors in a measuring system. For example, it may be based on a single bad reading. There are statistical methods and procedures to determine precision errors that include all the readings and provide a basis to eliminate certain bad data.

Span

All measuring systems are designed to operate over a specified range of measurands. The range of a measuring system describes the values of the measurand to which that measuring system will respond properly. Values of the measurand outside that range are not expected to produce useful results and output. For example, a flowmeter may have a range of 10 to 100 gpm (gallons per minute) and would not give a correct response to measurands of 5 or 130 gpm. The span of a measuring system is the difference between the upper and lower values of the range. Mathematically:

$$span = upper\ limit - lower\ limit \tag{7-4}$$

For example, a pressure transmitter with a range of –10 to 20 psig (pounds per square inch gauge) has a span of 20 – (–10) = 30 psig.

To summarize:

- *Accuracy* refers to the closeness with which the measurement approaches the true value of the variable being measured.
- *Precision* refers to the magnitude of random error.

Whenever a measurement is made, sources of precision or random error will add a component to the result that is unknown. However, with repeated measurements made under the same conditions, this result changes in a random fashion.[1]

1. Interestingly, *precision* is not a defined term in ISA 51.1-1979 (R 1993), *Process Instrumentation Terminology*, but it is used routinely by many measurement professionals.

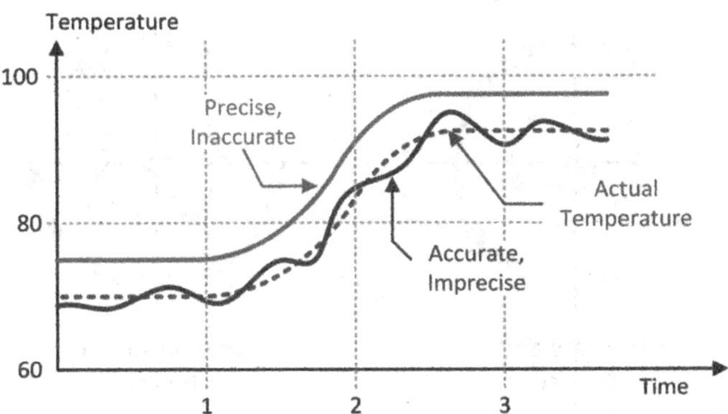

Figure 7-2. Accuracy and precision in a temperature measurement.

A more common term related to precision is *repeatability*, which is discussed in the next section. In matters of process control, precision and repeatability are more important than accuracy. To put this differently, it is usually more desirable to measure a variable precisely than it is to have a high degree of absolute accuracy. The distinction between these two properties of measurement is shown in Figure 7-2.

The dashed line in Figure 7-2 indicates the actual temperature of a fluid. The upper measurement, or solid line, illustrates a precise but inaccurate instrument. The lower line shows the measurement given by an imprecise but more accurate instrument. The first instrument (the upper measurement) has the greater "error."

Figure 7-3 provides an illustration, again using a target, to help in understanding the terms *accuracy* and *precision*.

Static and Dynamic Accuracy

When selecting and using measuring systems, it is also essential to distinguish between two types of accuracy: static or steady-state accuracy and dynamic accuracy. Static accuracy is the closeness of the approach to the true value of the measurand when that true value is constant. Dynamic accuracy, on the other hand, is the closeness of approach of the measurement when the true value is changing. These terms are illustrated in Figure 7-4. Note that in the figure, the dynamic accuracy increases as the actual value changes. Also, dynamic errors are usually larger than static errors, mainly because of the presence first-order lag behavior in measuring systems.

The numerical value of the dynamic accuracy depends on the nature of the dynamic change made by the true value of the variable being measured. The properties of the measuring system itself will also have an effect. For process control systems, a practical specification of the time variable is ramp forcing, like that shown in Figure 7-4, and one practical designation of dynamic accuracy is the dynamic error that results from ramp forcing.

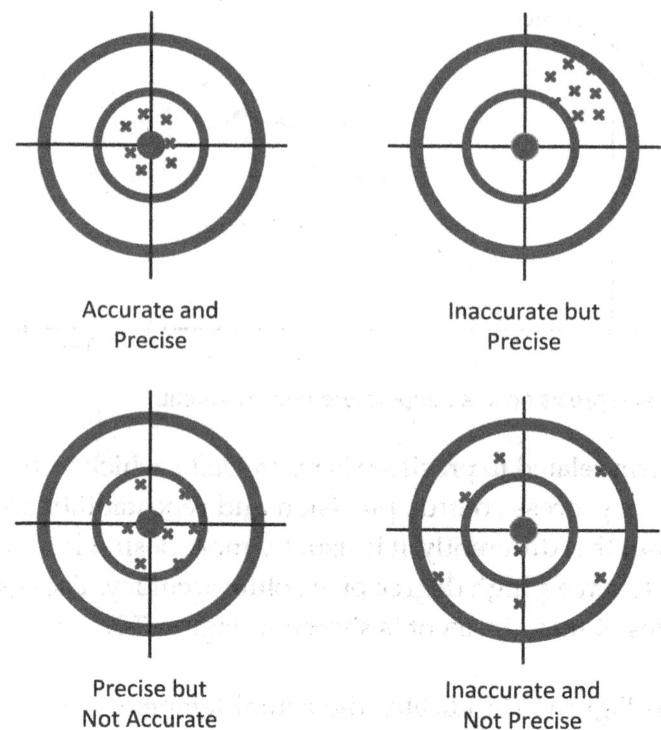

Figure 7-3. Illustration of accuracy and precision.

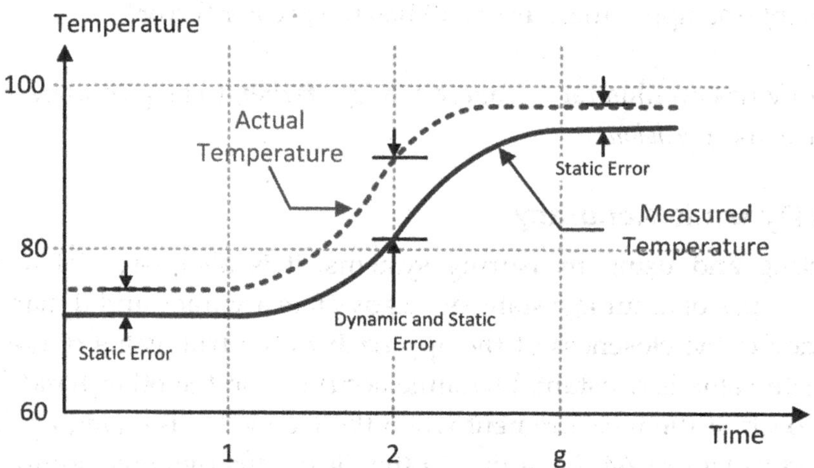

Figure 7-4. Dynamic and static error.

7.4 Sensitivity, Resolution, Repeatability, and Reproducibility

The *sensitivity* of a measuring device is defined by the ratio of the output signal change to the change in the measured variable. Sensitivity is calculated by dividing the engineering units of the output signal by the engineering units of the input signal. Clearly, the greater the output signal changes for a given input change, the greater the sensitivity of the measuring element. Sensitivity is a steady-state ratio and is in effect the steady-state characteristic of the element, or the steady-state transfer function of the element.

There is another kind of sensitivity that is important in measuring systems. This sensitivity is defined as the smallest change in the measured variable that will produce a change in the output signal from the sensing element. In many physical systems, especially those that contain levers, linkages, and mechanical parts, these moving parts have a tendency to stick and to have some free play, producing some hysteresis. The result is that small input signals may not produce any detectable output signal. Highly sensitive, well-designed, and well-constructed instruments are required so the control system has the ability to respond to small changes in the controlled variable.

Electronic amplification of sensor inputs can also exhibit limited sensitivity and some hysteresis.

The *resolution* of a measurement system is the smallest distinguishable increment in values and should not be confused with the accuracy of the instrument's measurement.

In digital measuring systems, analog signals are converted to a binary digital number with the use of an analog-to-digital (A/D) converter. The smallest difference between two numbers is then the resolution of the system. This means it is therefore always equal to 1 bit or, in the case of a digital display, 1 digit. It is also possible to express the resolution in units other than bits.

Example

A digital measuring system, such as a transmitter, has a 12-bit A/D converter. The transmitter is calibrated with a range from 200°C to 1000°C. The 12 bits represent $2^{12} = 4096$ different values. The resolution in degrees is:

$$\frac{1000 - 200}{4096} = 0.195°C$$

Repeatability is the closeness of agreement among a number of consecutive measurements of the output for the same value of the input, under the same operating conditions, approaching from the same direction, for full-range traverses. The degree of repeatability achieved between consecutive measurements is usually measured as nonrepeatability and is specified as a percentage of full scale or a percentage of span. Repeatability measured this way does not include hysteresis.

Reproducibility is the closeness of agreement among repeated measurements of the output, but are due to variations of the measurement system for the same value of input made under the same operating conditions over a period of time, approaching from both directions. Reproducibility includes hysteresis, deadband, drift, and repeatability.

7.5 Rangeability and Turndown Ratio[2]

Rangeability refers to the minimum and maximum measurable values of the process variable being measured. *Turndown ratio* is another way to determine rangeability, but in which the information is expressed differently. Turndown is the ratio of the maximum to the minimum measurable parameter. Mathematically:

$$turndown\ ratio = \frac{max}{min}$$

For example, if a given flowmeter has a 1:20. turndown ratio, the flowmeter is capable of accurately measuring down to 1/20th of the maximum flow. If the flowmeter has a full-scale rating of 25 L/min, then the flowmeter will only be able to measure down to 25/20 = 1.25 L/min of flow.

Consider a flowmeter where the maximum measurable flow is 100 gpm and the minimum measurable flow is 10 gpm. The rangeability of the flowmeter is then 10% to 100%. For the flowmeter, the turndown ratio is then 10:1.

7.6 Measurement Uncertainty Analysis

Measurement uncertainty analysis is a numerical, objective method for defining the potential error that exists in all data. It is fundamental to have some knowledge of the uncertainty in any measurement to understand either the state of a process or its performance. In measurement uncertainty analysis, errors are considered to be either random errors or bias errors. Random error adds a component to a measurement that is unknown, but with repeated measurements, it will change in a random fashion.

It is not enough to focus only on reducing precision errors. It is also essential to understand and reduce systematic or bias errors. Systematic errors or bias errors are constant in the sense that they affect every measurement of a variable by the same amount. They are not directly observable in the measurement itself.

---

2. Generally, the terms *rangeability* and *turndown* are associated with flow measurements and control valve sizing (see Chapter 8).

Measurement uncertainty analysis is a field that merits study by serious process control students. It gives clearer insight into a poorly appreciated, but most important, phenomenon.

7.7 Transmission Systems

When the sensor measures the controlled variable, the measured value must be transmitted in some way to the controller. This may be a distance of several feet or several thousand feet. In a similar fashion, it is necessary to get information from the controller to the final control element, such as the actuator of a control valve.

For years, many process control transmission systems used pneumatic tubing to transmit information as an air pressure signal. Although few of these pneumatic transmission systems are being installed today, many are operational in existing plants. These pneumatic systems introduce a time lag into the process dynamics of the control loop and thus influence the loop's performance, sometimes to serious disadvantage. For all these reasons, some discussion of pneumatic transmission systems is justified here.

The most common transmission medium is copper wire, either in the form of twisted pairs or in the form of coaxial cable. Twisted-pair wire has been used extensively for decades and are the most common transmission medium. Wireless technologies are becoming more and more prevalent, especially in mobile or temporary installations. Fiber-optic cable can also be used, and its installation is becoming increasingly common; but it is very expensive, and it is difficult to make multiple splices.

Many transmission systems are available and in use, and we will review some of the more common ones. Table 7-2 lists the most common transmission systems in use today (except for digital buses). We will give specific attention to the more established 3 to 15 psig analog pneumatic type and the 4 to 20 mA DC type.

Note that both the analog electrical and pneumatic transmission ranges have a nonzero value for the lowest value of the range. Any pressure below 3 psi and any current below 4 mA can be considered as dead zero and an alarm condition.

Typically, a 4 mA current or 3 psi pressure value represents 0% of scale, a 20 mA current or 15 psi pressure value represents 100% of scale, and any current or pressure value in between represents its corresponding percentage between 0% and 100%. Table 7-1 shows the corresponding current pressure values and percentage values for each 25% increment between 0% and 100%.

Table 7-1. Relationship between the different scales used.

Percentage %	Pressure psi	Pressure kPa	Current mA
0	3	20	4
25	6	40	8
50	9	60	12
75	12	80	16
100	15	100	20

Note that the pressure scales are nominal, meaning that the 3 to 15 psi scale does not precisely convert to the 20 to 100 kPa scale.

Table 7-2. Nominal transmission system signal ranges.[3]

Type	Medium	Values
Binary (on-off)	Electricity: alternating current (AC) direct current (DC)	 0 to 120 V 0 to 24 or 48 or 125 V
	Pneumatic	0 to 25, 35, 100 psig 0 to 170 kPa or 240 to 700 kPa
	Hydraulic	0 to 3000 psig 0 to 20,000 kPa
Analog (modulating)	Electricity: direct current (DC)	 –10 to +10 V, or 1 to 5 V, 4 to 20 mA, or 10 to 50 mA
	Pneumatic	3 to 15 psig or 6 to 30 psig 20 to 100 kPa or 40 to 200 kPa
	Hydraulic	0 to 3000 psig 0 to 20,000 kPa

Pneumatic Transmission

Pneumatic transmission may be used for distances up to several hundred feet. In a feedback transmission system, the controlled variable is measured and converted to an air pressure, and a transmitter, in effect, sends this air pressure signal through a single tube to a receiver where it is transduced to a position or force for use within the controller. A typical pneumatic transmission system is shown in Figure 7-5. The

3. George Platt, *Process Control: A Primer for the Nonspecialist and the Newcomer,* 2nd ed. (Research Triangle, NC: ISA [International Society of Automation], 1998).

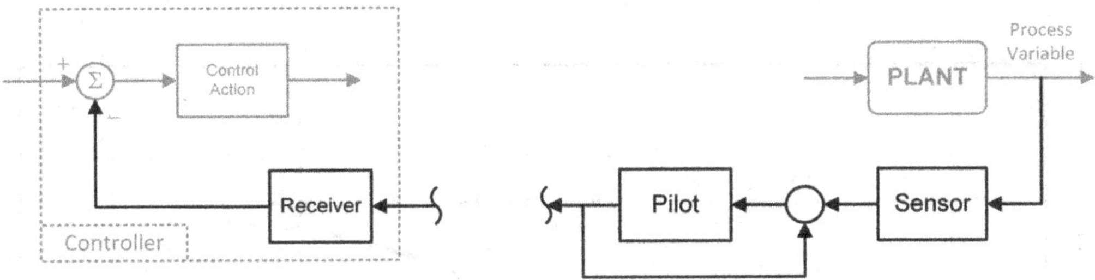

Figure 7-5. A pneumatic transmission.

controlled variable is sensed and converted to an air pressure, and the measured pressure is often used as a pilot signal to an amplifier. (An amplifying pilot is often employed in pneumatic transmission systems to increase the airflow capacity of the transmitter and to linearize the output signal.)

The connecting tube carries the transmitted pressure to a receiver (located in the controller case). This tube is almost always 1/4 in in outside diameter and may be copper, aluminum, or plastic. The receiver is simply a pressure-gauge element, and the transmitted air pressure is converted into the movement of a bellows or diaphragm. Thus, pressure is transduced into a position or force that is used within the controller.

One of the most impeding concerns about pneumatic transmission is the problem of transmitting a signal over significant distances, because there is a significant lag associated with the transmission of the pressure signal through the long connecting tube. The tube has volume distributed along its length, and there is resistance to fluid flow through this tube. If a step change in input signal of 3 to 15 psi is made to a pneumatic transmission tube, then a typical output signal would appear as shown in Figure 7-6.

As distances increase, the speed of response of pneumatic transmission systems becomes a problem, and alternate solutions must be found. As stated previously, very few pneumatic transmission systems are being installed today. However, pneumatic transmission systems are found in a range of applications from highly hazardous (explosive) locations to local garages.

Electrical Transmission Systems

The most common transmission systems use electrical signals, either voltage or current. Current is preferred over voltage because it is not affected by transmission line length and its associated electrical resistance drop in voltage, and it requires only two wires. There is an individual twisted pair of wires for each signal loop; unfortunately, in most manufacturing plants, this rapidly creates the need for literally thousands of such twisted pairs. They can be very expensive to install and maintain.

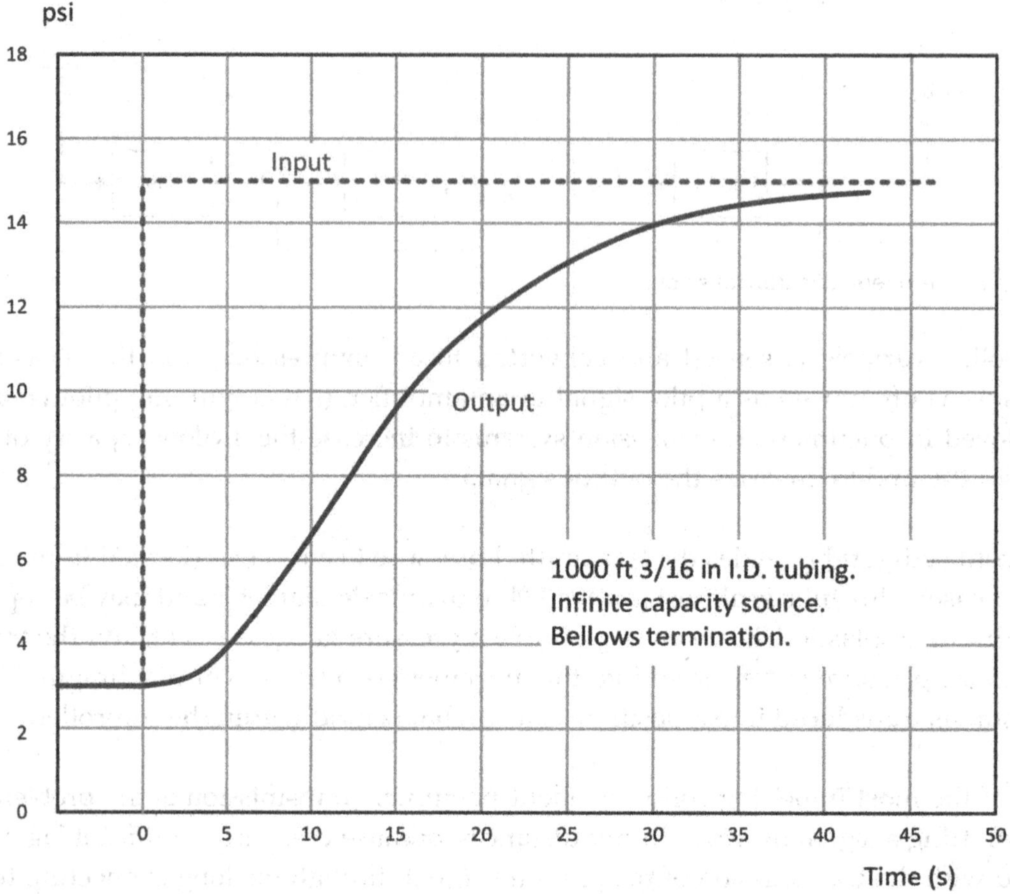

Figure 7-6. Input and output of a pneumatic transmission system.

Field wiring must provide for the galvanic isolation of signals from the process equipment to prevent ground loops. Field wiring is also often installed in hazardous locations that require explosion-proof or intrinsically safe media.

It is possible to consider using a sensor output directly as a signal for transmission, for example, using the mV output of a thermocouple as a direct signal. Problems with noise, signal weakness, and the like usually lead practitioners to use signal conditioning in the form of a transmitter to assist transmission, maintain signal integrity, and minimize signal noise.

As transmission systems become more complex and increasingly move into the digital world, it has become apparent that there is a need for a standard digital communication system for use in process control applications. The term *fieldbus* is used to describe equipment interconnected through the use of common wires, enabling communication between the connected units. Digital communication between process control elements and systems and digital fieldbuses will be discussed in more detail in Chapter 21.

7.8 The 4 to 20 mA Current Loop

One of the most popular signal transmissions used in industrial instrumentation systems is the 4 to 20 mA DC standard current loop.

The transmitter is the heart of the 4 to 20 mA current loop. It changes the measurand, such as temperature, humidity, or pressure, into an electrical signal. The signal is a current proportional to the physical parameter being measured: 4 mA represents the low end of the measurement range, and 20 mA represents the high end. The transmitter consumes between 7 and 15 VDC of loop voltage, depending on the model, to power itself. The lower voltage is the minimum voltage necessary to guarantee proper transmitter operation. The higher voltage is the maximum voltage the transmitter can withstand and operate to its stated specifications.

Current signals are standardized at 4 to 20 mA, and 0 mA is often used as a fault condition. The requirements of the transmitter are high output impedance, so that the output current does not vary with load, low offset drift, and low noise. The main advantage of a current signal is that there is no loss of signal amplitude over long wires, unlike voltage transmission. The internal resistance of the receiver should be low and is often set at 250 Ω, providing a 1 to 5 V signal internal to the receiver.

The 4 to 20 mA current loop is a robust and ideal standard for transmitting sensor signals and power. All the signaling current flows through all components; the same current flows even if the wire terminations are less than perfect. All the components in the loop have a voltage drop across them because of their inherent resistance, but the same signal current flows through them all.

The signal current is not affected by the component's voltage drop as long as the power supply voltage is greater than the sum of all the voltage drops around the loop at the maximum signal level of 20 mA.

Figure 7-7 shows a schematic of the simplest 4 to 20 mA current loop. There are four components.

1. DC power supply (usually 24 V)
2. Interconnecting wires

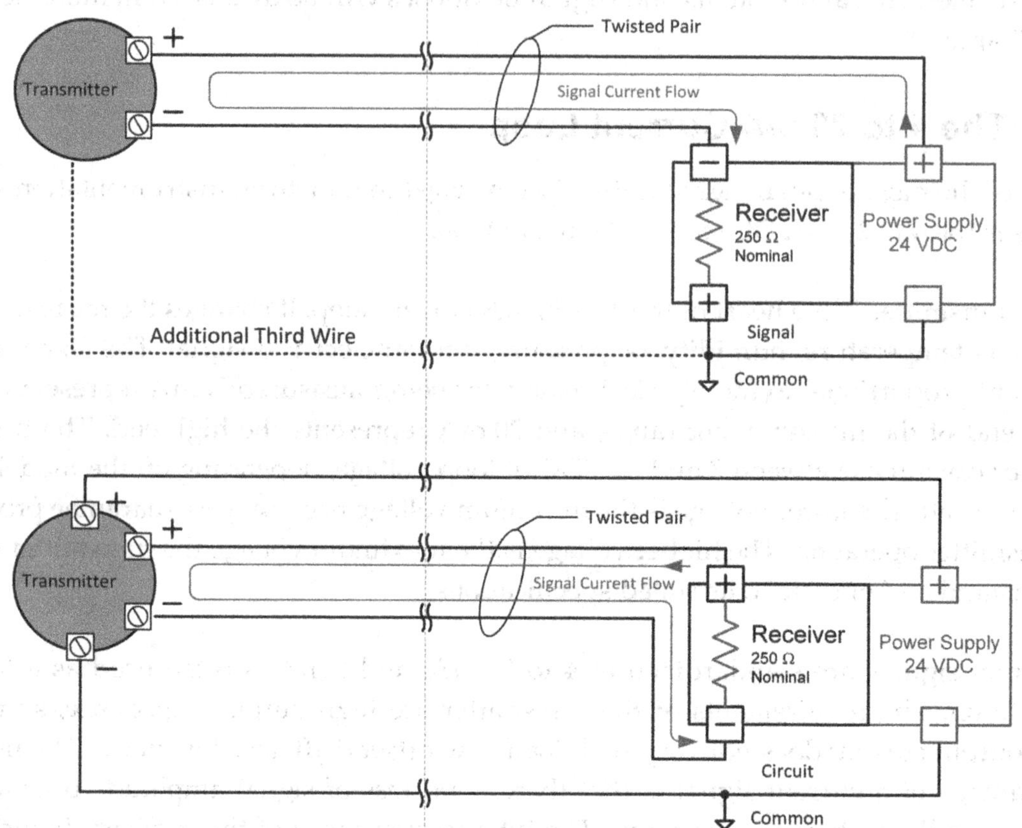

Figure 7-7. Two-wire, three-wire, and four-wire current loop configurations.

3. Transmitter
4. Receiver

The power supply is typically 24 VDC and is either part of the receiver or a separate unit. The receiver usually has a terminating resistor of 250 Ω that converts the 4 to 20 mA current signal to 1 to 5 V. Depending on the unit and application, a resistance of 100 Ω has become more prevalent, giving a voltage signal of 0.4 to 2 V.

There are three basic loop configurations used in industry: Type 2 for a two-wire configuration, Type 3 for a three-wire configuration, and Type 4 for a four-wire configuration.[4]

---

4. ANSI/ISA-50.00.01-1975 (R2002), *Compatibility of Analog Signals for Electronic Industrial Process Instruments* (Research Triangle, NC: ISA [International Society of Automation]).

In the two-wire system, the current that powers the sensor, conditioner, and transmitter also carries the signal, but the power/signal wires are floating from ground. In the three-wire system, both the transmitter and receiver are connected to a common ground. The four-wire type has separate wires for power and signals. See Figure 7-7.

Sending current through wire produces voltage drops proportional to the length of the wire. All wire has resistance, which is usually expressed in ohms per 1000 ft. The voltage drop across the wire length can be calculated using Ohm's law.

Ohm's law states that the voltage drop across the wire (or any resistive element) is directly proportional to the current through the wire and the voltage across the wire. The formula can be written as:

$$E = I \times R$$

where E is the voltage across the resistance in volts, I is the current flowing through the conductor in amperes, and R is the conductor resistance in ohms.

7.9 Smart Sensors

The term *smart sensor* refers to a device that has a microprocessor built into its internal operation. The signal sensed by the measuring device is brought into the transmitter and is immediately converted into a digital signal using an A/D converter. Most internal signal conditioning and processing is done digitally. Tremendous gains are made in accuracy, linearity, and versatility. The output of the transmitter can be in digital format, or it can be converted to a 4 to 20 mA signal. In addition to the advantages just cited, smart sensors can themselves be programmed by the computer that controls the loop where the sensor is used. The options for improvement are extensive and are just beginning to be fully exploited by hardware vendors.

7.10 Summary

- An electrical sensor is a device that detects a physical property (energy) and converts it directly to an electrical output signal.

- An electrical transducer is a device composed of a sensor, whose output is modified or conditioned. The transducer receives information in one physical form, modifies the information or its form or both, and sends an output signal.

- Transmitters are used to convert a signal format for the purpose of transmitting that signal.

- Accuracy is the amount of uncertainty in a measurement with respect to an absolute standard.
- Precision describes the reproducibility of the measurement.
- Sensitivity is the smallest amount of change that can be detected by a measurement.
- The 4 to 20 mA current loop has been the standard for signal transmission in control systems since the 1950s. In a current loop, the signal is proportional to the current flowing through the transmitter.

Exercises

7.1 Our entire presentation of sensors and transmission systems in this unit was structured around the idea of feeding a signal back to the comparator. What would be involved (how would the hardware be different) if the process variable being measured was not a controlled quantity but was instead simply being sent to a central control room for indication and recording?

7.2 A process chromatograph is used as a sensor to measure the composition of a liquid stream. It operates on a discrete basis, that is, it takes a sample and analyzes it and establishes a signal for feedback purposes. It then takes a new sample and starts its cycle over again. What effect does this mode of operation have on loop operation as compared to a continuous sensor such as a thermocouple or an orifice meter?

7.3 It was stated that sensor repeatability is more important than accuracy. How can this possibly be true?

7.4 How far can you transmit a pneumatic signal if you can tolerate a time constant lag of 3 s and a dead time of less than 1 s?

8 Actuators and Final Control Elements

Actuators are mechanical, electromechanical, pneumatic, and hydraulic devices that provide linear or rotational movement to position the final control element.

The *final control element* is that portion of the control loop that directly changes the value of the manipulated variable.

Learning Objectives

- Be familiar with the different types of actuators
- Be able to differentiate the actuator from final control element
- Understand the purpose and use of control valves, valve actuators, and valve positioners
- Be able to define rangeability and turndown ratio
- Know the meaning and use of valve coefficients and appreciate sizing considerations

8.1 Introduction

In previous chapters, we introduced the concept of the process control loop, and defined its three elements: measurement, comparison, and adjustment or manipulation. The final control element is perhaps the most important because it exerts a direct influence on the process. Final control elements contain the essential pieces of equipment to convert the control signal, generated by the process controller, into the action required to correctly manipulate the process to meet its desired value.

In this chapter, we will cover some of the fundamental principles of final control elements and actuators, such as control valves, motors, pumps, and solenoids. Because control valves are the single most common type of final control element in process control, we will discuss them first and in greatest detail. However, it should be understood that there is no attempt made here to be anything but a cursory overview of actuators and final control elements.

8.2 Control Valves and Actuators

In the vast majority of process control applications, the final control element is a valve. A control valve is in essence a device that regulates the flow of a process fluid. In a control valve, an actuator is connected to the valve's plug stem to move the valve between the open and closed positions to regulate flow. The valve body is mounted in the process fluid line and is used to control the flow of fluid in the process.

These valves are usually driven by pneumatic, solenoid or motor actuators. An example of a typical control valve and actuator is illustrated in Figure 8-1. This is a sliding-stem, single-seat control valve body with a pneumatic actuator.

There are essentially two types of control valves:

1. Linear-motion control valves that commonly have globe, gate, diaphragm, or pinch-type closures
2. Rotary-motion valves that have ball, butterfly, or plug closures

Each type of valve has its special generic features, which may, in a given application, be either an advantage or a disadvantage.

The control valve actuator is used to translate the specific output signal from the controller into a mechanical movement that drives the stem of the control valve. Control valve actuators may be pneumatic, electric, hydraulic, or manual. The most commonly used valve actuators in the process industries are pneumatic, and most of these have a diaphragm to move the stem. The diaphragm is spring loaded in opposition to the driving air pressure so that the valve stem position is proportional to air pressure.

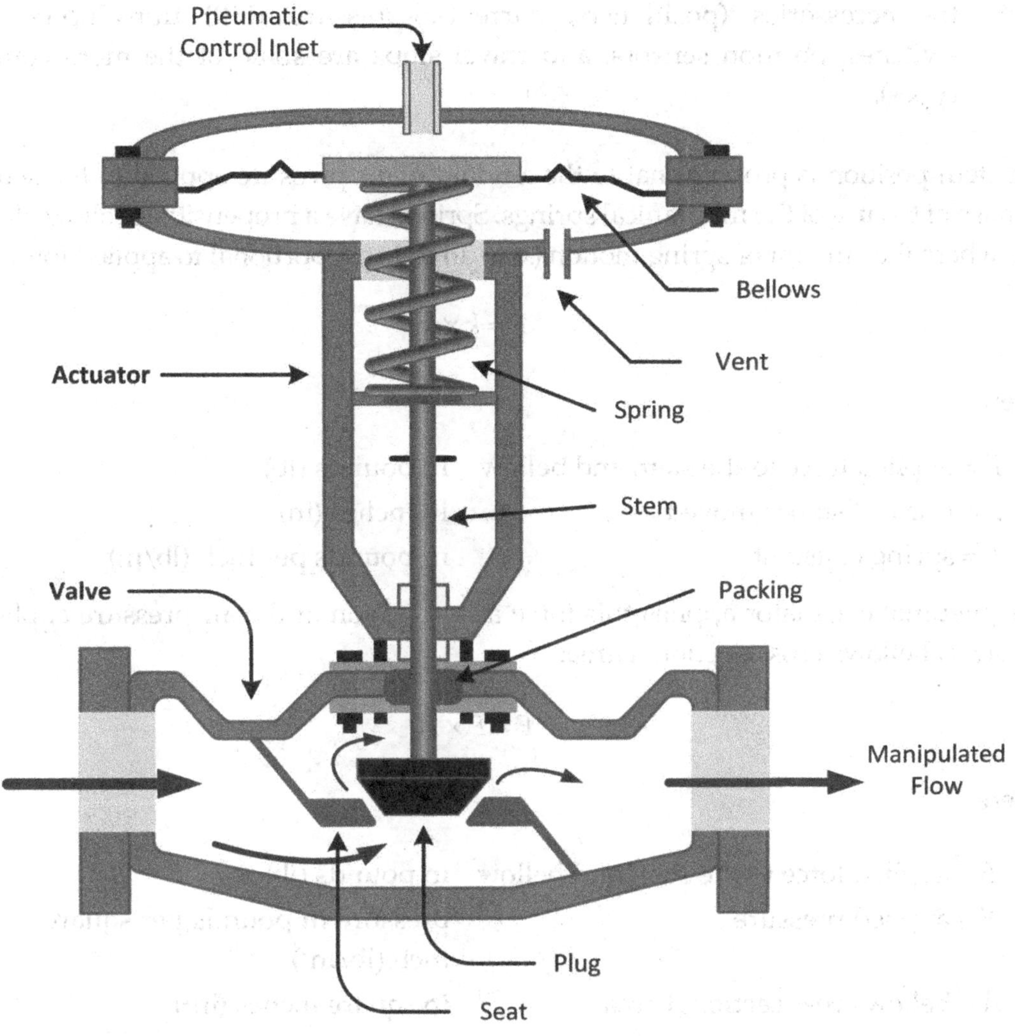

Figure 8-1. A typical sliding-stem control valve with a pneumatic actuator.

Usually, the diaphragm is made of a rubber-based fabric or other limp material and is supported by a backup plate. The maximum available stroke, or stem travel, is usually 2 to 3 in, but depends largely on the size of the valve. For longer strokes, the valve actuator may be a double-acting piston, or it may be a rotary pneumatic device. Electric actuators are less popular and cannot be used in a hazarded environment; electro-hydraulic actuators are used primarily in areas where no operating air is available.

Most installed control valves consist of three parts

1. the valve assembly (the working part)

2. the actuator (the part that positions the throttle inside the valve), and

3. the accessories (positioners, current-to-pressure [I/P] transducers, limit switches, position sensors, and travel stops are some of the more common types).

The stem position is proportional to the amount of air pressure applied to the actuator because of the use of the mechanical springs. Springs have a propensity to follow Hooke's law, where the amount of spring motion (x) is directly proportional to applied force:

$$F = k \times x \tag{8-1}$$

where

F = applied force to the stem and bellow	in pounds (lb)
x = linear distance moved	in inches (in)
k = spring constant	in pounds per inch (lb/in)

The pneumatic actuator applies this force as a function of the air pressure applied to the stem/bellows cross-sectional area:

$$F = P \times A \tag{8-2}$$

where

F = applied force to the stem and bellow	in pounds (lb)
P = applied pressure	pressure in pounds per square inch (lb/in^2)
A = bellow cross sectional area	in square inches (in^2)

Combining Equations 8-1 and 8-2 gives us the end result that actuator pressure linearly translates into valve stem motion:

$$F = k \times x = P \times A$$

$$x = \frac{P \times A}{k} \tag{8-3}$$

where

x = linear distance moved	in inches (in)
P = applied pressure	in pounds per square inch (lb/in^2)
A = bellow cross sectional area	in square inches (in^2)
k = spring constant	in pounds per inch (lb/in)

It is important to note that the force required to move the stem is not only the force opposed by the spring, but also the friction force of the padding and the force of the pressure differential of the valve. That is, the force of the intake against the plug.

These forces work against the desired positioning of the valve stem so the stem travel will not accurately correlate to actuating fluid pressure. A common solution to this problem is to add a positioner to the control valve assembly.

I/P Transducer

A *current-to-pressure* (I/P) transducer converts an analog signal usually 4 to 20 mA to a proportional linear pneumatic output signal in the range of 3 to 15 psig. Its function is to translate the electrical analog output from a process controller into a precise and consistent pressure value to control pneumatic actuators and other pneumatic equipment.

Typically, when a pneumatic activated control valve is controlled by an electronic controller, a current-to-pressure transducer is used. The I/P transducer will provide the control pressure needed to move the bellows. See Figure 8-2.

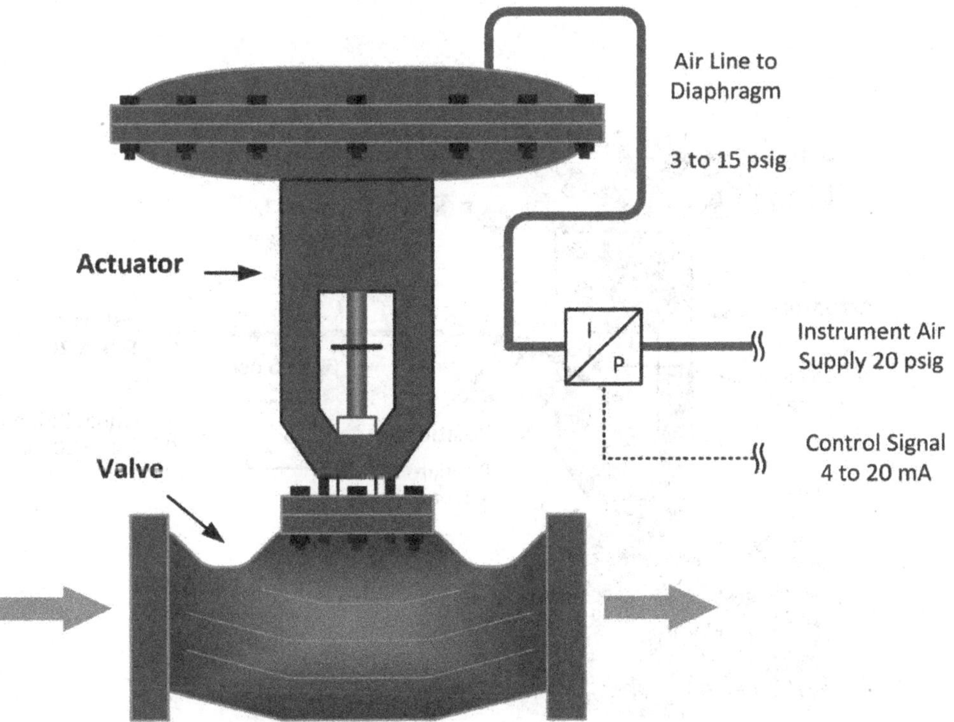

Figure 8-2. Control valve controlled by an I/P transducer.

Valve Positioner

A *valve positioner* is a control valve accessory that transmits a loading pressure to a pneumatic actuator in order to position the valve plug stem exactly as commanded by the signal from the controller. It is interesting to note, that the positioner is a position control loop in itself. The set point to the loop is the desired position of the stem. This signal is the output of the I/P transducer, which converted the controller output signal. It is often desirable to use a valve positioner to improve both the dynamic and the static behavior of the valve. Figure 8-3 shows a typical configuration.

Valve positioners are typically air pressure operated relays (air relays) that are located between the controller output and the valve diaphragm. A positioner usually has a separate air supply and a feedback signal from the valve stem that indicates the stem position. The positioner helps eliminate hysteresis, packing-box friction, and valve plug unbalance due to pressure drop of the manipulated flow. It also ensures that the valve stem is positioned in accordance with the controller output signal. The valve positioner is also useful for minimizing the lag associated with the valve response. Many positioners also may have built-in I/P transducers.

Positioners with a cam feedback mechanism are also used to linearize the control valve or provide a specific installed valve characteristic to the process under control.

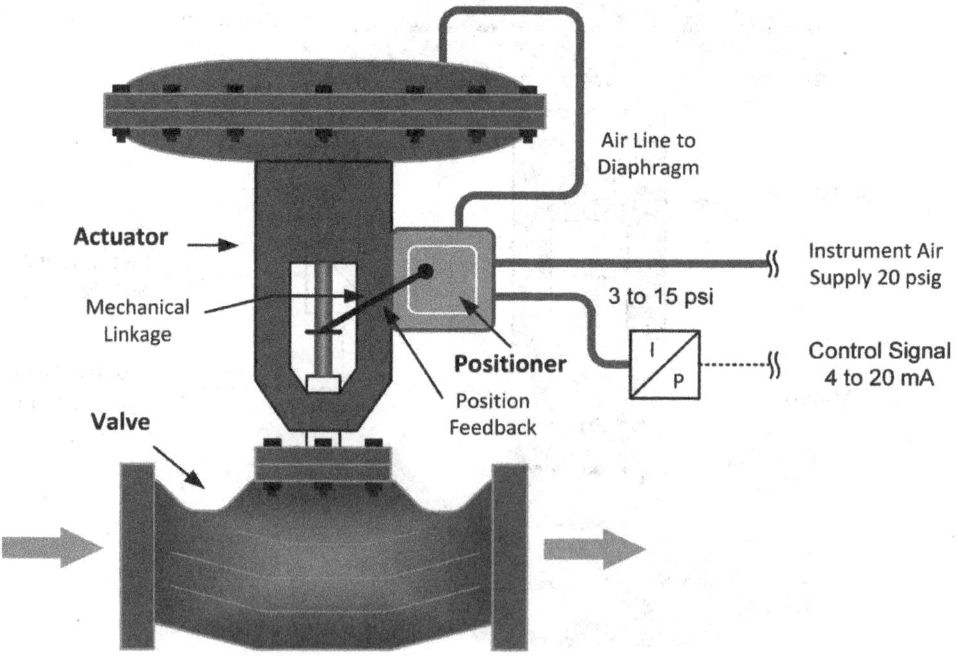

Figure 8-3. Control valve controlled by a positioner and an I/P transducer.

Concern should be exercised when applying cam linearization, given that it only works for slow reacting process loops such as temperature.

Control Valve Characteristics

Control valves are characterized by their rangeability, their turndown, and their flow characteristics or flow verses stem travel behavior.

Rangeability is the ratio of the maximum controllable flow through the valve to the minimum controllable flow through the valve. Mathematically:

$$rangeability = \frac{max\ controllable\ flow}{min\ controllable\ flow} \tag{8-4}$$

In a typical installed application, control valves do not close off entirely because that might damage the valve seat or cause the valve to stick. Flow in the closed position, therefore, is typically 2% to 4% of the maximum flow.

This corresponds to a rangeability of 50 ($\frac{100}{2} = 50$) to 25 ($\frac{100}{4} = 25$). Most linear-stem motion control valves show a rangeability of 20 to 70.

Turndown is defined as the ratio of the normal maximum flow through the valve to the minimum controllable flow. Mathematically:

$$turndown\ ratio = \frac{max\ system\ flow}{min\ controllable\ flow} \tag{8-5}$$

As a practical rule of thumb, a control valve should be sized so that the maximum flow under normal installed operating conditions is approximately 70% of the maximum possible flow. As a result, the turndown as a rule is about 70% of the rangeability.

Rangeability is the inherent valve characteristic, and the turndown ratio refers to the installed valve characteristic.

Flow through a control valve depends on the extent to which the valve is open as well as the pressure drop across the valve.

Control valve flow characteristics are the relationship of the change in the valve's opening to the change in flow through the valve. The most frequently used characteristics are quick-opening, linear, and equal-percentage, shown in Figure 8-4.

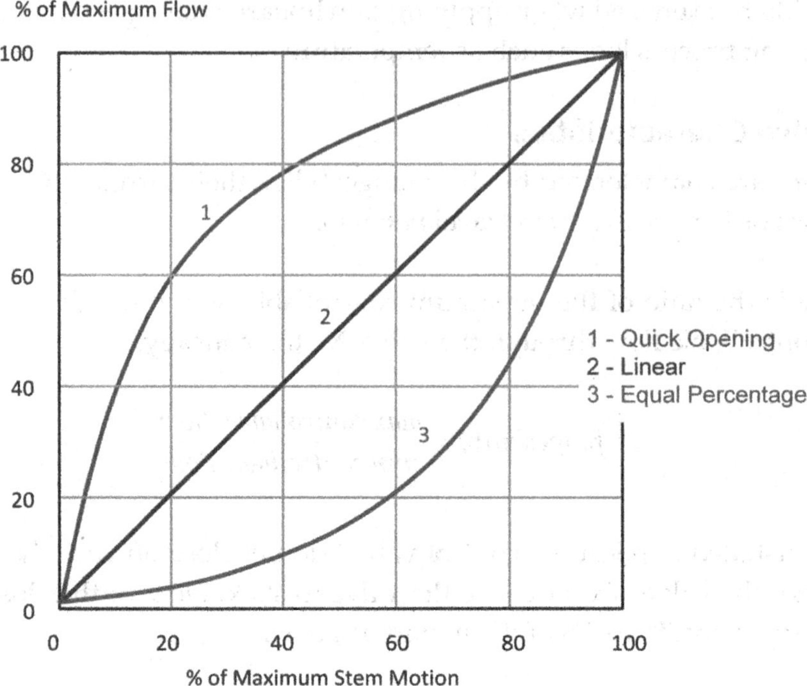

Figure 8-4. Valve flow characteristics.

Individual valve flow characteristics of valves that are available will not always fall into such orderly categories. These three types will, at least, provide the student with a systematic way to approach the many hundreds of valve trim arrangements that are available.

Valve Trim

The trim produces the different characteristics of the valve, especially the mating parts that throttle the fluid through the valve. Each process is unique and demands unique flow control characteristics. The characteristic of the valve can be achieved by reshaping the valve trim to get desired valve characteristics.

The operating parts of a valve that are usually exposed to the process fluid are called *valve trim* (see Figure 8-1), parts like the stem, plug, disk, and seating surface. The valve trim is the physical shape of the plug and seat arrangement. The shape of the valve plug determines the flow characteristics of the valve.

Different valve characterizations may be achieved by reshaping the valve trim. For instance, the plug profiles of a stem-guided globe valve may be modified to achieve the common quick-opening, linear, and equal-percentage characteristics. Figure 8-5 shows plug shapes that correspond to different valve flow characteristics.

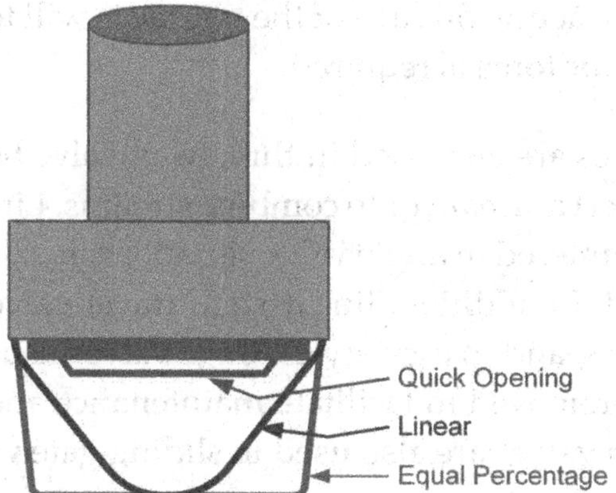

Figure 8-5. Valve trim modifications.

Valve characteristics can also be customized by changing the mechanical cam used by a positioner's feedback mechanism.

Linear-stem motion control valves have many different body styles. The most common takes the form of a globe. These valves may be either single-seated or double-seated; an example of a double-seated valve is shown in Figure 8-6. Single-seated valves are commonly employed in situations where tight shutoff is required or for valves with sizes of 1 in or smaller.

Double-seated valves generally experience leakage through the valve that is somewhat greater than in single-seated valves. This is because it is virtually impossible to close the two ports simultaneously, especially when thermal expansion and other factors are considered. The advantage of double-seated bodies, however, is that the hydrostatic

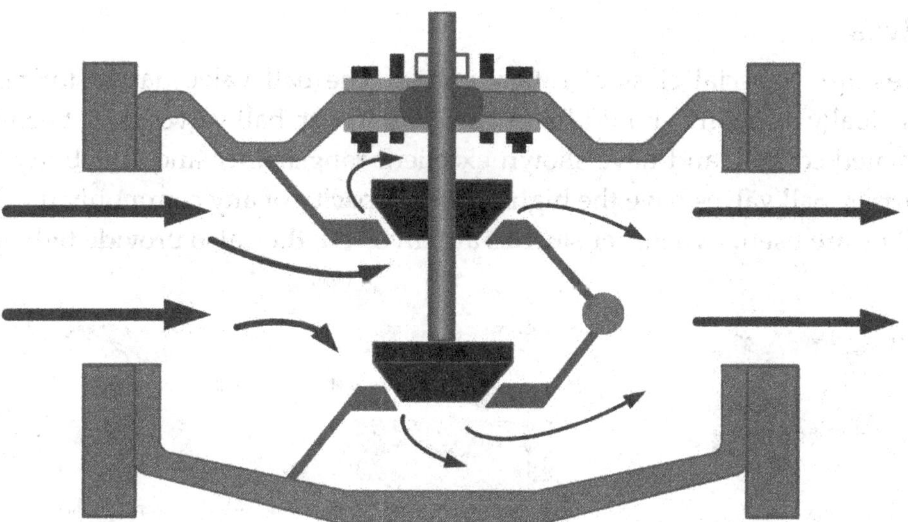

Figure 8-6. Double seated valves.

effect of the fluid pressure acting on each of the two seats will tend to cancel each other out, and much less actuator force is required.

Linear-stem motion valves are also used in three-way valve bodies where the control valve may be used to divert a stream or to combine streams. Linear-stem motion valves are also commonly encountered in angle valve situations. In such cases, the valve body is typically single-seated. In addition, linear-stem travel valves are built into Y-style bodies, in split-body styles, and in cage styles. Cage valves usually are designed so the valve trim can be easily removed to facilitate maintenance and replacement, if necessary. Linear-stem motion valves are also used in sliding-gate valve applications and in several similar, special applications.

Rotary Control Valves

Rotary-shaft control valves have the advantages of low weight, simplicity of design, relatively high flow rates, additional reliability and friction-free packing, and somewhat low initial cost. However, they cannot usually be used in sizes below 1 in.

Figure 8-7 shows several rotary valves configurations.

The most common rotary-shaft control valve is the butterfly valve, which is illustrated in Figure 8-7. Butterfly valves are used in sizes from 2 to 36 in, or larger. They are often used in applications that involve large flows at high static pressures but have limited pressure drop availability. Properly selected, the butterfly valve offers the advantages of low cost, lightness of weight, simplicity, and space-saving size. It also exhibits good flow control characteristics.

Ball Valves

Ball valves are a special class of rotary valves. The ball valve has historically been used principally as a tight shutoff hand valve. However, ball valves have been adapted for automated control, and have shown excellent rangeability and suitability for handling slurries. Ball valves have the highest flow capacity of any commonly used control valves. They are useful whenever slurries are involved; they also provide tight shutoff.

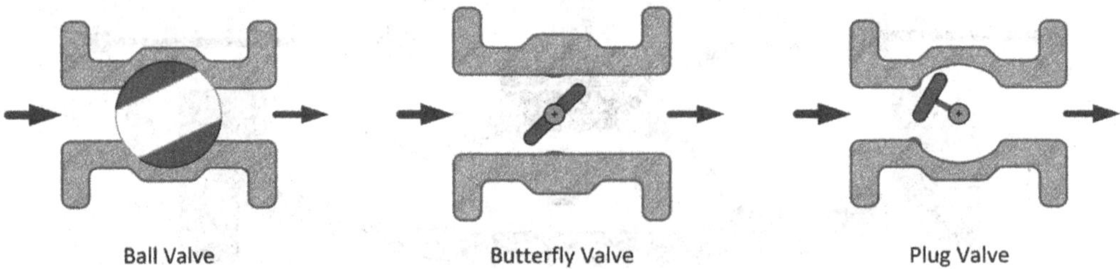

Figure 8-7. Rotary valve types.

8.3 Control Valve Selection and Sizing

There are many important factors to be considered in control valve selection. These include the following:

- The rangeability of the process and the maximum specific range of flows required by the process
- The range of operating loads to be controlled
- The available pressure drops at the valve location at maximum flow and at minimum flow
- The nature and condition of the fluid flowing through the valve

Control valve rangeability must exceed the rangeability of the process by some reasonable safety margin.

Not only must control valves have sufficient rangeability, but they must also be able to moderate flows that are well below the minimum requirement of the process. In addition, the maximum flow rate through the valve must exceed the peak requirement of the process. For many process plants, the normal range of operating loads is narrow, and in such cases, the control valve maintains nearly constant flow rates. Valves selected to operate at 60% to 70% of full capacity are usually used in such applications. There are many process units, however, in which the operating conditions vary considerably, that is, the load varies significantly. In such conditions, valves with equal-percentage flow characteristics are often used because they provide a more constant fractional sensitivity at all operating levels.

The nature and condition of the fluid significantly affect valve selection. The entire question of valve pressure drop enters not only into the sizing of the valve but also into the nature of the dynamic performance of the valve. The proper size of a control valve is most important to the overall operation of the feedback control loop. If the valve is oversized, it tends to operate only slightly open, and the minimum controllable flow is too large. In addition, the lower part of the flow characteristic curve is quite often non-uniform in shape. On the other hand, if the valve is undersized, the maximum flow needed may not be reachable.

In sizing control valves, it is standard practice to combine many terms from the basic orifice equation into the following general relationship for incompressible liquids:

$$m_L = C_v \sqrt{\frac{\Delta P}{G}} \tag{8-6}$$

where

m_L = liquid flow rate	in gallons per minutes (gal/min) (at the conditions where specific gravity G is taken)
G = specific gravity of liquid	(referred to water)
C_v = valve coefficient	in gallons per minutes (gal/min)
ΔP = differential pressure	in pounds per square inch (lb/in^2)

The *valve coefficient* (C_v) is defined as the flow rate of water in gallons per minute provided by a pressure differential of 1.0 psi through a fully opened control valve. The valve coefficient for any specific control valve must be determined by an actual test. The valve coefficient for a control valve of the linear-stem travel type is approximately equal to the square of the nominal valve size multiplied by 10.

For the flow of compressible fluids, the valve coefficient is employed with suitable conversion factors. The form of the equation is an approximation to the complete isentropic flow equation:

$$m_g = 63.3\, C_v\, Y\sqrt{\Delta P y_1} \tag{8-7}$$

where

m_g = flow rate of gas	in pounds per hour (lb/h)
C_v = valve coefficient	obtained for liquids
ΔP = differential pressure	in pounds per square inch (lb/i^2)
y_1 = specific weight	in pounds per cubic foot (lb/ft^3) at upstream conditions
Y = expansion factor	ratio of flow coefficient for a gas to that for a liquid at the same Reynolds number, varies from 0.667 to 1.0

The entire question of determining control valve size, flow rate, or pressure differential is made in most industrial applications by using special-purpose calculators or special-purpose computer programs. In actual practice, the sizing of control valves is much more complex than is illustrated here. The equations we have shown here do illustrate the basic principles involved, however. For specific details on the sizing of valves, the student should consult the books written on the subject or control valve vendors.

Power Failure

It is to be anticipated that power will fail intermittently at some time, and this may produce a loss of power or pressure to the final control element. In general, there are three potential patterns that might be encountered in power failure:

- The final control element may fail open.
- The final control element may fail in its last position.
- The final control element may fail closed.

In any given instance, it is possible that any one of these three situations may be desirable. The selection of a valve actuator must include considerable analysis to determine the failure mode.

One of the built-in characteristics of a spring-diaphragm-actuated control valve is its ability to fail-safe by the action of the precompressed actuator spring. Explicitly, to close or open the valve on air failure or the breaking of the signal wire or a controller malfunction. Cylinder-actuated valves do not fail-safe without additional equipment.

8.4 Motors

Motors find use as final control elements in position or speed-control applications. An electric motor is essentially a device that converts electrical energy into mechanical energy to produce motion. In many processes, an electric motor provides rotary or linear motion to a mechanical load. For example, motor-driven stirrers are used to mix chemicals. Electric motors find wide use in automatic process control systems because they can be accurately and continuously controlled over a wide range of speeds and mechanical loads.

Motors are electromechanical devices that use the interaction of a magnetic field and current-carrying conductors to convert the electrical energy into mechanical rotary energy. Motors are often used as discrete (on-off) control elements such as when driving a pump, conveyor belt, or other machine for the transportation of a substance.

Of all the available electrical motor types, the most common found in industrial applications is the three-phase AC induction motor.

Electric motors are generally rated in horsepower. One horsepower equals 550 ft-lb/s (foot-pound per second) or 745.7 W (watt). A wide range of designs are available to meet specific applications and are available in fractional horsepower to thousands of horsepower.

There are basically three types of motors:

1. AC motors
2. DC motors
3. Stepper motors

Replacing pneumatic actuators with electrical actuators such as a motor has not been very successful. There are however specific applications where electric actuators offer advantages over pneumatic actuators. For example, in applications where precise positioning is required; where no instrument air is available; or when the valve must fail in its *as is* position on power failure.

In most applications, a gearbox or gear-train is used to reduce rotational speed while at the same time increasing the output torque.

DC Motors

Direct current (DC) motors are used widely in low- and high-power control applications. DC motors are continuous actuators; meaning that they can produce continuous movement such as is required to move a conveyor belt or pump. The DC motor achieves this by producing a continuous angular rotation that can be used to rotate pumps, fans, compressors, and wheels.

DC motors consists of two parts, a *stator,* which is the stationary part, and a *rotor,* which is the rotating part. Four types of DC motors are available:

1. **Brushed motor** – This type of motor produces a magnetic field in a wound rotor by passing a current through a commutator and carbon brush assembly. The stator's magnetic field is produced by using either a wound stator field winding or permanent magnets. Generally, brushed DC motors are inexpensive, small, and easily controlled.
2. **Brushless motor** – This type of motor produces a magnetic field in the rotor by using permanent magnets attached to it, and commutation is achieved electronically. Brushless motors are generally smaller but more expensive than conventional brushed-type DC motors because they use Hall effect switches in the stator to produce the required stator field rotational sequence. However, they have better torque/speed characteristics, are more efficient, and have a longer operating life than equivalent brushed types.
3. **Stepper motor** – The stepper motor is a device that produces rotation through equal angles called *steps* for each digital pulse supplied to its input.

4. **Servomotor** – This type of motor is basically a DC motor with some form of positional feedback control connected to the rotor shaft. Servo motors are connected to and controlled by a pulse width modulation (PWM) type controller and are mainly used in positional control systems and radio-controlled models.

Most DC motors have a linear relationship between rotational speed and applied voltage, determined by the applied DC voltage and the output torque being determined by the current flowing through the motor armature windings. For every motor, there is a specific torque-speed curve and torque-current (power) curve (see Figure 8-8). Note that torque is inversely proportional to the speed of the output shaft. There is a trade-off between how much torque a motor delivers and how fast the output shaft rotates.

The stall torque (T_s) represents the point on the graph at which the torque is at its maximum, but the shaft is not rotating. The no-load speed (ω_n) is the maximum output speed of the motor when no torque is applied to the output shaft. The curve can then be approximated by connecting these two points with a line, whose equation can be written in terms of torque or angular velocity as follows:

$$T = \frac{T_s - \omega T_s}{\omega_n} \quad (8\text{-}8)$$

$$\omega = \frac{T_s - T}{\omega_n / T_s} \quad (8\text{-}9)$$

where

T = torque delivered in inch pounds (in-lb)
T_s = stall torque in inch pounds (in-lb)
ω = motor speed in radians per second (rad/s)
ω_n = no-load speed in radians per second (rad/s)

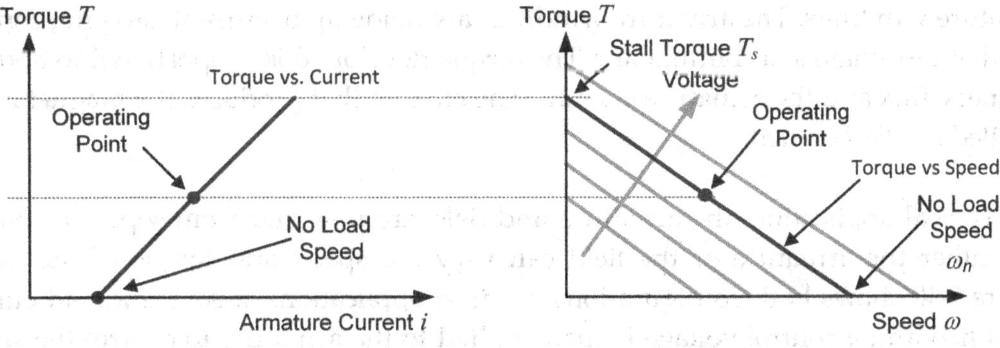

Figure 8-8. Torque-current and speed-torque curves of a typical DC motor.

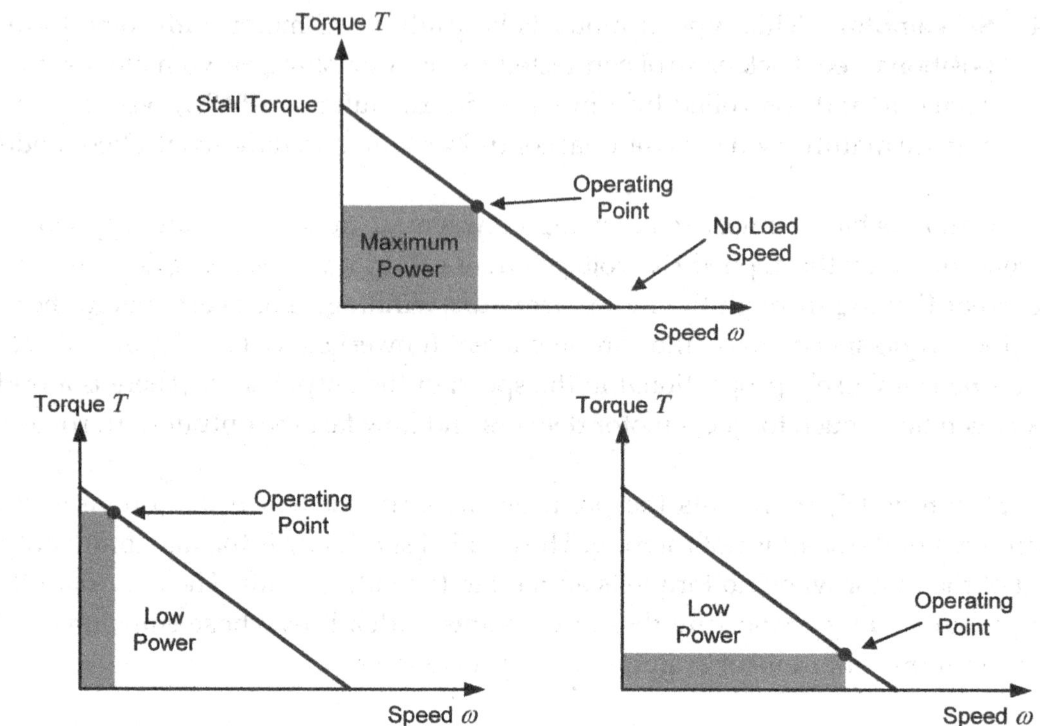

Figure 8-9. Maximum power in a DC motor.

This linear model of a DC motor is a practical approximation. If the specifications are given as two points, it is safe to assume a somewhat linear curve.

The power delivered by the motor is the product of torque and angular velocity. This corresponds to the area of a rectangle under the torque-speed curve with one corner at the origin and the second corner at the operating point on the curve (see Figure 8-9). Due to the linear inverse relationship between torque and speed, the maximum power occurs at the point where $\omega = (1/2)\omega_n$ and $\omega = (1/2)T_s$.

There are two electrical elements in a brushed DC motor: the field windings and the armature windings. The armature windings are made up of current-carrying conductors that terminate at a commutator. The torque developed is proportional to both the magnetic flux and the armature current. The current that produces the magnetic field is called the *field current*.

In a typical application, the armature and field are powered from separate sources, and either the armature or the field can vary the speed and torque of the motor. Figure 8-10 shows both configurations. In most applications, a constant field current is applied and a control voltage is then applied to the armature to control the speed.

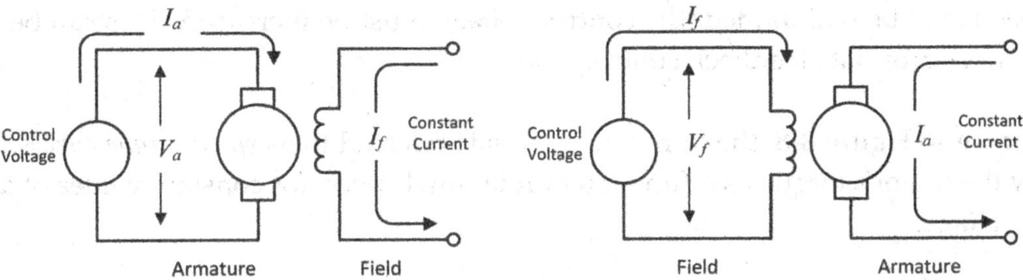

Figure 8-10. Armature and field control of a DC motor.

Using this method, the control source must be capable of supplying the motor's full power requirements.

DC servomotors used in a position control loop are based on the same principles as conventional DC motors. They are typically brushless and specifically designed for high torque and low inertia. Servo applications require that the motor operates smoothly without hesitation, so that the set point is reached smoothly, quickly, and without overshoot. This is usually achieved through high proportional gain.

Performance characteristics can be obtained from the motor's torque and speed specifications (see Figure 8-8). Keeping the field constant, the torque produced is closely proportional to the armature current. The intercept of the line and the horizontal axis is the value of current that is required to overcome the static-friction torque of the motor. The slope of the line is defined as the torque constant of the motor:

$$K_i = \frac{\Delta T}{I_a} \tag{8-10}$$

where

K_i = torque constant in inch per ampere (in/A)
T = torque in ounces per inch (oz/in)
I_a = armature current in amperes (A)

After the motor is rotating, a voltage proportional to the speed is generated in the winding. The relative rotating coil inside a magnetic field induces an electromotive force (emf). This emf acts opposite to the applied voltage causing the rotation and reduces the current flowing through the coils of the motor. The motor speed will be such that the back emf is just small enough in relation to the applied voltage to enable an armature current to drive the load. If the load on the motor is increased, the motor will slow down. As the motor slows, the back emf falls and a higher current is drawn from the source. This larger current will drive the increased load but at a lower speed.

If speed is to be maintained, the control voltage must be increased. This can be done with an appropriate feedback control loop.

Referring to Figure 8-8, the family of curves is referred to as *speed-torque curves*. They show the motor's torque as a function of rotational speed for constant values of armature voltage.

In practice, rather than specifying motor torque as a function of armature control, it is more convenient to define a torque constant in terms of armature voltage. When the motor is stalled (zero speed), no back emf is generated and the armature current is determined exclusively by the resistance of the armature. Subsequently, the following is true:

$$K_t = \frac{\Delta T_s}{\Delta V_a} \tag{8-11}$$

where

K_t = defined torque	in ounces-inch per volt (oz-in/V)
T_s = constant stall torque	in ounces per inch (oz/in)
V_a = armature voltage	in voltage (V)

The reduction in motor speed with an increasing torque is considered a damping constant and is defined as:

$$D_m = \frac{\Delta T_s}{\Delta \omega} \tag{8-11}$$

where

D_m = damping constant	in ounces per inch per radians per second (oz/in/rad/s)
T_s = stall torque	in ounces per inch (oz/in)
ω= rotational speed	in radians per second (rad/s)

The damping constant D_m represents the slope of the speed-torque curves in Figure 8-8.

When the motor is connected using field control, the armature current is held constant, and the armature torque and current are unaffected by the back emf. The torque will then be (theoretically) independent of speed, as illustrated in Figure 8-11. This is shown by the horizontal lines that illustrate torque versus speed for various values of field current.

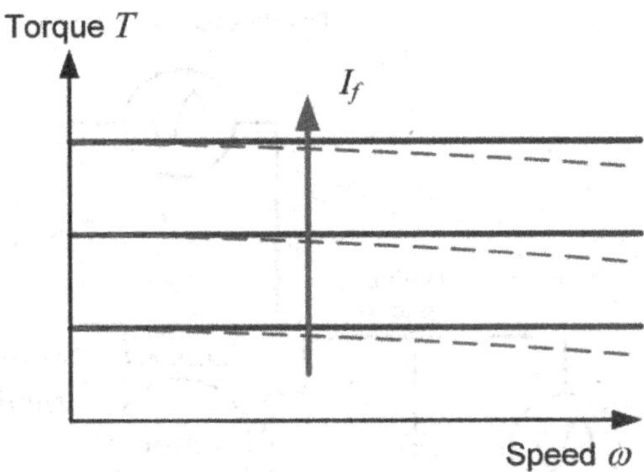

Figure 8-11. Field control speed-torque curves.

In practice, there will be some back emf as well as friction as the speed increases. As a result, armature current will drop off somewhat as the speed increases. This decrease in torque is generally advantageous in closed-loop control systems because the damping effect reduces the loop's susceptibility toward overshoot and instability.

AC Motors

Compared to DC motors, AC motors have the advantage of being less expensive, more rugged, more reliable, and practically maintenance free. Four types of AC motors are available:

1. Single-phase
2. Polyphase
3. Induction
4. Synchronous

The main advantage AC motors have over DC motors is that they are directly compatible with AC power. The most extensively used AC motor is the two-phase induction motor shown in Figure 8-12. This motor has two stator windings that are located at 90° with respect each other.

The principle of operation of the induction motor is shown in Figure 8-13. Both windings are operated from a sinusoidal source (generally 60 Hz). The reference winding is a fixed voltage, and the control winding is connected to an AC control voltage and is 90° out of phase with respect to the reference voltage. The currents through the two-stator windings produce lines of magnetic flux that pass through the rotor. These magnetic flux lines combine, and at any instant the effect of all the lines can be represented by a single vector that has some angular position relative to the axis of the rotor. As the

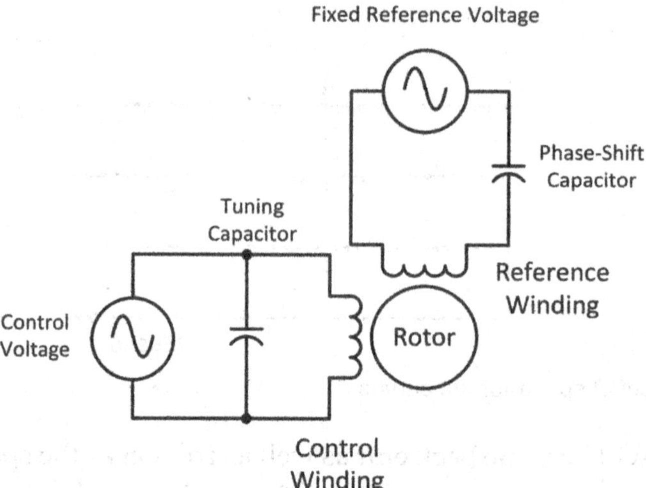

Figure 8-12. Two-phase AC induction motor.

amount of the currents change at successive instants of time, the vector angle changes. The result is a magnetic field that rotates at the frequency of the applied voltage.

As the lines of flux rotate, they cut across the rotor bars, which induce currents in the bars. These electric currents, in turn, produce a magnetic field that reacts with the stator magnetic field and causes the rotor to turn smoothly and continuously.

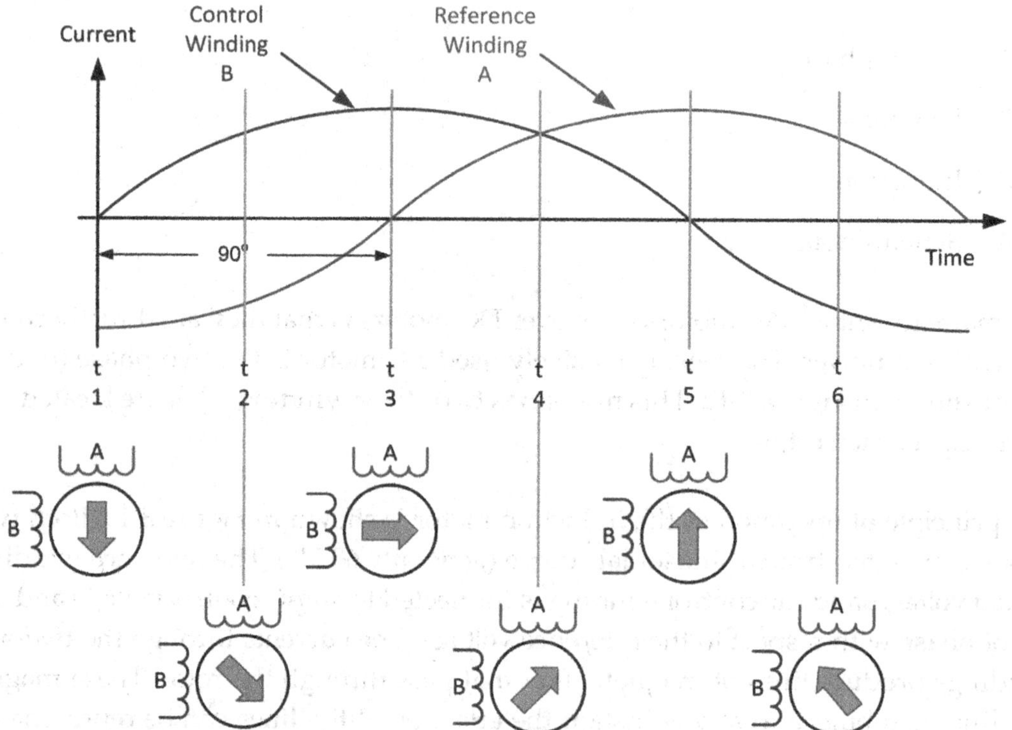

Figure 8-13. Rotation of a magnetic field in a two-phase motor.

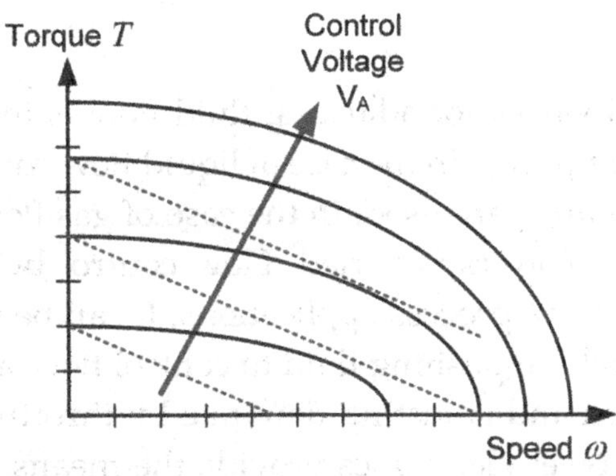

Figure 8-14. Speed-torque curves of an AC motor.

In a typical control application, the control winding voltage is variable. The resultant motor characteristics are shown in the speed-torque curves of Figure 8-14. These curves show that the stall torque is proportional to the control voltage.

The effect of load torque is to reduce the motor's speed, which causes a greater difference in speed between the rotating magnetic field and the motor. The induced magnetic field will increase with this difference, creating additional torque to pull the load but at a lower speed. Although the internal effects of rotor slowdown are somewhat more complex in the induction motor than in the DC motor, the external effects are about the same.

We can derive AC motor constants that are comparable to those of the armature-controlled DC motor. In two-phase AC induction motors, a tuning capacitor is added in parallel with the control winding (see Figure 8-12) to help match the output impedance of the control circuit to the input of the AC motor. The capacitor is usually chosen to produce a unity power factor, so the parallel resonant circuit will have a high input impedance, providing minimum power consumption on the control circuit. The equation for determining the input tuning capacitor is:

$$C = \frac{X_L}{2\pi f Z^2} \tag{8-12}$$

where

C = capacitance	in farads (F)
X_L = control winding inductive reactance	in ohms (Ω)
Z = control winding impedance	in ohms (Ω)
f = excitation frequency	in hertz (Hz)

8.5 Pumps

An alternative to control valves for adjusting fluid flow is to regulate the amount of fluid flow with the use of a pump. In the case of liquid flow control, this would take the form of variable-speed pumps are used. In the case of gas flow control, the rotational speed of compressors or blowers is varied. Flow control by machine speed control makes a lot of sense for some process applications. It can be more energy-efficient to vary the speed of the machine pushing fluid to control flow, as opposed to letting the machine run at full speed and adjusting flow rate by throttling the machine's outlet using a valve. Modern power electronics provide the means to electronically control the speed of almost any type and size of electric motor, using variable-speed motor drives.

There are certainly situations where using a variable-speed pump makes sense, but there are several reasons why this is not done in many situations:

- Variable-speed pumps often cannot increase or decrease fluid flow rates as rapidly as control valves can.
- Control valves can positively halt flow; a stopped pump or blower will not necessarily prevent flow from going through.

For these and other reasons, pumps tend not be used directly to control fluid flow.

Pumps can be divided into two main types:

1. Centrifugal
2. Reciprocating

Rotary pumps convert rotational energy of a motor into velocity or kinetic energy and then into pressure of a fluid that is being pumped.

Reciprocating pumps are commonly used to control the rate at which a volume of fluid is injected into a process stream or vessel, and they can be highly accurate.

Centrifugal Pumps

A centrifugal pump is the most common type of pump used today. It uses an impeller, which is a rotating blade, to generate suction that moves fluid through pipes. The rotating impeller creates what is known as *centrifugal force*, giving this pump design its name. The pump can be driven by an electric motor or engine (Figure 8-15).

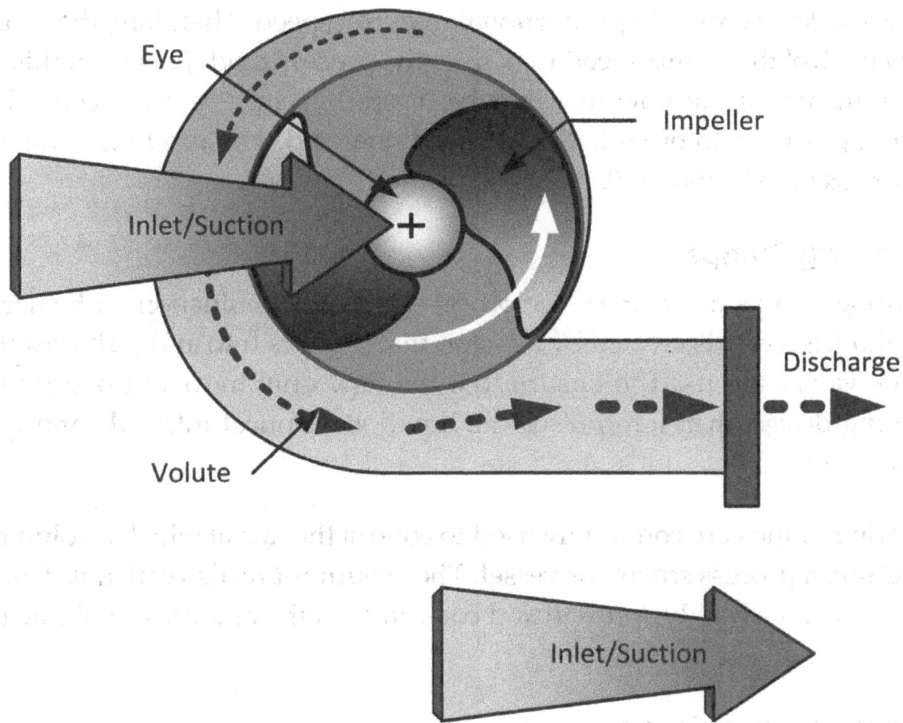

Figure 8-15. A typical centrifugal pump.

The energy changes occur in two main parts of the pump, the impeller and the volute. The impeller is the rotating part that converts energy into kinetic energy. The volute is the stationary part that converts the kinetic energy into pressure.

Centrifugal Force

Liquid enters the pump suction and then the eye of the impeller. When the impeller rotates, it spins the liquid sitting in the cavities between the impeller vanes outward and imparts centrifugal acceleration. As the liquid leaves the eye of the impeller, a low-pressure area is created at the eye, causing more liquid to enter the pump inlet.

Centrifugal pumps are classified into three general categories:

- **Radial flow** – A centrifugal pump in which the pressure is developed wholly by centrifugal force
- **Mixed flow** – A centrifugal pump in which the pressure is developed partly by centrifugal force and partly by the lift of the vanes of the impeller on the liquid
- **Axial flow** – A centrifugal pump in which the pressure is developed by the propelling or lifting action of the vanes of the impeller on the liquid

In most cases, flow is directly proportional to pump speed. Therefore, the control loop must only control the pump speed to deliver the process fluid. However, this does not guarantee the amount of flow that may be desired. In most process control applications, a pump is used to provide the required flow and pressure to the control valve, as was discussed in Section 5-9.

Reciprocating Pumps

Reciprocating pumps use a linear reciprocal stroke in combination with check valves to pump fluid. Suction is created by a piston that plunges into and pulls out the material. Check valves are used to ensure that the flow only moves in one direction. A reciprocating design therefore pulses the liquid at identical intervals and quantities (see Figure 8-16).

Reciprocating pumps are commonly used to control the rate at which a volume of fluid is injected into a process stream or vessel. These pumps are also called *metering pumps* because they can be highly accurate and consistent in the amount of volume they dispense per cycle.

Pump Performance Curve

The best way to select an industrial pump is by using a pump curve. Being able to read and understand a pump curve is a fundamental skill in the selection of a pump for a specific application. Pump performance is defined as the relationship between the flow rate and the total head developed by the pump.

Pump head is a measure of the power of a pump and is measured in meters (or feet). The greater the pump head, the greater the pressure that the pump can generate.

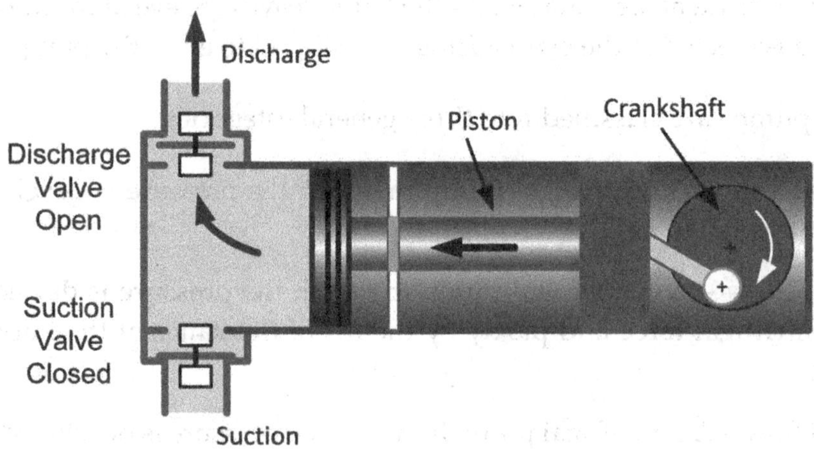

Figure 8-16. A typical reciprocating pump.

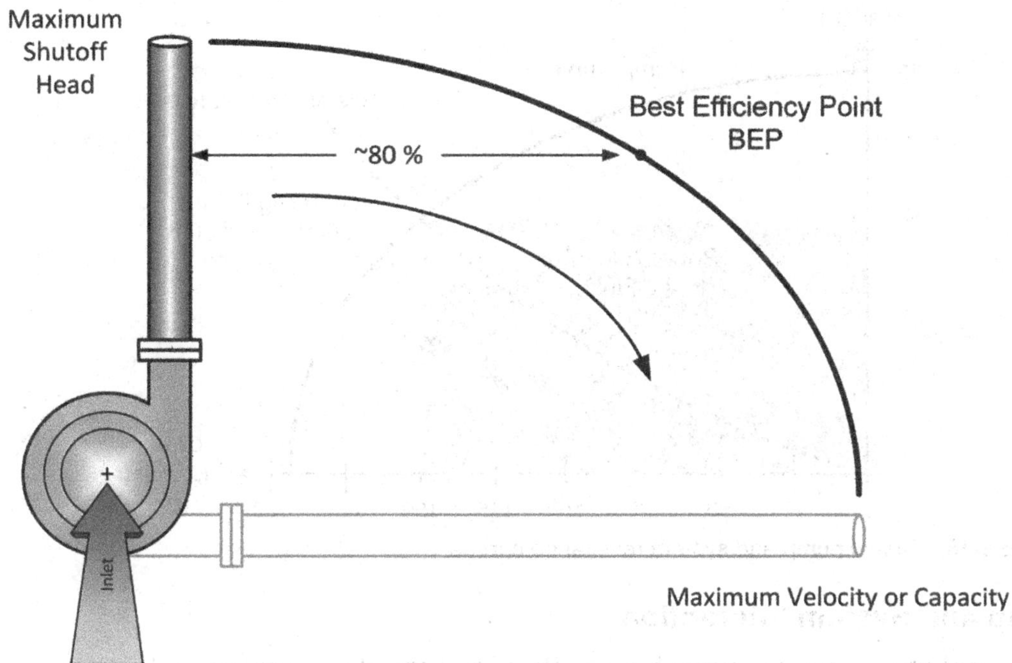

Figure 8-17. Construction of a pump curve.

Pump head can be measured by placing a tube on a pump's discharge and measuring the maximum height to which it can pump the water (see Figure 8-17). As the pump is rotated back to its horizontal position, it reaches its maximum velocity or capacity.

If the pump is operating as designed, it must operate at a position on the pump curve based on the resistance of the system into which it is pumping. An example of this was discussed in Section 5-10. The pump curve is a critical piece of information in reference to the pumping system.

Pump Efficiency

Efficiency is the ratio of the energy added to the liquid, called *water horsepower*, to the energy applied to the shaft, called *break horsepower*. This information is usually provided by the manufacturer, typically on the pump curve. The best efficiency point (BEP) occurs where there is the least amount of vibration of the pump, and the pump's efficiency decreases to the right and left of the BEP.

Power Added to the Fluid

The area formed by a rectangle that corresponds to the area under the pump curve with one corner at the origin and the second corner at the operating point on the pump curve represents the amount of power added to the fluid (see Figure 8-18).

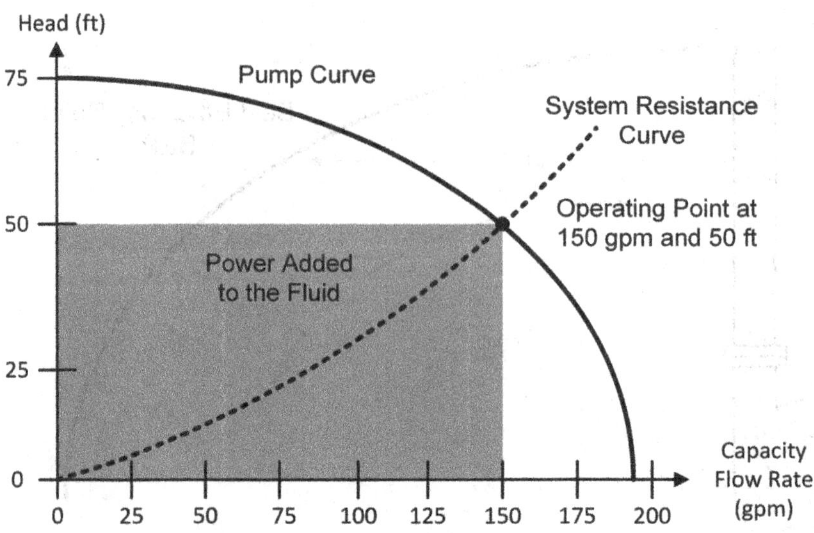

Figure 8-18. Typical pump and system resistance curves.

Pump and System Interaction

The pump must operate somewhere on its curve, and it will operate at the intersection of the pump curve and the system's resistance curve (see Figure 8-18).

8.6 Solenoids

Solenoid is the generic term for a coil of wire used as an electromagnet. The device creates a magnetic field from electric current and uses the magnetic field to create linear motion (see Figure 8-19). Common applications of solenoids are operating on-off valves.

A solenoid's component parts consist of a coil, to carry current and generate ampere-turns; an iron shell or case, to provide a magnetic circuit; and a movable plunger or pole to act as the moving working element. A main objective in the design of a solenoid is to provide an iron path capable of transmitting maximum magnetic flux density (B in Figure 8-20) with a minimum of energy input. A second objective is to get the best relationship between the variable ampere-turns and the working flux density in the air gap.

Magnetic flux lines are transmitted through the iron shell and the air gap between the shell and the plunger. An iron path is much more efficient than air, but the air gap is required to permit the movement of the plunger. The force of a given solenoid is inversely proportional to the square of the distance between the pole faces. The lowest force is generated when the distance is widest, and the strongest when the distance is smallest. The force is also directly proportional to the magnetic field and magnetic flux

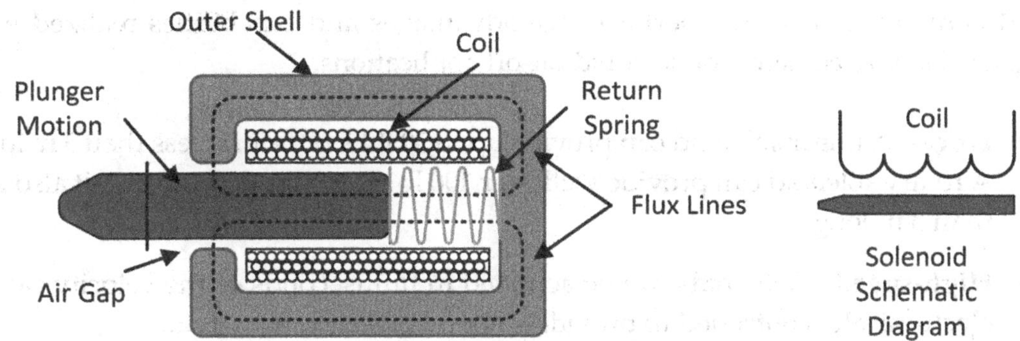

Figure 8-19. Construction and schematic diagram of a solenoid.

density generated through the iron path. However, increasing the magnetic field will not necessarily increase the magnetic flux density. Saturation of the iron path occurs as the magnetic field is increased. A point is reached at which the iron no longer is capable of carrying any increase in flux density (see Figure 8-20).

As the pole pieces are moved together or when input power is increased, the flux density of the magnetic circuit increases until the iron saturates. Beyond this point, any further increase in power only serves to add heat without a significant increase in force.

The number of copper wire turns, and the magnitude of the applied current are referred to as *ampere-turns*. The absolute value of the magnetic flux is determined by both the ampere-turn and the permeance of the magnetic circuit within the solenoid. The permissible temperature rise limits the magnitude of the power input. When using a constant voltage, the heat generated makes the coil less efficient because it reduces the ampere-turns and, therefore, the flux density and the force generated.

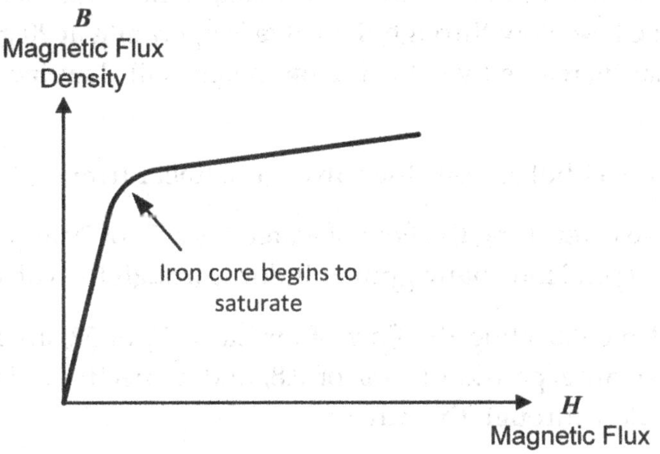

Figure 8-20. Saturation of the iron core.

Listed here are some of the performance advantages and capabilities realized with using a solenoid's actuator for solenoid on-off applications.

- **Force** – A linear solenoid can provide up to 30 lbf from a unit less than 3 in long. A rotary solenoid can provide well over 100 lb · in of torque from a unit also less than 3 in long.
- **High-speed** – Solenoids can be actuated in milliseconds or the velocity can be electronically controlled to provide smooth, noiseless actuation.
- **Minimum size** – Solenoids can pack more work per cubic inch than motors.

8.7 Summary

- This chapter provided a brief overview of some of the more common actuators and final control elements used in process control.
- Control valves are the single most used final control element used in process control. Most control valves are actuated using pneumatics actuators. Control valve characteristics can be customized by using different valve trim and with the use of a positioner. Positioners are also used to increase the required pressure needed for valves.
- Motors and solenoids are alternate means to actuate a control valve. Motors can provide more accurate control, and solenoids can provide faster valve closure.

Exercises

8.1 An equal-percentage control valve has 10 gpm flowing through it. When the air-to-valve signal increases by 0.5 psi, flow through the valve increases by 1.5 gpm. Much later, the base flow through the valve is operating at 20 gpm when the air-to-valve signal increases by 0.5 psi. How much will flow increase through the valve?

8.2 Repeat Exercise 8.1 but assume the valve has a linear trim.

8.3 A control valve controlling the flow of water has a C_v of 20 and an assigned pressure drop of 10 psi. How many gpm will flow through the valve?

8.4 A control valve controlling the flow of air has a C_v of 32, an assigned pressure drop of 10 psi, an expansion factor of 0.8, and a specific weight of 3.103 lb/ft_3. Calculate the flow through the valve.

8.5 Given an equal-percentage valve installed with 10% of the dynamic pressure drop across the valve, assume the valve operates about 50% to 70% open under normal conditions. Does the valve act more like an equal-percentage valve or more like a linear valve?

8.6 Consider several actual feedback loops, such as a home heating system, an oven, and the air compressor controller at a service station. For each loop, decide how you would like the final element to "fail" in the event of power loss.

9 Controllers

The controller is the element of the process control loop that processes information needed to perform the decision-making function. The controller can be considered the brain that enables automated systems to operate without human intervention.

The feedback controller determines the changes that are needed in the manipulated variable to compensate for disturbances that upset the process or for changes in the set point (SP). A clear understanding of the computational aspect of the controller is necessary for proper use and tuning of the process control loop.

Learning Objectives

- Understand the computational ability of a proportional-integral-derivative (PID) controller
- Explain proportional control action and its advantages and disadvantages
- Explain integral (reset) action and its advantages and disadvantages
- Explain derivative (rate) control action and its advantages and disadvantages

9.1 The PID Controller

Most process control is done using a PID controller. This is a special-purpose computer that uses the set point (SP) and measured value (MV) of the process as its input or forcing function, and calculates the changes needed in the manipulated variable to maintain the controlled variable at its desired set point despite changes in set point, load changes, or disturbances.

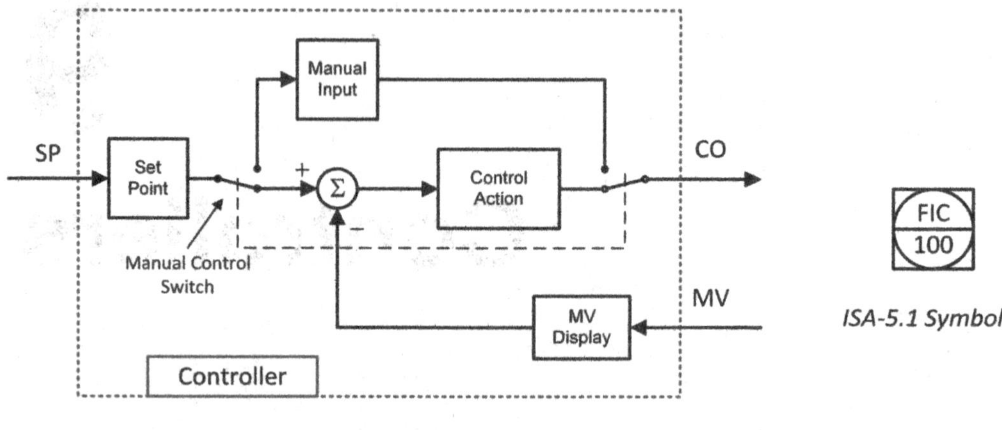

Figure 9-1. Block diagram and ISA symbol of a typical controller.

A PID controller is essentially a device or instrument that maintains a feedback control loop at its desired set point, whether that is the temperature in a room, a quadcopter, or the pressure in a vessel. Most (over 90%) controllers in industry are PID controllers.

Figure 9-1 is a simplified block diagram of a typical PID controller. The controller is symbolized using ISA-5.1[1] symbols as indicated.

The input applied to the controller is the MV and the required SP. The controller first calculates the error signal, which is the difference between the SP and the MV, which is the feedback signal. The controller then calculates the change needed in the controlled variable to compensate for disturbances that upset the process or changes in the SP. The controller responds to these changes by producing a controller output (CO) signal that drives the actuator to alter the process variable until the error signal is reduced to zero.

The controller may be as simple as a mechanical lever like the one in a common toilet tank as we saw in Chapter 1; or as complicated as a computer, a programmable logic controller (PLC) or a programmable automation controller (PAC). It may control one process or several simultaneously. The controller may be analog, digital, or pneumatic, or a combination of all three.

Most process controllers are designed as general-purpose industrial process controllers, although some have inputs that can be connected directly to thermocouples or

1. ANSI/ISA-5.1-2009, *Instrumentation Symbols and Identification* (Research Triangle Park, NC: ISA [International Society of Automation]).

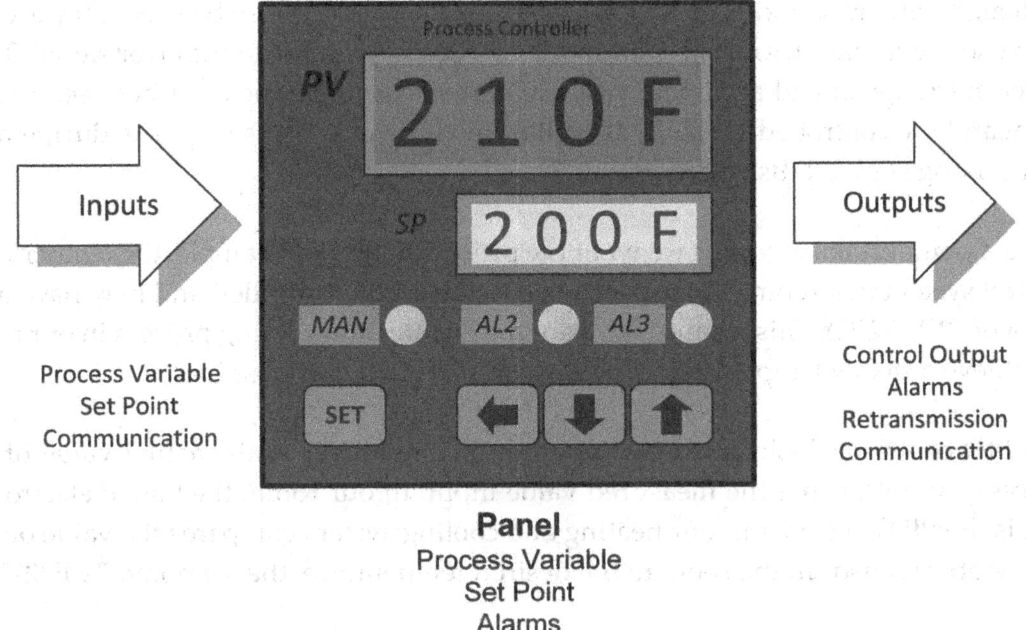

Figure 9-2. Front panel of a typical PID controller.

thermistors in addition to standard input ranges of 4 to 20 mA or 1 to 5 V. Its user interface includes custom programming of alarm messages in addition to all the standard required settings.

A typical PID controller offers a programmable universal process input and a variety of optimal outputs that allow flexibility in different applications. Many controllers also provide serial communications such as RS-485 or TCP/IP. Many offer sophisticated control algorithms in addition to standard PID control, including adaptive tuning. A typical PID controller is shown in Figure 9-2.

The controller inputs and outputs signal are all functions of time. The time required for the controller to respond to the error signal depends on the control mode used.

There are four control modes of operation that are commonly performed by the controller section of a closed-loop system:

1. On-off
2. Proportional
3. Proportional-integral
4. Proportional-integral-derivative

Although derivative and proportional-derivative modes can also be used, they are seldom used without integral. All four modes of control respond to the error signal. They differ in the speed and accuracy with which they eliminate the error between the set point and the controlled variable, as well as the type of process response during a set point change or load disturbance.

The set point is the value that we want the process to be. For example, the temperature control system in a room can be controlled using a PID controller, and may have a set point of 72°F (22°C). This means "we want the heating and cooling process in our room to achieve a steady temperature of as close to 72°F (22°C) as possible."

The PID controller looks at the set point and compares it to the actual value of the process variable, using the measured value input. In our room, the box of electronics that is the PID controller in our heating and cooling system compares the value on the temperature sensor in the room to the desired temperature, the set point, 72°F (22°C).

If the set point and the process value are the same, then no action is required of the controller; it will set its output to zero or to some value that maintains the process variable at its desired value.

However, if there is a discrepancy between the set point and the process value, an error is generated and corrective action is needed. In our room example, this will either be cooling or heating depending on whether the process value is higher or lower, respectively, than the set point.

This process is repeated until the room has reached 72°F (22°C) and there is no error. Then a disturbance hits the system and the controller must take action again. In our room, the disturbance may be the sun beating down on the roof, raising the temperature of the air inside, or doors or windows opening.

Unfortunately, if we want peak performance from our process control loops in the real world, we need a controller that is a bit more complicated than the one described here.

9.2 On-Off Control

On-off control is the most basic of control modes. In this mode, the system has only two states, on and off. One state is used when the process value (e.g., temperature, level, or flow) is above the desired value, and the second state is used when the controlled variable is below the desired value.

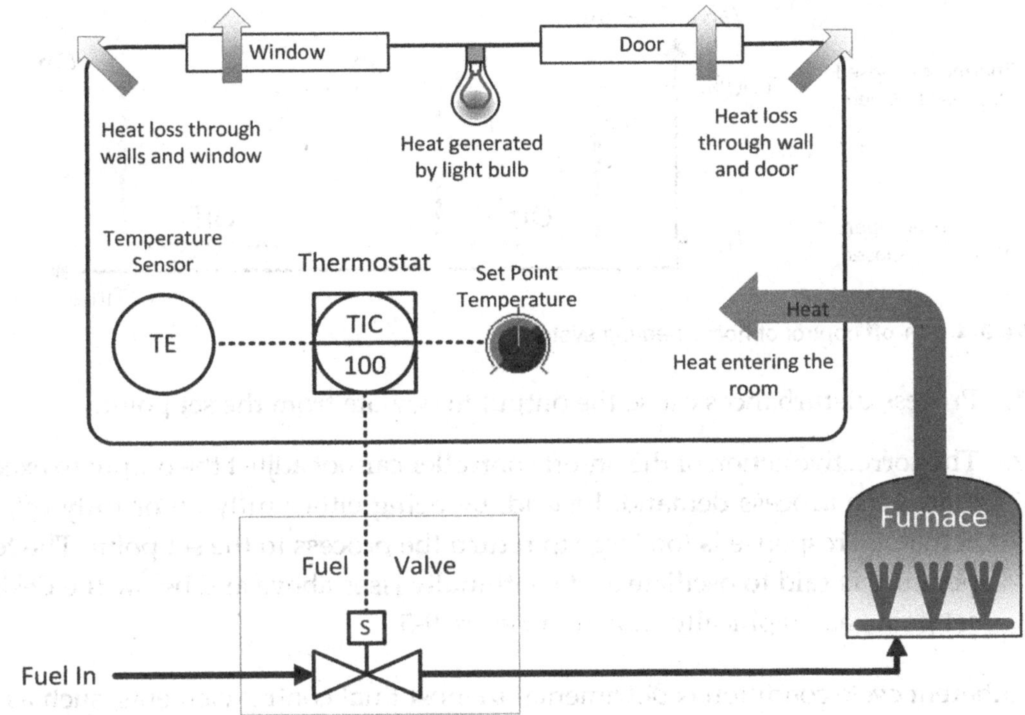

Figure 9-3. Home heating system.

The desired value is called the set point, and the process variable is the parameter we are controlling (e.g., temperature, level, or flow). The on-off control mode is also referred to as *two-position control* or *bang-bang control*.

The home heating system shown in Figure 9-3 illustrates this mode of control.

The thermostat is the measuring device. When the room temperature (the process variable) falls below the desired setting (the set point), the thermostat closes a switch that is connected to a solenoid fuel valve in the furnace room, as shown in Figure 9-4. With the switch closed, the valve is fully opened. The furnace turns on and begins to generate heat. When the room temperature rises above the desired setting, the thermostat switch opens the switch connected to the furnace fuel valve. An open switch closes the fuel valve to extinguish the flame. With the furnace off, the temperature in the room continues to rise before it begins to fall. When the temperature is low enough, the furnace turns back on as shown in Figure 9-4.

Because the process variable (temperature) must deviate from the set point to cause control action, the process response will continually cycle. The cycling occurs because of two factors:

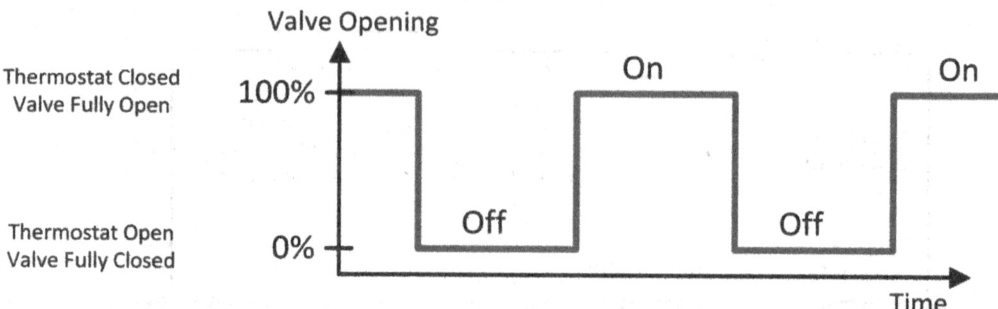

Figure 9-4. On-off control of home heating system.

1. Process disturbances cause the output to deviate from the set point.
2. The corrective action of the on-off controller cannot adjust the output to exactly match the process demand. Instead, by being either fully on or fully off, the actuator's response is too large to return the process to the set point. The temperature is said to oscillate as it continually rises above and below the desired set point, as graphically shown in Figure 9-5.

The inherent cycle condition is detrimental to most final control elements, such as the fuel valve, pumps, and relays. By turning the output on and off frequently, the rapid oscillation wears down equipment and shortens its life cycle. But the main impact is that it becomes almost impossible to be on the set point.

Differential Gap

To prevent this rapid cycling, the time between the oscillations can be lengthened by adding an on-off differential gap to the controller function. The differential gap is also referred to as the *deadband,* or *hysteresis.* The differential gap forces the controlled variable to move above or below the desired set point by a specified amount before the controlled action will change again. Most room thermostats have a built-in differential gap of approximately ±3°F (±2°C)[2].

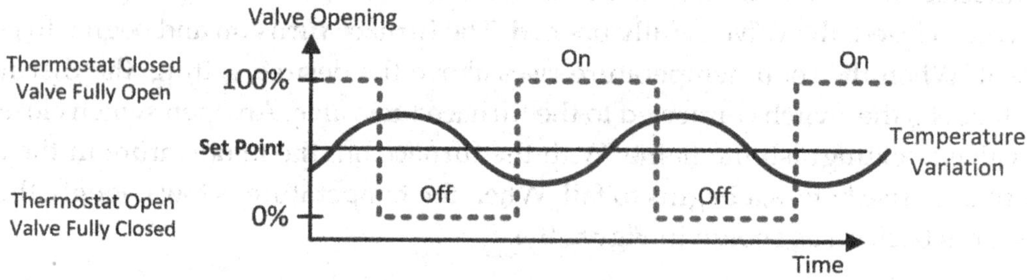

Figure 9-5. Graph illustrating temperature oscillation above and below the desired set point as the fuel valve is opened and closed.

2. These are typical nominal values used in home thermostats.

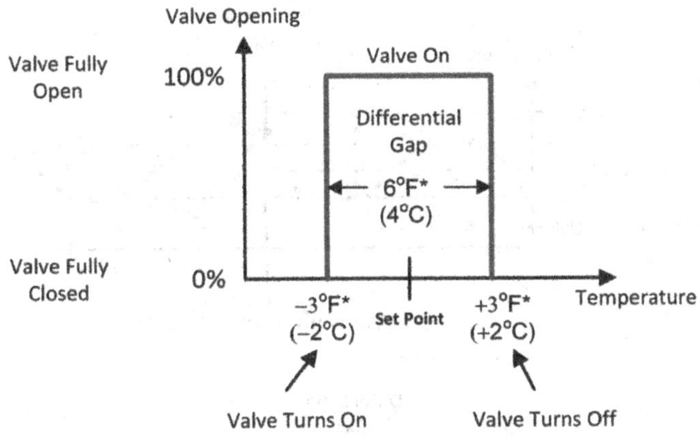

Figure 9-6. Differential gap of an on-off thermostat.

Figure 9-6 illustrates the differential gap function added to the thermostat device. In the example, the temperature must rise 3°F (2°C) above the desired SP before the furnace will turn off. To turn the furnace back on, the temperature must fall 3°F (2°C) below the desired SP. The *differential gap* is defined as the smallest change in the controlled variable that causes the value to shift from on to off, or from off to on. Therefore, the differential gap for the thermostat in this example is 6°F (4°C).

The differential gap can also be expressed as a percentage of the full range of the controlled device.

$$\%\ differential\ gap = \frac{differential\ gap}{total\ control\ range} \times 100\% \tag{9-1}$$

Example

The temperature range on a typical home thermostat with a differential gap of 4°C is between 5°C and 30°C.

What is its percentage differential gap?

Solution

The thermostat's full range is (30 – 5)°C = 25°C. Therefore

$$\%\ \text{differential gap} = \frac{4°\text{C}}{25°\text{C}} \times 100\% = 16\%$$

This means a temperature variance of 4°C represents 16% of the full control range.

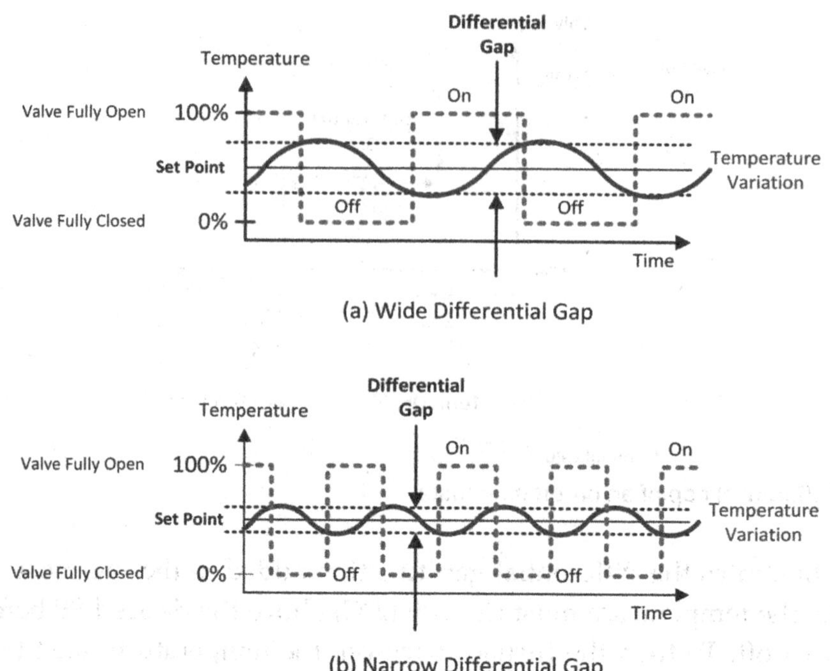

Figure 9-7. Comparison of systems with wide and narrow differential gaps.

The graph in Figure 9-7(a) illustrates the operation of the thermostat, the fuel valve in the furnace, and the resulting room temperature. It shows that the room temperature does not respond instantly after the fuel valve is turned on or off. For example, after the thermostat turns the fuel valve off, the furnace and ducts contain enough heat so that the temperature in the room will not immediately begin to fall. Likewise, the temperature in the room does not rise as soon as the furnace turns on. A certain amount of time is required before the heat generated inside the furnace travels to the location, heats up the room, and is measured by the thermostat. This lagging effect of the temperature behind the thermostat switching action is called hysteresis.

Figure 9-7(b) shows the effects of narrowing the differential gap. The narrow gap causes an increase in cycling with a smaller deviation from the desired set point. The wider differential gap shown in Figure 9-7(a) causes less frequent cycling, but at the expense of a larger deviation from the desired set point. A compromise must be made between frequency of cycling and deviation from the set point.

9.3 Proportional Control

Proportional (P) control action provides better control than on-off control, primarily because proportional output operates linearly anywhere between fully on and fully off. As its name implies, its output changes proportionally to the input error signal.

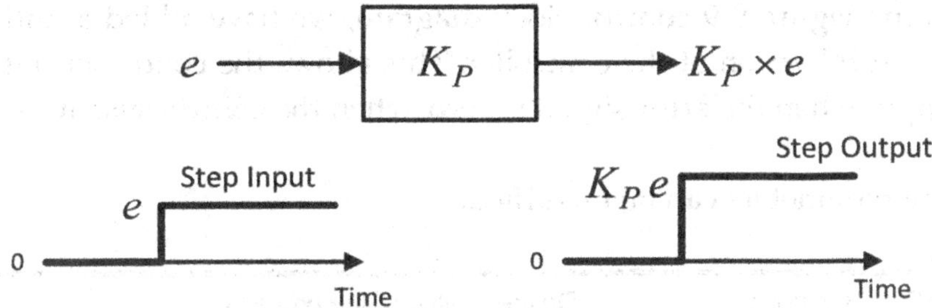

Figure 9-8. Proportional action.

The greater the error, the more the output responds. This action returns the controlled variable to the desired set point value without the rapid cycling of on-off control, provided that any initial offset was removed at the set point.

A proportional controller changes its output in proportion to the error signal. In block diagram form, proportional control action is represented as shown in Figure 9-8.

A proportional block amplifies or attenuate (reduces) the input in relation to the output. This can be represented graphically using a step input as shown in Figure 9-8. For proportional-only control, K_P is simply a proportionality constant.

Typically, a process connected to a proportional controller is configured as shown in Figure 9-9. Because we are concentrating on the controller, we are making the assumption that the controller output is directly connected to the plant input. This means we are assuming that the transfer functions of the actuator, the plant, and the measuring device are all equal to 1, meaning they do not contribute to the analysis of the controller. In practice, of course, this is a gross simplification. However, it will help us to fully understand the controller's contribution to a process control system.

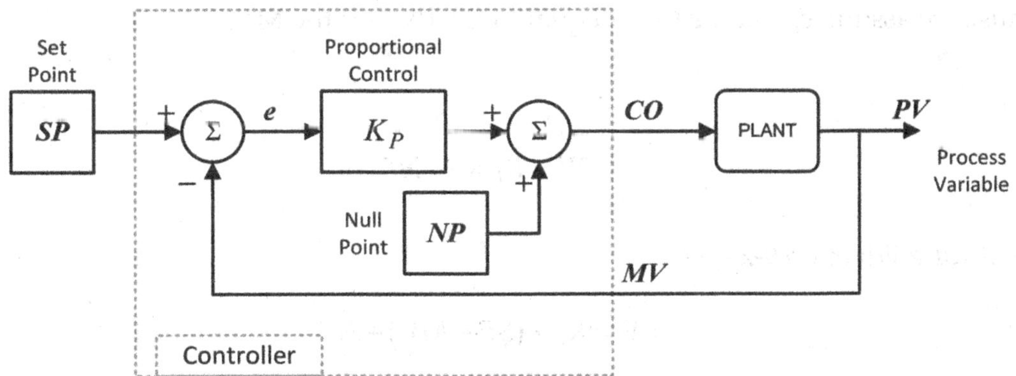

Figure 9-9. Proportional controller block diagram.

Note that in the Figure 9-9 control block diagram, we have added a null point (NP) block to the control action of the controller. This allows the controller output to have a specific output when the error signal is zero, when the control system is balanced.

The following parameters can be identified:

SP	set point	Desired value of the process
MV	measured value	Measured value of the process
e	error	Deviation of the desired value from the measured (actual) value
NP	null point	Adjustment to set the nominal controller output and to reduce offset
CO	output	Controller output
PV	process value	Actual value of the process
K_P	proportional gain	Controller gain

Based on our block diagram:

$$e = SP - MV \tag{9-2}$$

In a proportional controller, the CO is fed back through the process as the MV to the controller input to adjust the output for changes in the SP or to adjust for load changes or disturbances.

The proportional controller can be mathematically described from the block diagram in Figure 9-9 as:

$$CO = K_P \times e + NP \tag{9-3}$$

Because we assumed that the CO is equal to the PV and the MV:

Then

$$PV = K_P \times e + NP$$

Substituting Equation 9-2 gives:

$$PV = K_P \times (SP - MV) + NP$$

But $$MV = PV$$

Then

$$PV = K_P \times (SP - PV) + NP$$
$$PV = K_P SP - K_P PV + NP$$
$$PV + K_P PV = K_P SP + NP$$
$$PV(1 + K_P) = K_P SP + NP$$
$$PV = \frac{K_P SP + NP}{1 + K_P} \qquad (9\text{-}4)$$

Equation 9-4 describes the contribution the proportional controller makes to the control system and can be used to calculate the actual steady-state process value when changes in the SP are initiated. However, the equation does not account for possible disturbances added to the process.

Another point to notice is that we did not solve for the proportional controller's transfer function because the block diagram includes a null point.

Proportional control is used primarily in systems where manual adjustment of the null point is possible, or disturbances are not permanent.

Example

Assume the following conditions:

$$SP = 50\%$$
$$K_P = 1$$
$$NP = 50\%$$

Then when the *SP* changes from 50% to 60%, calculate the new output:

$$PV = \frac{K_P SP + NP}{1 + K_P} = \frac{1 \times 0.6 + 0.5}{1 + 1} = 0.55$$
$$= 55\%$$

So, the new MV also changes to 55%.

The output is now not equal to the new SP but has an offset of (60 – 55)% = 5% of full scale.

Note that the desired level of 60% has not been reached.

Next, assume that we now increase the proportional gain (K_P) from 1 to 4, giving us:

$$SP = 50\%$$
$$K_P = 4$$
$$NP = 50\%$$

Then when the SP changes from 50% to 60%, we get:

$$PV = \frac{K_P SP + NP}{1 + K_P} = \frac{4 \times 0.6 + 0.5}{1 + 4} = 0.58$$
$$= 58\%$$

Note that the desired level of 60% is still not reached but it is much closer.

This example shows that for a proportional controller, a change in set point will be corrected but will never reach the desired response for the new set point without readjusting the null point. Increasing the K_P will improve the response. This inability to reach the set point is referred to as *offset error.*

This example has ignored the response of the plant, actuator, and sensor/transmitter.

In practice, in a system with a large capacitance (e.g., a liquid tank or a heated room), increasing the K_P will result in system oscillation and most likely an unacceptable response. Figure 9-10 shows an example of a typical temperature system response with an increase in the SP from 15°C to 20°C, with a K_P of 4.

Usually, the NP would be adjusted so that the controller output is the same as the desired SP.

A better way to adjust for offset correction is needed. This is integral control and will be discussed in Section 9-4.

The K_P of the controller is also referred to as the *proportional sensitivity* of the controller. It indicates the change in the manipulated variable per unit change in the error signal. In a sense, the proportional sensitivity or gain is an amplification or attenuation and represents a parameter on a piece of hardware that must be adjusted by an operator. In effect, the K_P is a knob or a number to be adjusted in the controller.

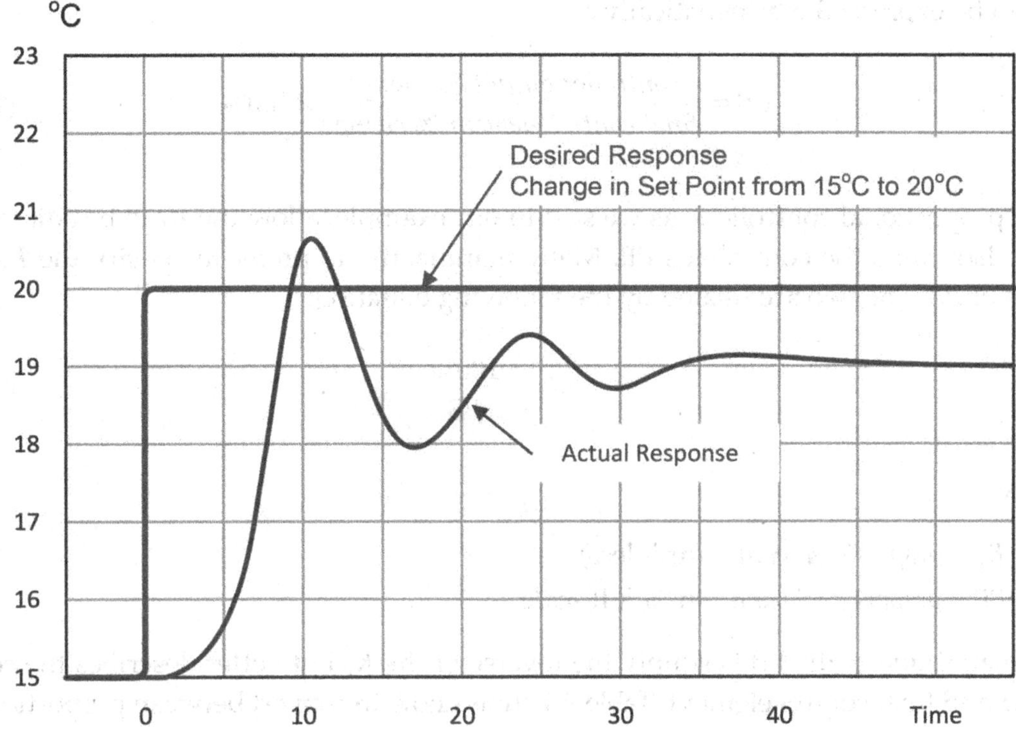

Figure 9-10. Typical response of a change in the SP for a large capacitance system.

Proportional Band

A proportional controller only operates within a specific range of error and proportional gain (K_P). The boundaries of this range correspond to the zero and the maximum value of the controller output. When expressed in normalized units, the range is from 0% to 100%.

The K_P (amplification or attenuation) can also be expressed as *proportional band* (PB).

PB is defined as:

The percentage change in error (input) that is required to produce a 100% change in output as a result of proportional control action.

This can be restated as:

The percentage change in the controller output, which causes the final control element to go through 100% of its range.

PB can be expressed mathematically as:

$$PB = \frac{controller\ output\ \%\ change}{final\ control\ element\ \%\ change} \times 100\% \qquad (9\text{-}5)$$

Most proportional controllers, as we saw in our example, allow the K_P to be adjusted. This also limits the controller's PB. Many manufacturers prefer to specify the K_P in terms of PB. The two are related by the following equation:

$$K_P = \frac{100\%}{PB} \qquad (9\text{-}6)$$

where

$K_P = proportional\ gain$ (unit less)
$PB = proportional\ band$ in % full scale

The magnitude of the PB is simply the inverse of the K_P but better describes the controller and final control element. Table 9-1 shows how to convert between proportional gain and PB values.

A wide proportional band (high percentages of PB) correspond to a less aggressive response, and a narrow proportional band (low percentages of PB) corresponds to a more aggressive response. This is graphically illustrated in Figure 9-11, which shows the effects of varying PBs.

The width of the PB setting on a controller determines how much controlled variable change is required to cause a given amount of movement by the final control element.

Table 9-1. Relation between K_P and PB.

$PB(\%) = \frac{100\%}{proportional\ gain}$	
PB	K_P
1%	100.0
10%	10.0
50%	2.0
100%	1.0
500%	0.2
1000%	0.1

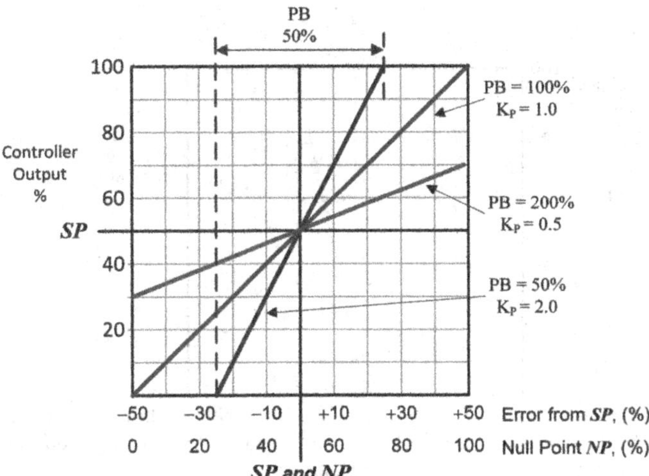

Figure 9-11. Effect of the PB on the final control element (or valve opening).

For example, to cause a final control element to move through 100% of its range, a controller with a PB setting of 100% requires that the controlled variable change twice as much as it does in one having a PB setting of 50%.

In a system that has a narrow PB, the response to a disturbance or set point change is more aggressive but may take longer to stabilize. The response to a system with a wider PB will take longer.

It would appear that the system with the narrow PB is better because the set point would be restored more quickly. However, the characteristic of a narrow PB is that a system has a tendency to oscillate. When the system responds quickly, it tends to overshoot the SP. The system tries to correct itself by providing a controller output in the opposite direction. However, the system overshoots again in the opposite direction. The oscillations usually die out, at which time the system becomes stable. If the PB is too small, oscillations will not stop. A PB of 0% will cause the system to operate almost the same as the on-off control. When the system continues to oscillate, it is unstable.

The K_P determines the amount of control action that will be generated by a given error signal: A small gain value will result in little control output change for a given change in the error signal, whereas a large gain value will result in a large change in the control output for the same change in the error signal.

The PB term clearly identifies the region in which the controller is acting in a proportional manner, and it is for this reason that we use the PB term rather than K_P term. Figure 9-12 shows the same concept when the NP is moved from its 50% center point to 60%.

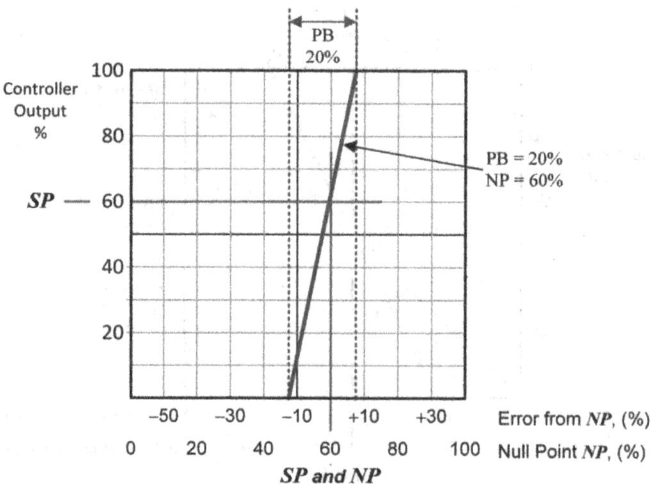

Figure 9-12. SP and NP moved to 60%.

Understanding Controller NP (Bias or Offset)

Proportional control is somewhat simple and the simplest of the continuous controllers to tune, that is, there is only one parameter to adjust. It also provides rapid response, and is relatively stable dynamically. Proportional control has one major disadvantage, however. At steady state, it exhibits offset, that is, there is a difference at steady state between the desired value or SP and the actual value of the controlled variable when the SP is changed, as was illustrated previously.

The proportional-only control mode computes a controller output (Equation 9-3) signal as:

$$CO = K_P \times e + NP$$

where

CO = controller output signal
NP = controller null point (also called the *bias* or *offset*)
K_P = proportional controller gain

Controller error was computed (Equation 9-2) as:

$$e = SP - MV$$

where

SP = set point
MV = current value of the measured PV

The controller's K_P describes how the controller output signal changes given a change in the controller error. A larger K_P means the controller output will change more for a given error.

Suppose the following is the equation or algorithm used for a cruise control in an automobile and *CO* is the flow of gas to the engine computed by the controller. Further suppose that the velocity SP for the car is 45 mph (70 km/h) and the current measured velocity is also 45 mph (70 km/h). Because *CO* equals *SP*, the error equals zero and the equation becomes:

$$\begin{aligned} CO &= K_P \times e + NP \\ &= K_P \times 0 + NP \\ &= NP \end{aligned}$$

If *NP* is zero, then the equation says that when the SP equals measurement, the flow of gas to the engine is zero. This, of course, makes no sense. Clearly, if the car is traveling at 45 mph (70 km/h), then some flow of gas is going to the engine. This baseline value of the controller output is the null point, bias, or null value. In this example, the bias is the flow of gas that, in open-loop control causes the car to travel the desired velocity of 45 mph (70 km/h) when the disturbance variables are at their normal or expected values.

Consider a second example, a gravity drained tank. Here suppose the measured level equals the SP value, so the error equals zero. If no liquid is flowing into the top of the tank, the tank will empty. Therefore, to maintain the liquid level in the tank at the SP, some baseline flow rate of liquid, or *NP*, must always be entering the tank.

The *NP* value determines what control output will be generated when the error is zero. This is often set to 50% so that the control device will be 50% open with zero error. The primary problem with proportional control is that some nonzero error signal is usually required to generate the control output necessary to stabilize the process at the SP, so the process cannot be controlled precisely. The control output required to achieve the SP would have to be exactly equal to the offset value (50% in the preceding example) for the control to be accurate at the SP. In any other situation, a nonzero error signal would be required to generate the appropriate control output, and this would result in a process that is always off its SP. This phenomenon is called *steady-state error.*

Then, when a process is under proportional-only control and the set point equals the measured value, some baseline or bias value of the controller output must exist, or the measured PV will drift from the SP. This bias value of the controller output is determined from the design level of operation of the process to be controlled. Specifically:

> NP is the value of the controller output that, in open loop (manual mode), causes the measured PV to maintain steady state at the desired level of operation when the process is at its desired or expected values.

In a proportional-only controller, *NP* is assigned a value as part of the controller's initial SP and the value remains fixed once the controller is put in automatic.

Reverse-Acting, Direct-Acting, and Control Action

If a positive error increases the controller output, the controller is said to be direct-acting. If the opposite is the case, a positive error decreases controller output, the controller is said to be reverse-acting.

For a process with a positive gain, the process variable increases when the controller output increases. So, in a closed loop, if the process value is too high, the controller must decrease the controller output signal to correct the error. The controller action is the reverse of the process gain. When a process has a positive gain, then the K_P of the controller must be reverse-acting. Conversely, when process gain and K_P are negative, the controller must be direct-acting, or:

- K_P and process gain are positive then reverse-acting
- K_P and process gain are negative then direct-acting

In most commercial controllers, a positive value of the controller gain is always entered. The sign (or action) of the controller is then assigned by specifying that the controller is either reverse- or direct-acting to indicate a positive or negative process gain, respectively. If the wrong control action is entered, the controller will quickly drive the final control element to full-on open or full-off closed and remain there until a proper control action entry is made.

Whether a controller must be reverse- or direct-acting is determined by the configuration of the process. In the case of a tank level control, for example, it depends on the placement of the control valve. If the valve controls the flow out of the tank, we would want a positive error (level too high) to increase the control output, opening the valve and letting more fluid out of the tank; thus, a direct-acting controller would be used.

If the valve controls the flow into the tank, however, a reverse-acting controller would be used, which would respond to a high level by closing the valve and reducing the flow into the vessel.

9.4 Integral (Reset) Control

An *integral* (also called *reset*) controller changes its output in proportion to the integral of the error signal. In block diagram form, integral action is represented as shown in Figure 9-13.

An integral block accumulates (integrates) the input error over time. The input in relation to the output can be represented graphically using a step input as shown in Figure 9-13.

Typically, a process connected to an integral controller is configured as shown in Figure 9-14.

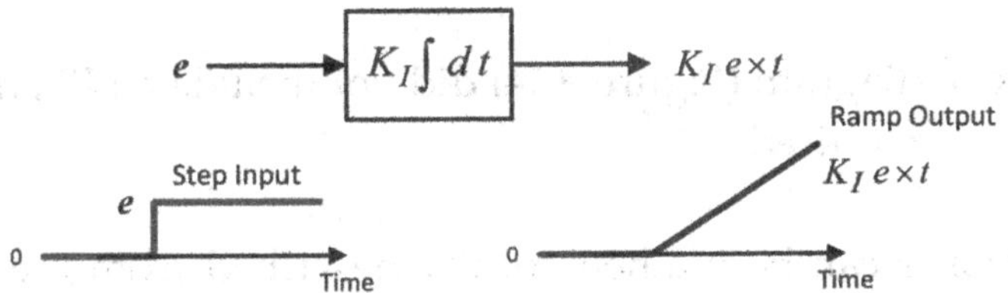

Figure 9-13. Integral action.

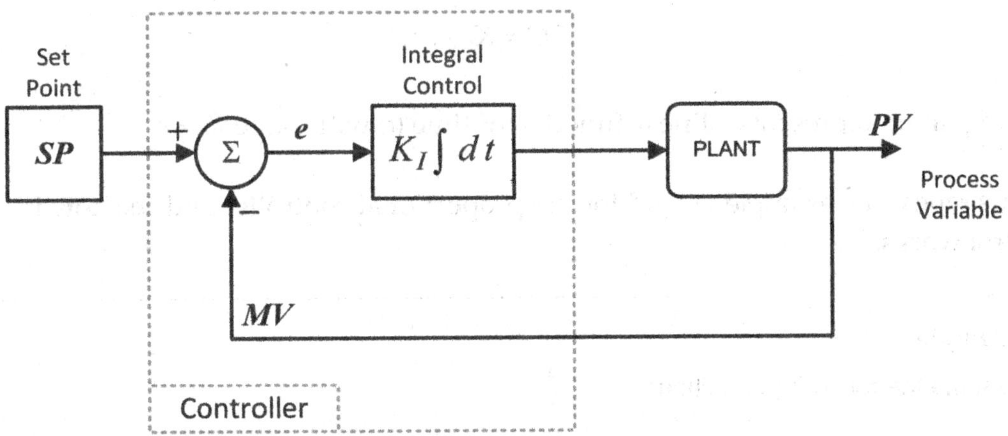

Figure 9-14. Integral controller block diagram.

Note: For clarity, the actuator and measuring device connected between the controller output and the process and between the process output and the MV are not shown.

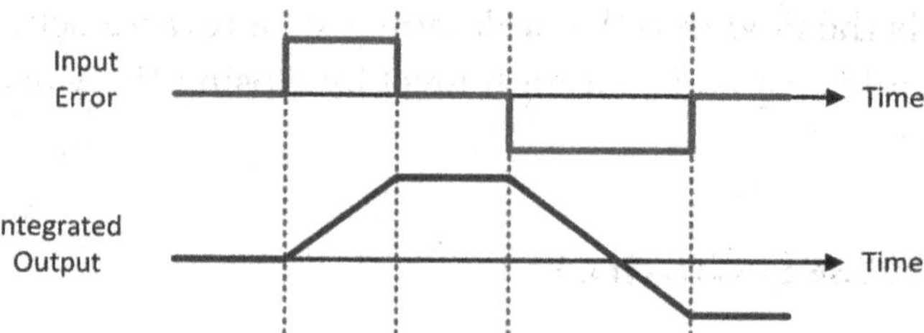

Figure 9-15. Integral action response.

The action of the integral control mode is to generate a controller output that is proportional to the accumulated error over time. When the error is positive, the controller output will increase; when the error is zero, the controller output will stay at the same level; and when the error is negative, the controller output will decrease. The graph in Figure 9-15 shows a typical integral action response curve.

When we analyzed the system response of the proportional controller, we noticed one major disadvantage: it exhibits an offset at steady state (after a change in the SP). This means there is a difference between the desired output and the actual value of the controlled variable.

Notice that the block diagram (Figure 9-14) did not include a NP adjustment. We will see later why it is not required.

The integral controller can be mathematically described from the block diagram in Figure 9-14 as:

$$CO = K_I \int e\,dt$$

$$CO = K_I \times e \times t \tag{9-7}$$

This equation represents a linear function of time (t) with slope $K_I \times e$.

Let us repeat the example we did for the proportional controller and see how integral control works.

Example

Assume the following conditions:

$$SP = 50\%$$
$$K_I = 1$$

Note that we did not specify an NP in this example. Also, assume that the system is balanced at 50%.

Then when the *SP* changes 10%, say from 50% to 60%, we can calculate the new output.

Solution

When the *SP* changes from 50% to 60%, the new error signal is:

$$\begin{aligned} e &= SP - MV \\ &= 60\% - 50\% \\ &= 10\% \end{aligned}$$

This example shows that the integrator is now accumulating this error over time, and the output is, in effect, increasing at a rate of 10%. As the controller output increases, the process value increases, the measured value increases, and the error will start decreasing. The decrease in the error will reduce the slope of the integrator output, effectively producing an exponential response until the error has reached 0%, at which time the process value has reached its desired value, as shown in Figure 9-16.

The controller output will now reach the final value, unlike the proportional controller. However, it now takes more time to reach the final value than the proportional controller with a calibrated NP.

An integral controller acts the same as though an operator has *reset* the NP.

The rate of change, or the slope of the integrator output, is proportional to the integral gain K_I. The inverse of K_I is the reset rate and is in units of time. The meaning of the term *reset rate* will become clear in the next section.

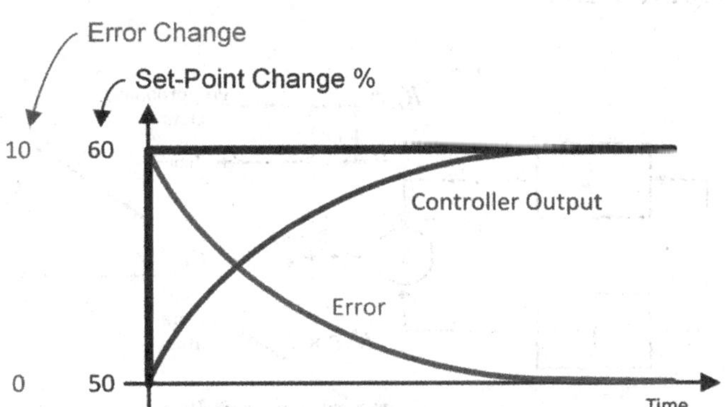

Figure 9-16. The relationship of error and controller output for integral control.

Table 9-2. Comparison of controller integral settings.

$T_I = \frac{s}{Repeat}$	$K_I = \frac{Repeat}{s}$	$T_I = \frac{min}{Repeat}$	$K_I = \frac{Repeat}{min}$
1	1.00	0.0167	60.0
5	0.20	0.0833	12.0
10	0.10	0.1667	6.0
60	0.0166	1.0	1.0
120	0.0083	2.0	0.5
240	0.00417	4.0	0.25
480	0.00208	8.0	0.125
1000	0.00100	16.667	0.06

Mathematically:

$$K_I = \frac{1}{T_I} \tag{9-8}$$

The actual units of K_I and T_I depend on the controller being used. Common units are repeats per minute or per second, and minutes or seconds per repeat, respectively. Their relationships are tabulated in Table 9-2.

Proportional plus Integral Control

We saw that the proportional controller responded instantly (not taking into consideration the plant response) but did not reach the final value, leaving what we called an offset error. The integral controller did reach the final value but had a slowed-down response. By combining both proportional and integral (PI) control, we have an initial (almost) instantaneous proportional response followed by the integral action to reduce (reset) the remaining offset. Proportional plus integral action is represented in block diagram form in Figure 9-17.

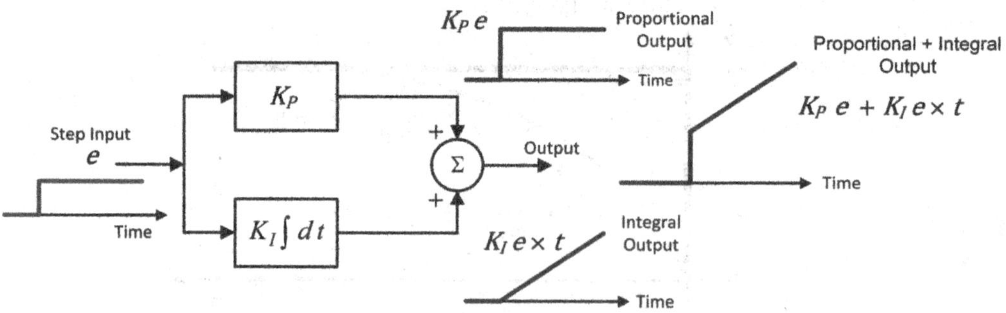

Figure 9-17. Proportional-plus-reset controller's response to a step change in error.

The proportional block response to the step input is instantaneous and proportional, and the integral response accumulates the input error over time. The combined output in relation to the input is also shown Figure 9-17.

Typically, a process connected to a proportional-integral controller is configured as shown in Figure 9-18.

It is now a good time to clearly define the terms *repeats per minute* and *repeats per second.*

The integration constant K_I (repeats per time) or T_I (times per repeat) specifies the number of times per minute or second that the proportional part of the control action is duplicated. In other words, how long (in minutes or seconds) it takes for the integrator to reach the same amplitude (when connected in open-loop control) as the K_P. This is graphically illustrated in Figure 9-19. Note, however, that in practice the controller is set in closed-loop mode and the response shown below is not directly applicable.

The advantage of including integral control with proportional control is that the integral action will eliminate the offset error. However, typically there may be some decrease in stability due to the integral action if the integral time is set too low. One important exception to this is in liquid flow control. Liquid flow control tends to be

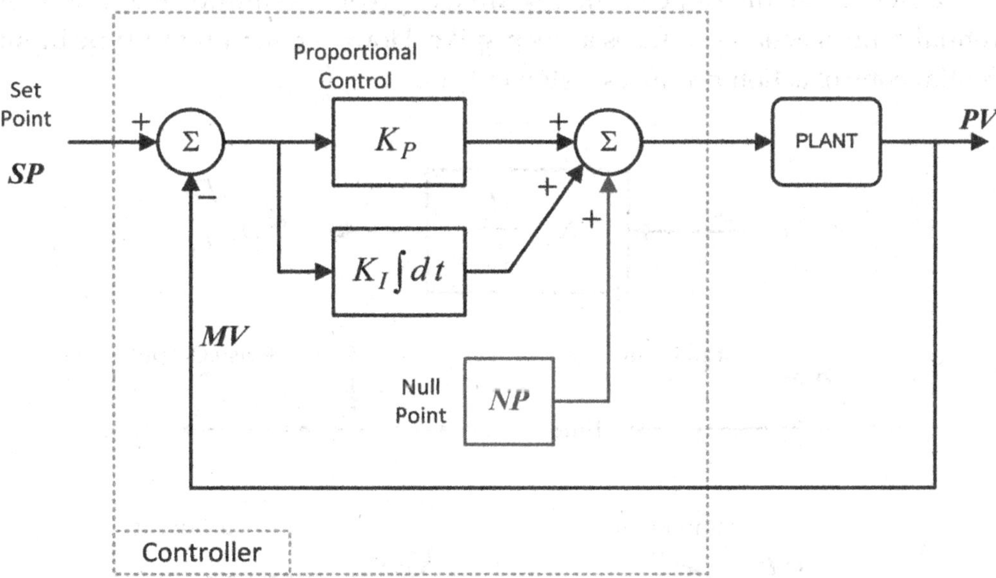

Figure 9-18. Proportional-integral controller block diagram.

Note: For clarity, the actuator and measuring device connected between the controller output and the process and between the process output and the MV are not shown.

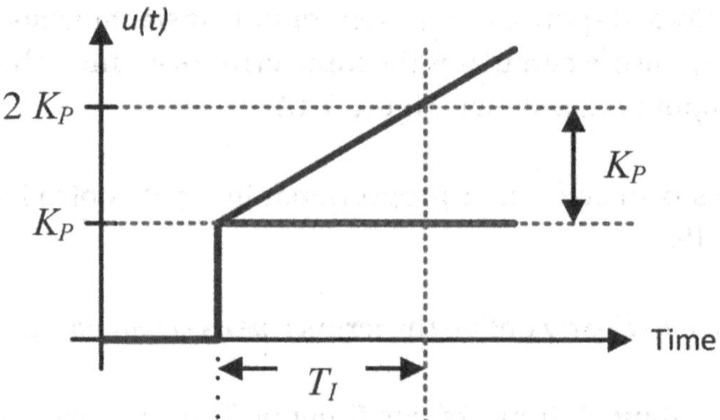

Figure 9-19. Graphical illustration of T_I.

extremely fast and often very noisy. As a result, integral action is often added to provide damping and a filter effect is added to remove the noise.

9.5 Differential (Rate) Control

Differential control provides a fast, temporary response to a change in the set point or disturbance.

A differential controller changes its output proportional to the rate of change or differential of the error signal. In block diagram form, differential control is shown in Figure 9-20. For a step input, the slope of the rise time approaches infinity. For that reason, a differential control action produces a steep spike. However, for a ramp-type input, the differential control action produces a step output.

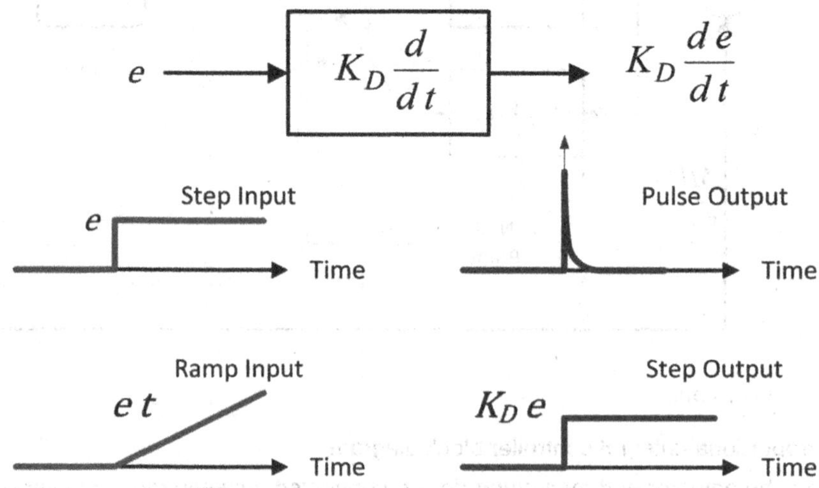

Figure 9-20. Differential control.

A differential block reacts to the rate of change of the input signal. This can be represented graphically using a step or ramp input as shown.

Differential control action is never used by itself (although theoretically possible) because if the error does not change, the output would be zero. Differential action is based on the rate of change of the error and is therefore referred to as *rate* control.

By adding derivative action to a controller, lead is added to the controller, compensating for any lag around the loop. This lag is caused by the plant's capacitance and/or the actuator/transmitter.

Derivative action is a very difficult control action to implement and adjust. Its usage is usually limited to cases in which there is an extensive amount of lag in the process. This is often the case with temperature control problems.

It is conceivable to have control that is based exclusively on the rate of change of the error signal. Although this is theoretically possible, it is not practical because, while the error might be huge, if it were fixed and constant, the controller output would be zero. For this reason, differential control is usually found in combination with proportional control.

Because differential control is rarely used by itself, to better understand derivative control, we will look at a proportional differential controller next.

Proportional and Differential Control

In block diagram form, proportional plus differential (PD) action is represented as shown in Figure 9-21.

The proportional block responds proportionally to the input ramp and the derivative block responds to the rate of change of the input ramp. Because the rate of change is constant, the differential block will produce a step of constant magnitude proportional to the rate of change. The combined output in relation to the input can be represented graphically as shown in Figure 9-21.

Note that the response is based on a ramp input rather than a step input because differential control of a step input provided a pulse at the rising edge and zero elsewhere as shown in Figure 9-20.

The differentiation constant (K_D) is also called the *derivative time* (T_D) or *rate time.* The magnitude of the differential controller output is proportional to the rate of change of the actual error signal.

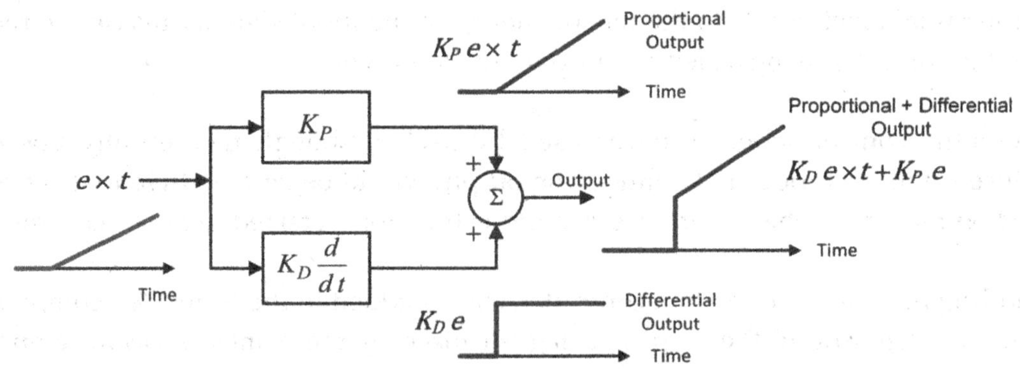

Figure 9-21. Combined proportional and differential control action.

T_D is defined as the time interval by which the rate action advances the effect of the proportional control action. T_D represents the amount of time that the differential control action is ahead of the proportional-only action. This is graphically illustrated in Figure 9-22. Note, however, that in practice the controller is set in closed-loop mode, and the response shown in the figure is not directly applicable.

By inspecting the combined output of both the proportional and derivative control modes, it is possible to see that the derivative time (T_D) which is adjusted in the controller, is really an adjustment of the amount of lead that is introduced in the controller. The addition of differential control to the controller makes the loop more stable if it is appropriately tuned. Because the loop is more stable, the K_P may be slightly higher, and thus it can decrease offset more than proportional action alone. But adding the differential does not eliminate offset.

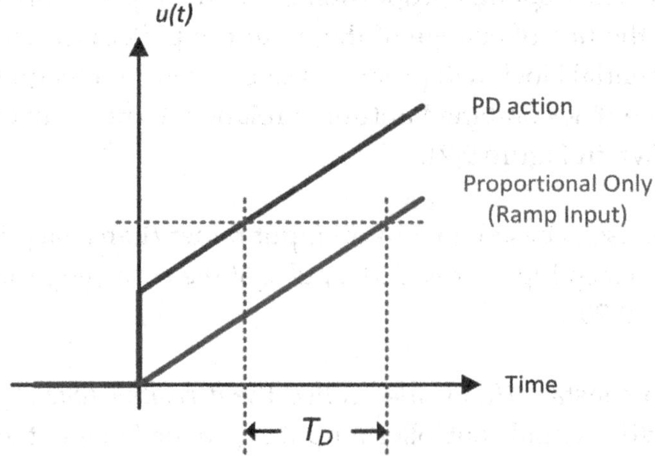

Figure 9-22. Graphical illustration of T_D.

9.6 PID Control

Proportional-plus-integral-plus-derivative control, or *three-mode control*, is the most sophisticated, continuous controller available in feedback loops. Figure 9-23 shows a typical block diagram for such a PID controller.

In three-mode control, we apply all three control actions to the process under control. When properly tuned, it can give rapid response and exhibits no offset, and minimum overshoot and any oscillation will fade away quickly. But it can be very difficult to tune. As a result, it is used only in a very small number of applications, and it often requires extensive and continuing adjustment to stay properly tuned. It does offer very good control when good tuning is implemented.

If we regard the three-mode control action as a transfer function, we can develop the following equation:

$$\mathrm{CO} = K_P e(t) + \frac{1}{T_I}\int e(t)\,dt + T_D \frac{d}{dt}e(t)$$

Taking the Laplace transform gives:

$$\mathrm{CO} = K_P\, e + \frac{1}{T_I} e \frac{1}{s} + T_D\, e\, s$$

$$= e\left(K_P + \frac{1}{T_I\, s} + T_D\, s\right)$$

$$\frac{\mathrm{CO}}{e} = K_P + \frac{1}{T_I\, s} + T_D\, s$$

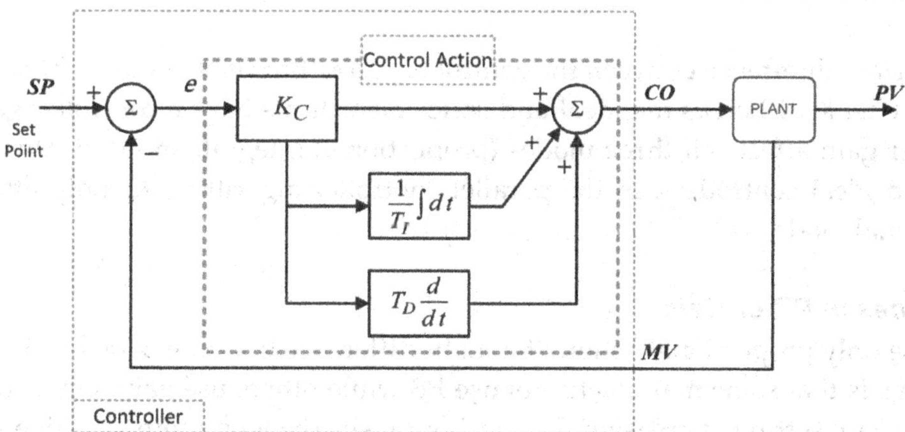

Figure 9-23. PID controller with three-mode control.

Note: For clarity, the actuator and measuring device connected between the controller output and the process and between the process output and the MV are not shown.

This form of the equation is found in many texts on process control automation.

The PID Control Algorithm

Most feedback control loops in industrial process control employ some type of PID control algorithm. Several forms of the PID equation are in use.

There are three major classifications of the PID algorithm that most manufacturers' algorithms fit under: series, ideal, and parallel. The most common of the three PID algorithms is the ideal form (also called the *ISA form*). Manufacturers vary on their names for these categories. To confirm which algorithm is used, one must look at the equation of the controller. In our explanations of the different control actions, we have used the parallel form of the equations. The three equations in simple form are:

1. Ideal: $CO = K_C\left(e(t) + \frac{1}{T_I}\int e(t)\,dt + T_D\frac{d}{dt}e(t)\right)$
2. Parallel: $CO = K_P e(t) + \frac{1}{T_I}\int e(t)\,dt + T_D\frac{d}{dt}e(t)$
3. Series: $CO = K_C\left(e + \frac{1}{T_I}\int e(t)\,dt\right) \times \left(1 + T_D\frac{d}{dt}e(t)\right)$

The series controller's strange-looking form makes it act like an analog type of controller, and it can be constructed with only one pneumatic (or electronic) amplifier. Thus, pneumatic controllers and early electronic controllers often used the series form to save on amplifiers, which were expensive at the time. Some manufacturers also use the series form in their digital algorithms to keep tuning similar to electronic and pneumatic controllers.

An important difference between the controller algorithms is that the parallel controller has a true K_P, whereas the ideal and series algorithms have a controller gain (K_C). Controller gain affects all three modes (proportional, integral, and derivative) of the series and ideal controllers; in the parallel controller algorithm, K_P only affects the proportional mode.

Differences in PB or Gain

If you use only proportional action, the main difference between series and the ideal algorithms is that some manufacturers use PB while others use gain. On a controller using the gain setting, increasing this setting makes the loop more sensitive and less stable, while decreasing PB on controllers using it will have the same effect.

Parallel Controllers

With parallel controllers, controller gain is not multiplied by the error signal only. Integral and derivative actions are independent of the controller gain. At first, it might seem that the parallel controller is easier to work with because of this independence. But parallel algorithms require very different integral and derivative tuning parameters than other controllers do.

Reset Windup in Integral Mode

Because the integral control mode accumulates the error over time, there could be situations where an error persists due to the inability of the final control element to bring the process to its desired value. For example, when heating a room, it is possible that the amount of heat required to maintain the room at its desired temperature exceeds the amount of heat loss in the room. In our definition of integral control action, there is no limit to the growth of this accumulation of error. For this reason, the computed controller output could become significantly greater than 100% under some conditions. However, the controller will not saturate. In addition, the final control element, like a valve, cannot be more than fully open (or closed) but the controller output would be increasing. If the error were finally reversed, requiring the valve direction to reverse, the controller would not be able to do so, at least not until the integral mode contribution had been reduced by the persistent error of the opposite sign.

This condition is referred to as *reset windup.* Most PID controllers and algorithms include windup protection to prevent this from happening when integral mode is being used.

Derivative Spikes

Derivative mode was defined in terms of the time derivative of the error. However, derivative mode does not necessarily need to be engaged in response to a SP change, thereby avoiding the spikes associated with a SP change. This can be implemented by applying derivative mode only to changes in the measured variable as opposed to the error, as shown in Figure 9-24. Another approach is to filter the SP change to avoid sharp changes.

Filtering the Derivative Mode

By reacting to the rate of change of the controlled variable, the derivative mode is subject to noise. Applying derivative mode to a noisy signal can introduce disturbances by needless variation of the manipulated variable. The effect of noise can be reduced by first low pass filtering the derivative mode, also shown in Figure 9-24.

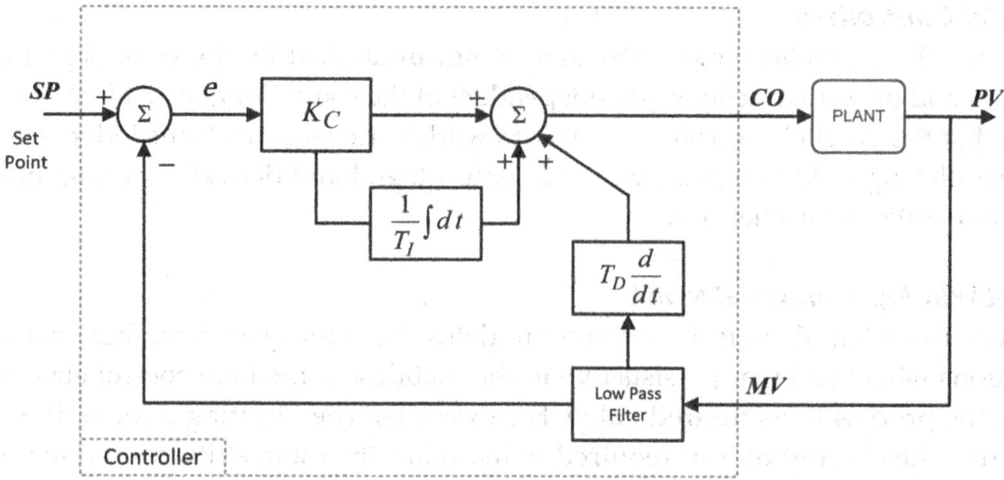

Figure 9-24. Proportional control with derivative mode filtering.

The equation will be as follows:

$$CO = K_C \left[e(t) + \frac{1}{T_I} \int e(t)dt + T_D \frac{d}{dt} MV(t) \right] \tag{9-9}$$

In other words, the derivative term is calculated directly from the MV, representing the actual PV. The filtering of the MV is then often also applied to the proportional and integral terms.

It is important to recognize which variables will be affected when the SP is changed versus when the load or PV is changed.

Continuous processes usually have PV load changes, while batch processes tend to have more SP changes. Depending on how the controller is being used, how the controller reacts to SP and PV changes, and how tuning constants are determined or calculated, it is possible to have a controller perform better one way than another.

9.7 Summary

By way of review, we will summarize the different PID control modes presented.

Table 9-3 provides descriptions of the basic control modes. Table 9-4 illustrates the response action of these various control modes to different input-forcing functions that may be produced by the error signal. Table 9-5 provides a summary of the basic characteristics of these controllers. The application of each control mode and proper tuning procedures are discussed in the next chapter.

Table 9-3. Basic controller modes.

Symbol	Description	Mathematical Expression
P	Proportional	$CO = K_C \cdot e$
I	Integral (Reset)	$CO = \frac{1}{T_I}\int e\,dt$
PI	Proportional plus integral	$CO = K_C\left(e + \frac{1}{T_I}\int e\,dt\right)$
PD	Proportional plus derivative	$CO = K_C\left(e + T_D\frac{de}{dt}\right)$
PID	Proportional plus integral plus differential	$CO = K_C\left(e + \frac{1}{T_I}\int e\,dt + T_D\frac{de}{dt}\right)$

Table 9-4. Characteristics of controller modes.

On-Off, Two-Position	Simple
Proportional	• Simple • Inherently stable when properly tuned • Relatively straightforward to tune • Experiences offset at steady state
Proportional plus integral	• Removes steady state offset • Improved dynamic response than integral alone • Potential instability possible due to the introduction of lag
Proportional plus differential	• Very stable • Less offset than proportional alone (use of higher K_P possible) • Reduces lags providing a more rapid response
Proportional plus integral plus differential	• Most complex • Rapid response • No offset • Difficult to tune • Best control if properly tuned

Table 9-5. Basic graphical characteristics of control action.

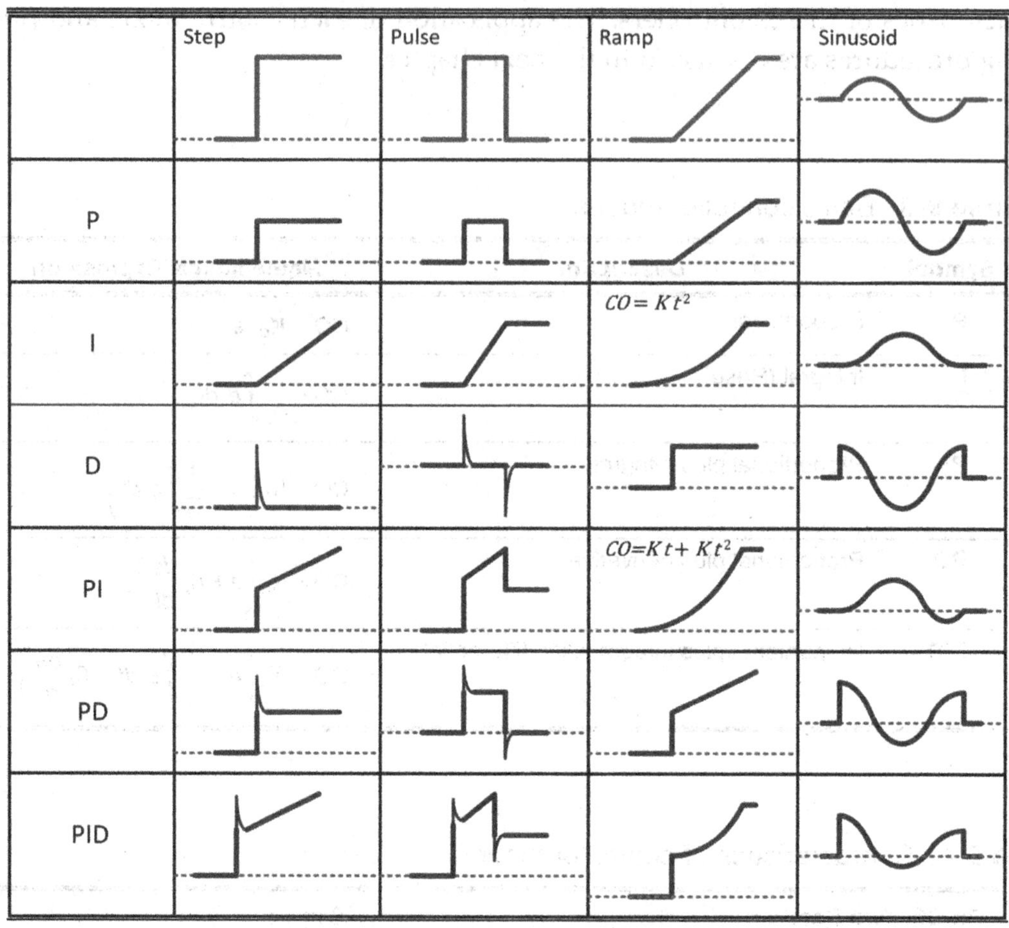

Exercises

9-1 For a proportional controller, answer the following questions:

A. What gain corresponds to a proportional band of 200%?

B. What proportional band corresponds to a gain of 0.2?

9-2 Proportional-only controllers exhibit offset at steady state. How is this possible? (Offset implies an error; therefore, why doesn't the controller output eliminate the error?)

9-3 In this chapter, the error is shown as the input to the controller. Does it make any difference if the error is caused by a change in the SP in one case or by a disturbance in another?

9-4 Proportional control is often described by the equation $m = K_C \times e + M_r$, where M_r is an adjustable "manual reset" that can be used to bias the output so as to eliminate offset at a specific operating point. Will this eliminate offset at other operating points? Explain your answer.

9-5 It has been proposed that a controller might be designed to check the magnitude of the error. If the error is large, the controller would operate as a proportional-only controller; if the error is small, it would operate as a reset-only controller. What would be the advantage(s) of this type of controller?

9-6 In practice, all controllers include some filtering of the error signal before or within the rate mode. Why is this necessary?

10

Tuning Control Loops

A feedback control loop is of little value if it is improperly tuned. A properly tuned control loop maintains its process value at its desired set point, regardless of external disturbances. Chapter 9 explained the different control modes available on most proportional-integral-derivative (PID) controllers. This chapter will explain how to adjust them.

Tuning a feedback control loop involves setting the proportional, integral, and derivative values to get the best possible control for a specific process control loop. Controller tuning is at best an educated trial-and-error process that requires experience more than anything.

It is important that you understand how the controller in a feedback control system should be tuned.

Learning Objectives

- Develop a fundamental understanding of tuning a feedback controller
- Calculate the tuning parameters for a process control loop using the Ziegler-Nichols ultimate method
- Calculate the tuning parameters for a process control loop using the process reaction curve method

10.1 Know Your Controller

An important aspect of tuning a process control loop is to understand exactly what type of controller you are using, what it is doing, and how it is doing it. Controllers come in many different flavors. Make sure to read the manual for the controller you are using. All PID controllers are tuned by adjusting the following three parameters:

1. K_C – Controller gain (or proportional band)
2. T_I – Reset rate (or integral gain)
3. T_D – Derivative time

How to adjust these tuning parameters on a feedback control loop is one of the least understood and most poorly practiced aspects of process control, yet it is very important.

Proportional

The proportional gain adjustment (K_P) on many PID controllers is not expressed in terms of proportional gain, but in terms of proportional band (PB). PB is defined as the span of values of the input that corresponds to a full or complete change in controller output. This is usually expressed as a percentage. Note that proportional gain is equal to controller gain (K_C). K_C is related to K_P and PB by the following equations:

$$K_P = K_C$$
$$PB = \frac{1}{K_P} \times 100\% \quad (10\text{-}1)$$

The larger the PB, the greater the area around the set point in which the proportional action takes place. Note that although PB and K_P are the same, proportional gain does not directly provide information about the controller's actual range as does PB.

Because most controllers have a scale that indicates the value of the final controlled variable, the PB can be conveniently expressed as the range of values of the controlled variable that corresponds to the full operating range of the final control element. This full operating range of the final control element is often inferred to be the operation of a final control valve through a full stroke.

Integral

Integral gain, also known as *reset,* is a function that adjusts the proportional bandwidth with respect to the set point to compensate for offset from the set point; that is, it adjusts the controlled variable to the new set point after the system stabilizes. Integral gain and reset time are related as:

$$K_I = \frac{1}{T_I} \quad (10\text{-}2)$$

Derivative

Derivative gain, also known as *rate,* senses the rate of rise or fall of the controlled variable and adjusts the PB to minimize overshoot or undershoot.

$$K_D = T_D \quad (10\text{-}3)$$

Estimating controller tuning parameters requires considerable knowledge about the process and the controller used. As mentioned in Chapter 9, different PID algorithms are used in industry.

These differences in the algorithms have a major impact on tuning a controller. All the popular tuning rules assume the controller does not have a parallel structure and as a result has controller gain. To tune a parallel controller using any of these rules, the integral time must be divided by and derivative time multiplied by the calculated controller gain.

Another difference between the controller algorithms is the interaction between the integral and derivative modes of the series (interactive) controller. This, of course, is only of significance if the derivative mode is used. In most PID controller applications, derivative mode is seldom used.

Parallel algorithms are generally found only in textbooks and help in understanding the parameters, as discussed in Chapter 9. The advantage of the parallel configuration is that a change in one of the values does not affect the other two. A disadvantage is that they are more difficult to tune because changing the K_p value will require retuning the K_I and K_D values. Series control algorithms are the most commonly used in most analog controllers. Ideal control algorithms are used in most digital controllers.

The tuning methods described in the following sections are based on the ideal algorithm, which is:

$$CO = K_P\left(e(t) + K_I\int e(t)dt + K_D\frac{d}{dt}e(t)\right)$$

$$\text{or} \quad CO = K_C\left(e(t) + \frac{1}{T_I}\int e(t)dt + T_D\frac{d}{dt}e(t)\right) \quad (10\text{-}4)$$

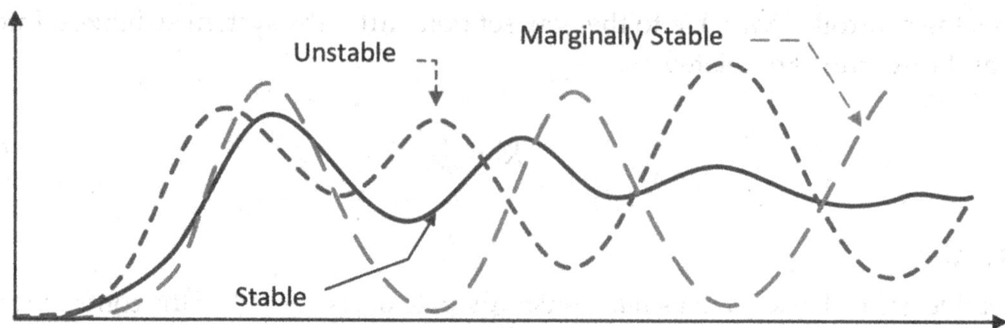

Figure 10-1. Stable, marginally stable, and unstable responses.

Tuning methods depend on being able to accurately determine the behavior of the process variable, specifically, whether the process variable is stable, unstable, or marginally stable. A stable process is one in which the process variable is lagging behind the controller output, and an unstable process is one in which the process variable is leading the controller output, causing increased oscillation. A marginally stable process is when the process variable is 180° out of phase with the controller output, causing sustained oscillation. Figure 10-1 illustrates all three cases.

The goal of process tuning is to always end up with a stable response. Many procedures have been developed with good results. The following sections describe some of the most widely used procedures, and when it is best to use each.

10.2 Criteria of Good Control

The first predicament encounter when tuning a control loop is determining what is meant by good control. Perfect control would mean that the controlled variable follows the set point exactly and that any disturbances have no effect on maintaining the process variable at its desired value. However, due to the dynamic behavior of the process, final control element, and instrumentation used, this is not possible. The definition of *good control,* as might be expected, will differ from one process to the next. A reference compromise criterion that many tuning procedures are based on for determining good control is to adjust the controller so that the process response curve has an amplitude ratio or decay ratio of 1/4. A decay ratio of 1/4 means that the ratio of the overshoot of the first peak in the process response curve to the overshoot of the second peak is 4 to 1. This is illustrated in Figure 10-2.

There is no mathematical justification for specifying a decay ratio of 1/4, but it represents a compromise between a rapid initial response to a change in set point or load disturbance and a relatively fast settling time. In many cases, this criterion is not sufficient to specify a unique combination of controller settings. For example, there are

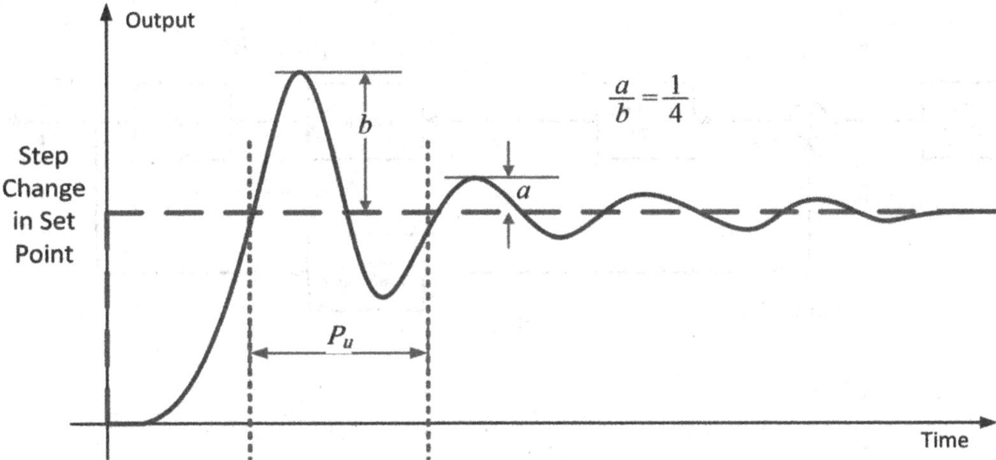

Figure 10-2. Process response curve for a 1/4 decay ratio.

an infinite number of settings that will yield a decay ratio of 1/4, each with a different period. This illustrates the difficulty of defining what constitutes good control.

In some cases, it is important to tune the process loop so that there is no overshoot. In other cases, you may want a slow and smooth response. Some control loops need a fast response, and significant oscillation may not be a problem. The point is, good control must be determined based on the requirements for each specific process control loop.

10.3 The Tuning Concept

The feedback controller is only one piece of hardware in the entire process control loop; many other hardware devices are connected to form the balance of the loop. From the controller's point of view, all components outside the controller are seen as one cluster. For the purposes of adjusting the feedback controller, it is convenient and sufficient to view the cluster as everything else within the feedback control loop (see Figure 10-3).

There is one parameter to adjust in a single-mode controller, for example, the controller gain (K_C) in a proportional-only instrument. There are two parameters to adjust in a two-mode controller, for example, the controller gain (K_C) and the reset time (T_I) in a proportional-integral (PI) controller. There are three parameters to adjust in a three-mode controller, for example, the controller gain K_C for the proportional mode, the reset time (T_I) for the integral mode, and the T_D for the derivative mode. The increase in the number of parameters also increases the difficulty in finding appropriate tuning parameters. When you adjust the controller, the gains around the loop tend to dictate what should be the optimum gain in the controller. Similarly, the time constants and

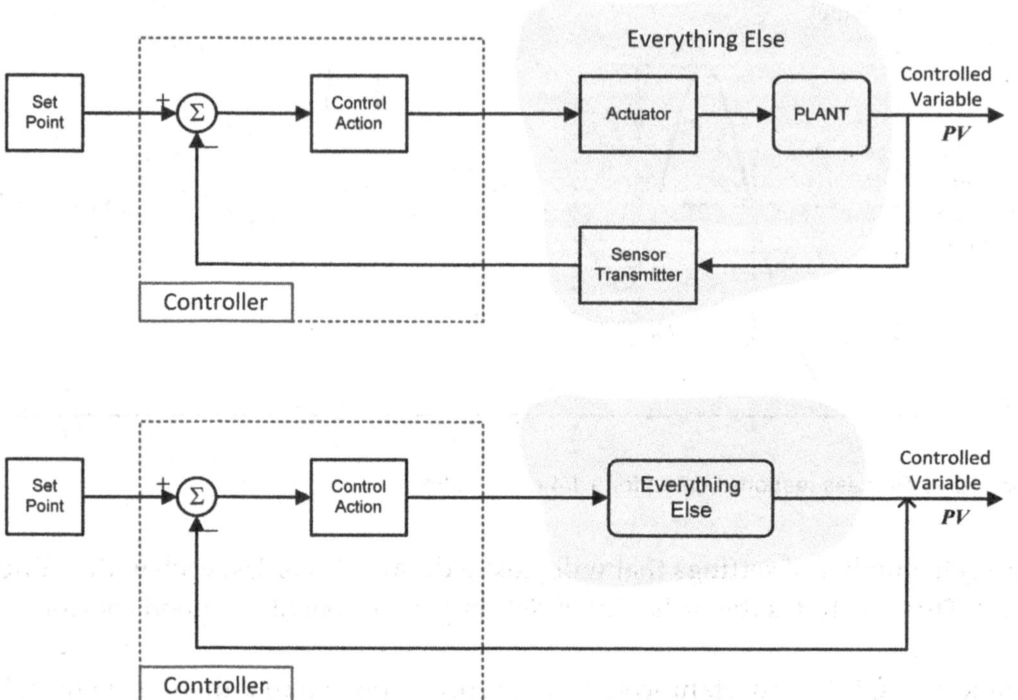

Figure 10-3. The balance of the loop as "seen" by the feedback controller.

dead times that characterize the lag dynamics for everything else around the loop tend to dictate what will be the optimum value of the reset time and what the derivative time in the controller should be. Stating this differently, before you can calculate or select the best values for the tuning parameters in the controller, you must get some quantitative information about the overall gain and the process lags that are present in the balance of the feedback loop. This illustrates quite clearly why controllers cannot be preset at the factory but instead must be individually tuned for each individual loop.

It is also helpful to establish a mathematical model of a process control loop and to appreciate the role of the controller as a mathematical equation within that model. Each individual block around the feedback control loop represents a transfer function and is a mathematical statement for each piece of hardware because all these blocks are coupled together. The intent of the controller is then to cancel out all other contributions in the loop. If you can determine what you think good control should be, then you can, in effect, specify the overall solution to determining the controller settings. Basically, the tuning of the controller represents the adjustment of the individual parameters in the equation that represents the controller. As you adjust these parameters for the controller equation, you modify the solution. You change the response of the overall system toward a point that represents good responses or good control.

Table 10-1. Effect of controller parameters on the process loop response.

Response	Rise Time	Overshoot	Settling Time	Steady-State Error
K_C	Decrease	Increase	Undefined	Decrease
K_I	Decrease	Increase	Increase	Eliminate
K_D	Undefined	Decrease	Decrease	Undefined

Mathematically, in an ideal situation, the transfer function of the control action PID(s) times the transfer function of everything else equals some fixed gain. In other words, all time-dependent time constants cancel out.

The PID parameters affect the dynamics of the control loop in different ways, but there are four characteristics that should be considered:

1. **Rise time** – The amount of time for the process value to reach its desired value.
2. **Overshoot** – How much the peak level is higher than the steady-state, normalized against the steady-state.
3. **Settling time** – The time it takes for the system to converge to its steady state.
4. **Steady-state error** – The difference between the steady-state output and the desired output.

Each PID parameter affects these considerations, as shown in Table 10-1.

10.4 Closed-Loop Tuning Methods

Techniques for adjusting controllers may be classified as either open-loop or closed-loop methods. One of the first methods proposed for tuning feedback controllers was the ultimate method proposed by Ziegler and Nichols[1] in 1942, and it is still in use today. The term *ultimate* was attached to this method because to use it, you have to determine the ultimate gain (sensitivity) and ultimate period for the loop. The ultimate gain is the maximum allowable value of gain (for a controller with only the proportional mode in operation) where the closed-loop system is still stable.

For any feedback control system, if the loop is closed (if the controller is on automatic) and under proportional gain only, you can increase the proportional controller gain,

1. J. G. Ziegler, and N. B. Nichols, "Optimum Settings for Automatic Controllers," *Transactions ASME* 64 (November 1942): 759.

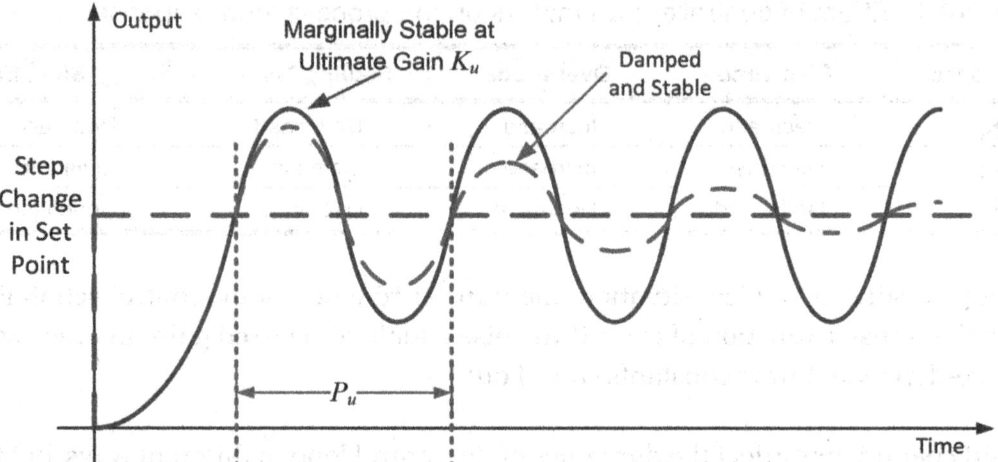

Figure 10-4. Responses to illustrate ultimate gain and ultimate period.

and as you do so the loop will tend to oscillate more and more. As you continue to increase the proportional gain further, you will observe continuous cycling or continuous oscillation in the controlled variable. This is the maximum gain at which the system may be operated before it becomes unstable; it is referred to as the *ultimate gain.* The period of these sustained oscillations is then the ultimate period. If you increase the gain further, the system will become unstable. Ultimate gain and ultimate period are illustrated in Figure 10-4.

To determine the ultimate gain and the ultimate period, perform the following steps:

1. Tune out the reset and derivative action from the controller, leaving only the proportional mode.
2. Set T_I equal to infinity.
3. Set T_D equal to zero (or as close to these values as is possible on the controller).
4. Maintain the controller on automatic; that is, leave the loop closed.
5. With the gain of the proportional mode of the controller at some low value, impose an upset on the process and observe the response. One uncomplicated method for imposing the upset is to move the set point for a relatively short moment and then return it to its original value. A second method is to introduce a disturbance or load change.
6. If the response curve from Step 3 does not damp out (Marginally Stable in Figure 10-4), the gain is too high (the PB setting is too low). The gain should be decreased (the PB setting should be increased) and Step 5 repeated.

7. If the response curve in Step 5 damps out (Damped and Stable in Figure 10-4), the gain is too low (the PB is too high). The gain should be increased (the PB setting should be decreased) and Step 5 repeated.

8. When you obtain a response curve similar to the Marginally Stable curve in Figure 10-4, note the values of the ultimate gain (or ultimate PB) setting and the ultimate period of the associated response curve. This ultimate gain where the sustained oscillations are encountered is the ultimate gain (K_u) and the ultimate period is P_u.

The ultimate gain and the ultimate period are then used to calculate controller settings. In the case of proportional controllers, Ziegler and Nichols correlated that a value of gain equal to 1/2 of the ultimate gain would often give a decay ratio of 1/4. By comparable reasoning and testing, the following equations were established to represent acceptable rules of thumb for controller settings for more complex controller modes. Ziegler and Nichols therefore proposed the following tuning rules of thumb (Table 10-2).

Remember that the equations in Table 10-2 are empirical, so exceptions are expected. The equations are intended to achieve a decay ratio of 1/4, which is the presumed definition of good control. Modern controllers, which are the type most commonly encountered today, have gain adjustment provisions that are, as gain should be, dimensionless. The gain setting typically represents the ratio of milliampere (mA) output to the change in milliampere input to the controller. Controllers with sensitivity dials have gone out of use because of their lack of universality, but Exercises 10-3 and 10-4 at the end of this chapter are based on sensitivity and are included in this book simply to emphasize the difference between sensitivity and gain.

Table 10-2. Ziegler and Nichols closed-loop tuning method.

Proportional	$K_C = 0.5\ K_u$	(10-5)
Proportional-Integral	$K_C = 0.45\ K_u$	(10-6)
	$T_I = \frac{P_u}{1.2}$	(10-7)
Proportional-Derivative	$K_C = 0.6\ K_u$	(10-8)
	$T_D = \frac{P_u}{8}$	(10-9)
Proportional-Integral-Derivative	$K_C = 0.6\ K_u$	(10-10)
	$T_I = 0.5\ P_u$	(10-11)
	$T_D = \frac{P_u}{8}$	(10-12)

Note that the closed-loop ultimate method presented in this section is not very practical for slow processes, such as large kilns, fractionating units, or boilers. This is because it can take days for a pattern to develop, and in some cases, it would not be safe or economical to have the process oscillate.

Example

For a temperature control system whose ultimate gain (K_u) is 0.4 and ultimate period (P_u) is 2 min, determine settings for the various controllers using the Ziegler-Nichols ultimate method.

Proportional

$$K_C = 0.5K_u = 0.5 \times 0.4 = 0.2$$

Proportional plus Reset

$$K_C = 0.45K_u = 0.45 \times 0.4 = 0.18$$

$$T_I = \frac{P_u}{1.2} = \frac{2 \text{ min}}{1.2} = 1.67 \text{ min}$$

Proportional plus Reset plus Derivative

$$K_C = 0.6K_u = 0.6 \times 0.4 = 0.42$$

$$T_I = 0.5P_u = 0.5 \times 2 \text{ min} = 1.0 \text{ min}$$

$$T_D = \frac{P_u}{8} = \frac{2 \text{ min}}{8} = 0.25 \text{ min}$$

10.5 The Process Reaction Curve

There are processes, such as integrating processes, where no sustained oscillation will ever occur and this method cannot be used. A second problem inherent in the closed-loop ultimate method is that you must cause the loop to oscillate to determine the ultimate period and the ultimate gain. In some cases, this can cause operational problems. On the other hand, open-loop methods usually require only that a single upset be imposed on the process. These methods provide more precise data about the dynamics of the process, and often give slightly better tuning results, though the variation in satisfaction from loop to loop can be quite significant.

Not only may a process reaction method be applied to all processes, but it can also be applied to processes that are still on the drawing board by making engineering calculations about the flow and energy transfer rates as well as the mass and energy storage capacities of the processing unit to be controlled.

The process reaction curve is the reaction of the process to a step change in the input signal. The process reaction curve is the reaction of the "Everything Else" shown in Figure 10-3. It is important to have a complete picture of exactly what this process

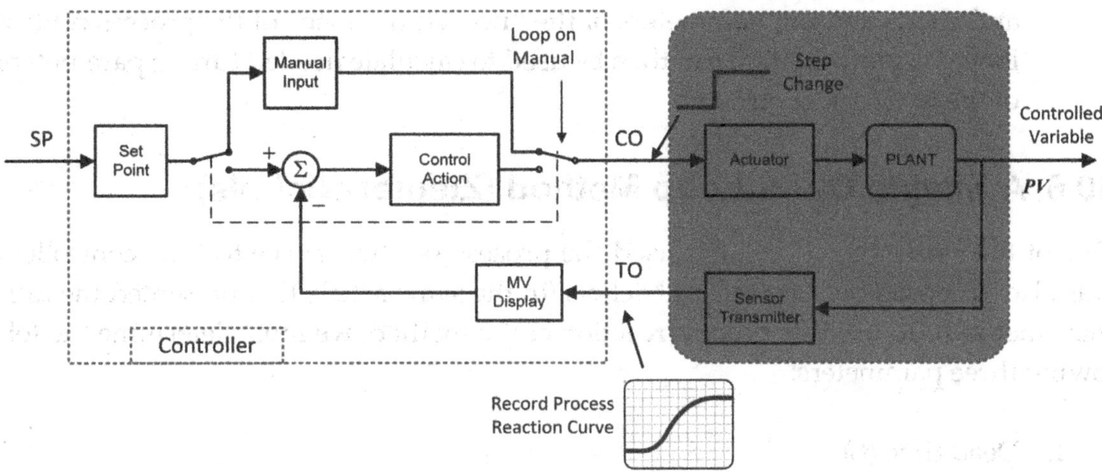

Figure 10-5. Determining the process reaction curve.

reaction curve might represent in a feedback control loop, and this is shown in some detail in Figure 10-5. Placing the loop in manual mode enables the operator to introduce a step change to the overall process. The dedicated trend recorder (analog or digital) available on most controllers or an external trend recorder can then be used to record the process reaction curve.

In general, a process reaction curve can be determined as follows:

1. Let the system come to a steady state.
2. Place the controller on manual operation, that is, remove it from automatic operation.
3. Let the system reach its steady state.
4. With the controller still in manual mode, impose a step change to the process. The size of the step change is significant because the process may exhibit nonlinear characteristics. Typically, using a unit step change of the manipulated variable is a common starting point.
5. Record the response of the controlled variable. (The response may be recorded by a dedicated trend recorder for the loop, but it is often desirable to have a supplementary recorder available that has a faster chart drive. This is because dedicated trend recorders tend to be very slow.)
6. Return the controller output to its previous value and return the controller to automatic operation. After this has been done, the recorded process reaction curve may be used to give significant information about the overall dynamics

and characterizing parameters of the "Everything Else" of the process control loop. This information can then be used to calculate needed tuning parameters of the feedback controller.

10.6 A Simple Open-Loop Method (Ziegler-Nichols)

One of the earliest methods that used the process reaction curve to tune controllers was also proposed by Ziegler and Nichols (in the same article that presented the ultimate method). To use this process reaction curve method, we must determine the following three parameters:

1. Dead time (t_d)
2. Time constant (τ)
3. Process gain (G_P) in percent of span

Details for finding these parameters were discussed in Chapter 5.

Figure 10-6 illustrates a sample determination of these parameters for a specific temperature control loop. The approach that Ziegler and Nichols used is identical to the *tangent method* in Chapter 5.

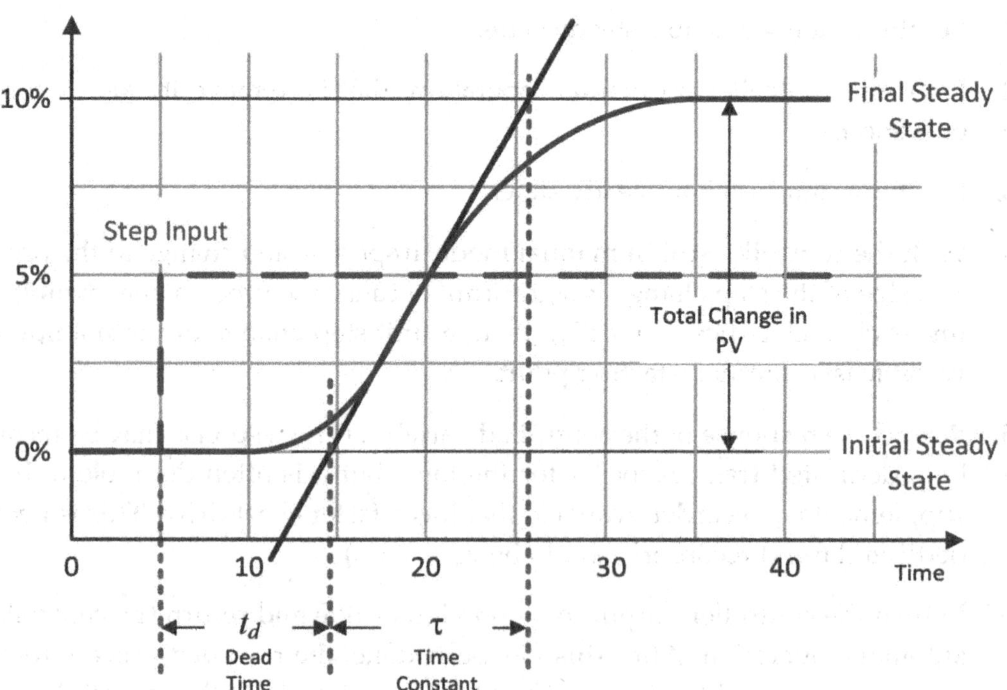

Figure 10-6. Process reaction curve for a sample feedback control loop.

Table 10-3. Ziegler and Nichols open-loop tuning method.

Proportional	$K_C = \frac{\tau}{t_d \times K_P}$	(10-13)
Proportional-Integral	$K_C = 0.9\frac{\tau}{t_d \times K_P}$	(10-14)
	$T_I = \frac{t_d}{0.3}$	(10-15)
Proportional-Integral-Derivative	$K_C = 1.2\frac{\tau}{t_d \times K_P}$	(10-16)
	$T_I = 2\,t_D$	(10-17)
	$T_D = 0.5\,t$	(10-18)

To obtain process information parameters in the process reaction curve method, draw a tangent to the process reaction curve at its point of maximum slope.

The process gain can be calculated as:

$$G_P = \frac{\%\ change\ in\ output}{\%\ change\ in\ input}$$

This slope (R_r) is the process reaction rate. Where this tangent line intersects the original initial steady-state baseline gives an indication of the process dead time (t_d). Using this information, Ziegler and Nichols proposed a series of rules of thumb (Table 10-3) that can be used to calculate appropriate controller settings.

In summary, to use the process reaction curve method, it is necessary to obtain a specific process reaction curve with a known input step size. Then, using the three parameters you graphically determined from the process reaction curve, calculate the needed tuning parameters.

Example

Consider a temperature control system with sensitivity of $S_P = 2.5$°F/% that underwent a step input change of 10% and produced the reaction curve shown in Figure 10-7.

From the curve, we determine the following:

$$t_d = 1.2\ \text{min}$$
$$\tau = 2.1\ \text{min}$$

Next, we calculate the process gain (G_P):

$$G_P = \frac{\%\ change\ in\ output}{\%\ change\ in\ input} = \frac{25°F}{10\% \times 2.5°F/\%} = 1$$

Using Ziegler and Nichols's open-loop method, calculate the required controller settings:

Proportional	$K_C = \frac{\tau}{t_d \times K_P} = \frac{2.1\,min}{1.2\,min \times 1} = 1.75$
Proportional-Integral	$K_C = 0.9\frac{\tau}{t_d \times K_P} = 0.9\frac{2.1\,min}{1.2\,min \times 1} = 1.57$ $T_I = \frac{t_d}{0.3} = \frac{1.2\,min}{0.3} = 4\,min$
Proportional-Integral-Derivative	$K_C = 1.2\frac{\tau}{t_d K_P} = 1.2\frac{2.1\,min}{1.2\,min \times 1} = 2.1$ $T_I = 2\,t_d = 2 \times 1.2\,min = 2.4\,min$ $T_D = 0.5\,t_d = 0.5 \times 1.2\,min = 0.6\,min$

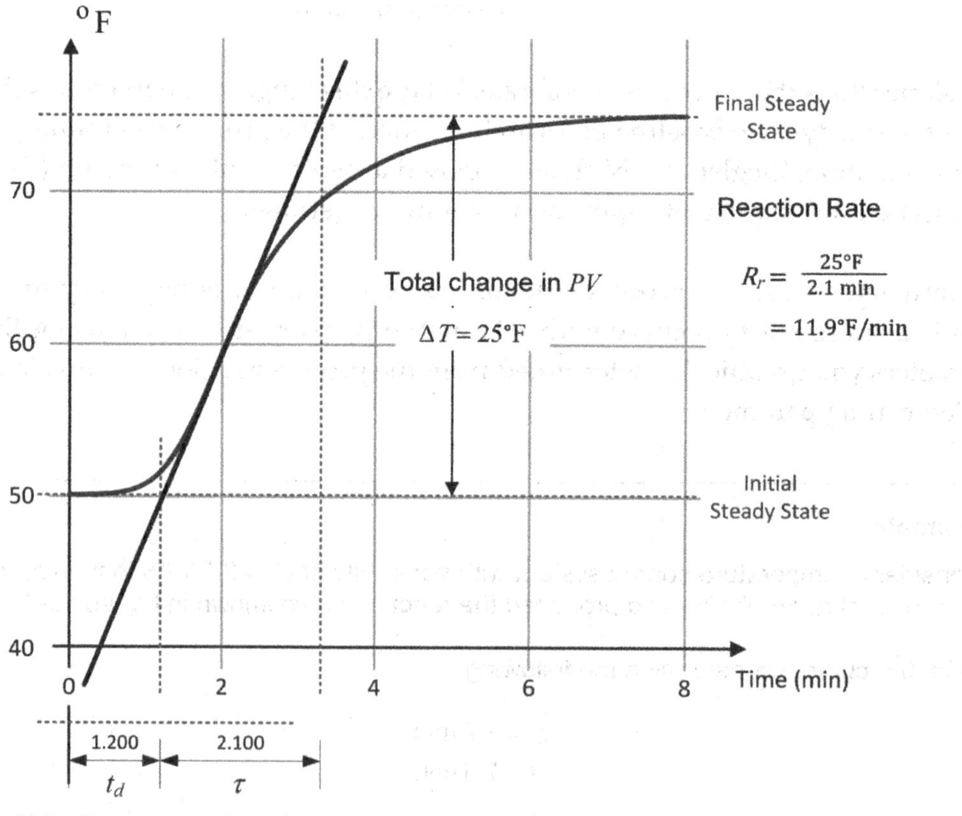

Figure 10-7. Process reaction curve for a temperature feedback control loop.

Ziegler and Nichols indicated that these rules of thumb should give a decay ratio of 1/4, which is their inherent definition of good control.

10.7 The Need to Retune and Adaptive Tuning

Feedback control loops may be mathematically classified as linear or nonlinear. Most industrial processes are nonlinear; that is, they will exhibit different responses at different operating points. If a control loop is linear, it will exhibit the same response at different operating points. Similarly, if an input change is a mirror image of a previous input change, the output or response will be a mirror image. If a control loop is linear, it may be tuned once, and it will stay tuned forever as long as the process conditions do not change. But if a loop is nonlinear, its tuning may require adjustment (or it will be poorly tuned). This phenomenon is a major obstacle to keeping control systems well-tuned.

It does not matter whether any one individual component of the control loop is linear or not. What matters is whether or not the "Everything Else"—everything except the controller—is, in its aggregate form, linear or not. Sensors and transmission systems are usually designed by vendors to be linear.

One exciting aspect of the advances being made in the tuning of industrial control systems involves the concept of self-tuning or adaptive tuning. This idea is based on the concept that one should be able to design and implement controllers that are capable of tuning themselves or of being tuned automatically. Such control systems and individual controllers do exist and are increasingly used for initial tuning of a process. The design and installation of systems and controllers capable of adaptively tuning themselves, which means adapting themselves to the process situation that is manifest at a given moment, has been practical only for such situations as aircraft, special reactors, and paper machines. But microprocessors and digital capabilities have changed all this. Today, the computational capabilities required to do adaptive tuning are relatively inexpensive, and there is a strong movement toward the implementation of such systems.

The natural tendency when designing a self-tuning controller is to design an expert system that automatically performs the steps that an experienced control operator would follow. This approach is basically one of "pattern recognition" in the closed-loop response of a control loop. To say this differently, the controller inspects the response pattern, measures overshoot and damping, and notes the times of successive peaks (thus establishing the period of oscillation). Based on these measured parameters, adaptive tuning recalculates the PID tuning parameters.

This sounds simple, but it is not. One of the most popular, and the first commercially successful, single-loop, adaptive-tuning systems for the fluid process industries was a rule-based expert system. It had more than 200 rules, most of which were to keep the pattern recognition algorithm from being confused by peaks that were not caused by controller tuning parameters.

An alternative approach to adaptive tuning is to use linear recursive regression to estimate the parameters of a discrete linear model of the process from the sampled values of the controller output and the controlled variable. Once the discrete process parameters have been calculated, they can be used to calculate controller parameters. Several commercially available controllers use this approach for adaptive tuning.

10.8 Summary

There are numerous different tuning techniques available, and they vary considerably in terms of thinking and implementation. They also vary in terms of how they define good control, whether they are open-loop or closed-loop, and how mathematically complex they are. Different methods can give widely varying results. It is impossible to say that one method is clearly superior to all others.

It is fair to say that the tuning of control systems is as much an art as it is a science. It requires a great deal of practice and experience to develop a sensitive and professional feel for the tuning of systems. But more importantly, it requires a good understanding of the process being controlled.

Exercises

10.1 How is it physically possible in two-mode and three-mode controllers (such as proportional-integral or proportional-integral-derivative) for there to be more than one set of controller settings that give a 1/4 decay ratio?

10.2 Why is it that all liquid-level control loops do not have the same tuning parameters, all flow control loops do not have the same tuning parameters, and so on?

10.3 For a level control loop whose ultimate sensitivity is 0.3 psi/ft and whose ultimate period is 3 min, calculate the settings for the various types of controllers using the ultimate method. (Sensitivity is a gain with dimensions or engineering units; gain is dimensionless.)

Section 2

Control Strategies

This section (Chapters 11 through 17) discusses different control strategies.

Control strategy refers to a plan or methods applicable to a specific control requirement that cannot be satisfied using a simple control loop. It usually consists of combining multiple control loops to realize specific control outcomes, or applying a nontraditional approach in combination with a standard process control loop.

Chapter 11 addresses cascade control. Cascade control involves the use of two controllers with the output of the first controller providing the set point for the second controller. This form of control is used to improve responses to disturbances.

Chapter 12 focuses on feedforward control. Feedforward control applies corrective action by measuring disturbances and anticipating corrective action before the disturbance enters the process. It is usually applied in conjunction with feedback control.

Chapter 13, "Multivariable Control," explains the process for controlling two or more interdependent process variables.

Chapter 14 explains how ratio control is used to maintain a constant flow rate of two or more streams in a process at a specified proportion relative to one another.

Chapter 15 discusses special control strategies such as three-element boiler control, selective control, override control, duplex and split-range control, and auto-selector or cutback control.

Chapter 16, "Statistical Process Control|," explains the application of statistics to process control. This type of control applies statistical methodology to measure and control product quality in real time during the manufacturing process.

Chapter 17, "Fuzzy Logic Control," provides an introduction to applying set theory and fuzzy logic to process control problems.

11 Cascade Control

Up to this point, whenever feedback control was discussed, attention was focused on a single feedback control loop. This chapter will expand that concept to several variations of this single-loop approach to include a secondary control loop using the principle of cascade control.

Learning Objectives

- Recognize the basic principle of cascade control
- Understand the basic concepts that govern the selection of a secondary controlled variable
- Understand the general guiding principles for cascade control mode selection and tuning

11.1 The Concept of Cascade Control

Cascade control enables the reduction of disturbances that are external to the process. The process variable, which is to be maintained at a fixed set point, is called the *primary variable*. A secondary variable that may introduce disturbances to the primary variable is identified. The output of the primary controller is then used as the set point of the secondary controller.

For example, in a home heating system (Figure 11-1), room temperature is maintained by controlling the amount of fuel to a furnace. The temperature transmitter (TT 100) reacts to disturbances in the room, and the controller calculates the required change

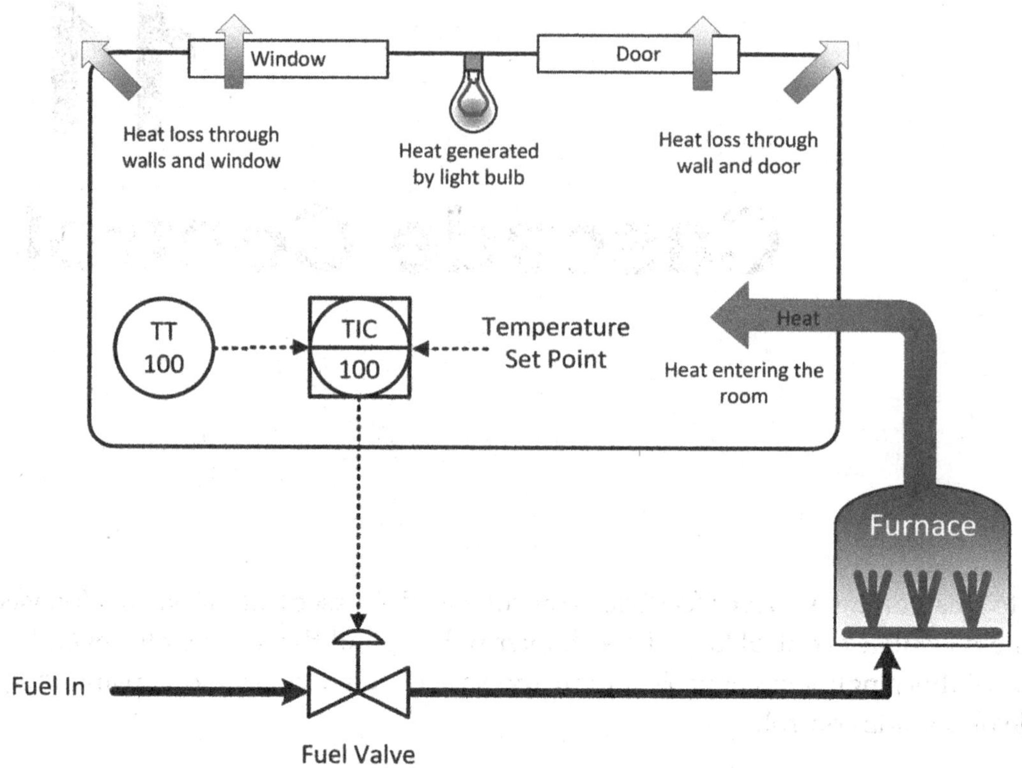

Figure 11-1. Feedback control of a home heating system.

needed to the fuel valve to maintain the desired room temperature. However, the fuel line supplying the fuel may also add disturbances, such as variations in fuel supply pressure, which in turn changes the amount of fuel supplied to the furnace. The temperature controller only responds to temperature changes measured by TT 100. It is not able to respond to fluctuations in fuel pressures because the time lag and dead time of this type of disturbance with respect to temperature is much larger than the fluctuations in fuel pressure. The temperature controller, not the valve opening, must determine the required fuel flow rate. This means we must also control the fuel flow rate.

The fuel flow rate can be controlled by using cascade control. A cascade control system consists of a primary (outer) control loop and a secondary (inner) control loop. In the home heating system example, the primary control loop is room temperature, and the secondary control loop is the fuel flow rate. Figure 11-2 illustrates how cascade control implements this arrangement.

The temperature controller (TT 100) controls the temperature not by controlling the valve opening, but rather by controlling the actual fuel flow rate to the furnace. The flow controller (FC 101) in turn controls the fuel valve and maintains the fuel flow as directed by the temperature controller. This arrangement will minimize any

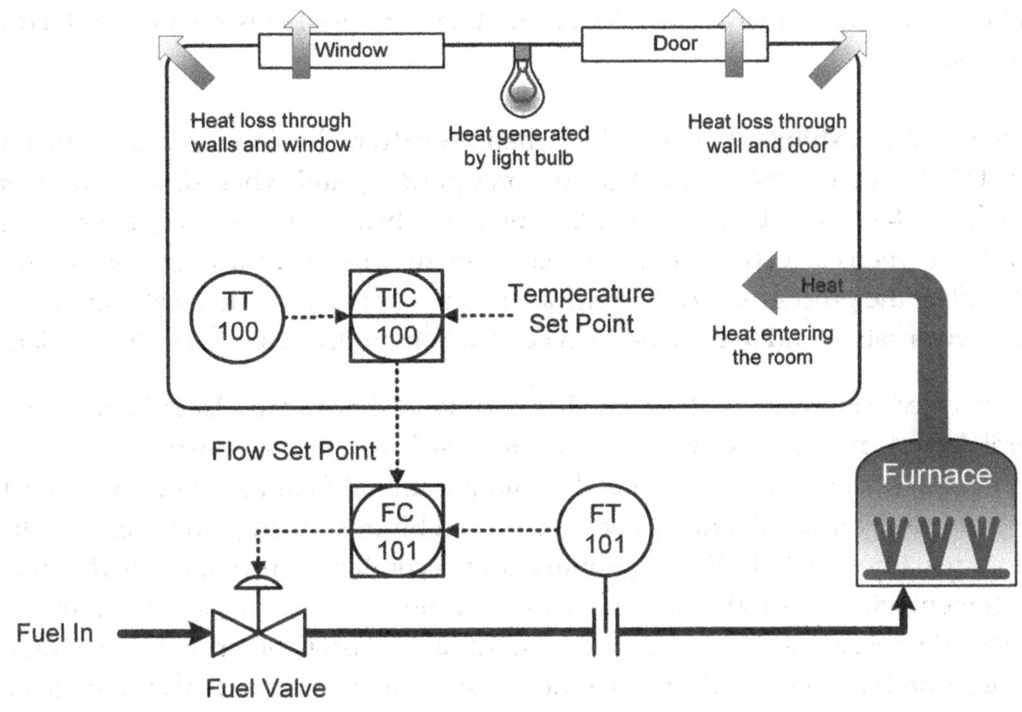

Figure 11-2. Cascade control of a home heating system.

disturbance caused by the fuel pressure supply and provides for a room temperature control.

The general concept of cascade control is to nest one feedback control loop inside another feedback control loop. Cascade control uses the output of the primary controller to direct the set point of the secondary controller as if it were the final control element. The heating system example is again illustrated in Figure 11-3 in block diagram form, clearly showing how the secondary control loop is nested inside the primary control loop.

In effect, you take the process being controlled and then identify an intermediate variable that can cause disturbances to the process. This intermediate variable can

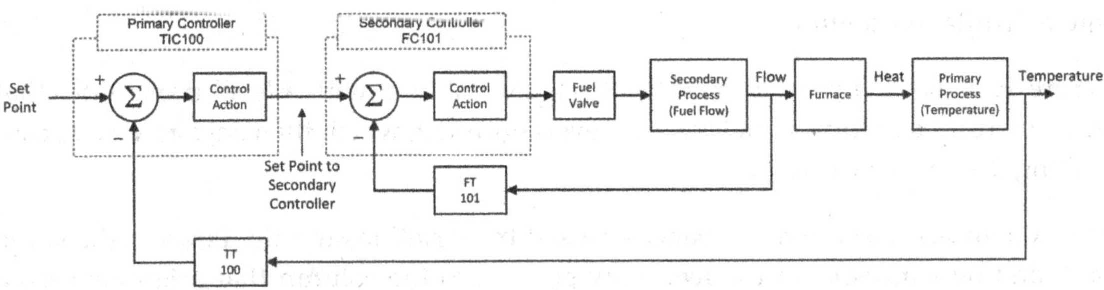

Figure 11-3. The concept of cascade control.

then be used as the inner control loop. The total process lag is split or divided into two parts.

Cascade control exhibits its real value when an extremely slow process is involved. When this happens, errors can exist for long periods, and when disturbances enter the process, there may be a significant time delay before any corrective action takes effect. Also, when corrective action is taken, there may be a significant delay before any results affect the process. Cascade control offers the opportunity to locate intermediate controlled variables and to provide corrective action on disturbances with less delay.

The use of cascade control does involve additional hardware. As evident from the general layout, it requires an additional feedback controller, sensor, and feedback transmission system. There is no need for an additional final control element such as a control valve. These general appearances are a bit misleading, however. Hardware vendors usually supply both the primary and secondary controllers for the cascade arrangement within a single controller case, and the total cost for the two controllers is not twice the cost of a single controller. In addition, it is probably true that the intermediate or secondary controlled variable was considered to be important enough to be set as one of the variables that is sensed and transmitted back to the central control room for indicating and/or recording purposes. If this is the case, there is little to no incremental cost associated with the secondary controlled variable's sensor and feedback transmission system. The additional cost in most cascade control strategies usually is not a significant factor, especially considering its benefits.

In general, cascade control offers significant advantages to the user, and it is perhaps one of the most underutilized feedback control strategies because of the difficulty in tuning cascade control. Most plants could increase their use of cascade control to significant advantage.

11.2 Applications

There are many applications for cascade control. Figure 11-4 shows a second example. In this case, several liquid streams flow together into a surge tank and, in turn, are fed into a distillation column.

Figure 11-4 shows a conventional feedback control loop with a level transmitter that measures and transmits tank level to a level controller, which then adjusts a valve controlling feed into the column.

The system is subject to disturbances caused by variations in inlet flows to the surge tank and by variations in the feed flow going into the column that originated from

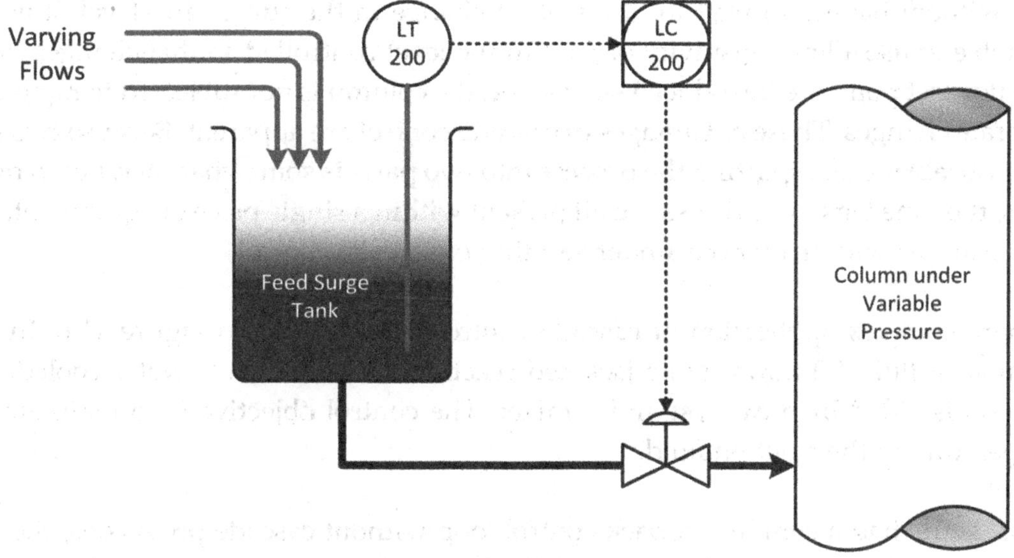

Figure 11-4. Column feed control system, conventional feedback.

pressure drop variations across the control valve due to pressure variations at the column.

Figure 11-5 shows a cascade arrangement of the same system in which the feed into the column is flow controlled and the set point on this flow control feedback loop is the manipulated variable of the primary level control feedback loop.

In this case, when the column pressure changes, pressure drop across the control valve changes, feed flow changes, and it is sensed and controlled quickly by the flow control

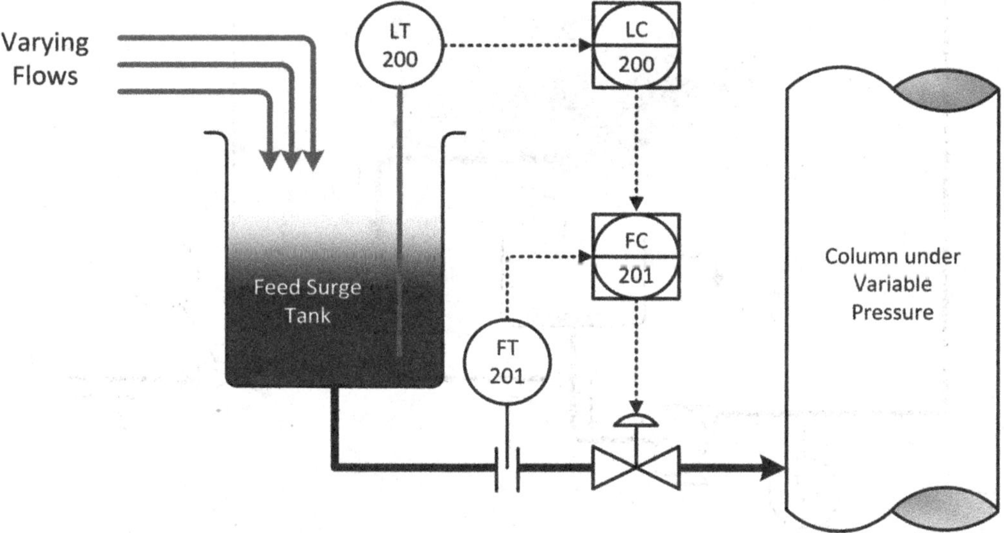

Figure 11-5. Column feed control system, cascade control.

loop without having to wait for a resultant change in the surge tank level. It is also desirable to use a low sensitivity or gain in the level controlled so that changes in the feed flow set point are introduced slowly and the column is not subjected to rapid feed flow rate changes. These advantages of cascade control are apparent. Because two separate variables exist, splitting the process into two parts is somewhat easy, but in many cases, the time lags in a process are all present within a single piece of equipment. The next example will further our understanding of cascade control.

A more complex application of cascade control is illustrated in Figure 11-6. In this example, a fluid (*F*) flows into a jacketed reaction kettle, which is water cooled. The reaction is aided by a catalyst and a mixer. The control objective is to maintain the temperature of the reaction fluid.

If this kettle had a simple feedback control loop without cascade provisions, the sensor on the primary controlled variable (*T*, the reacting mass temperature) would go

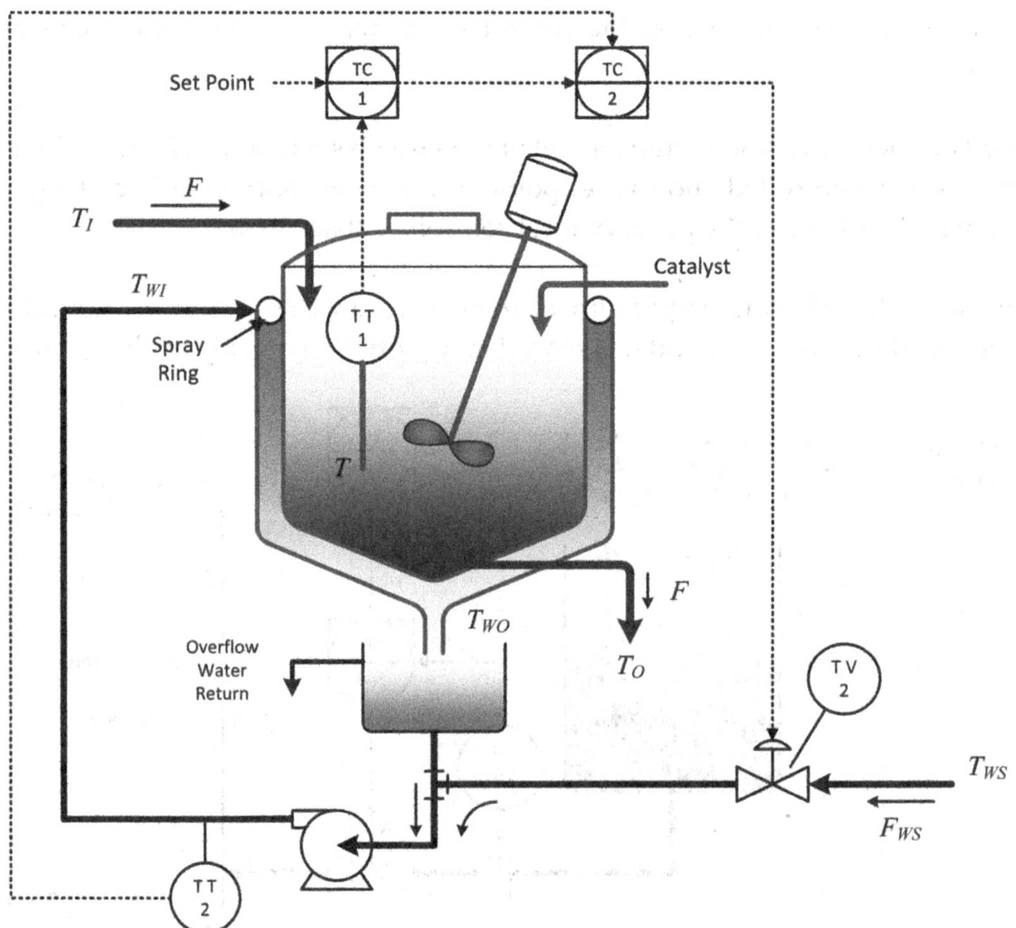

Figure 11-6. Cascade control on a jacketed kettle.

to feedback controller TC 1 and the output of the feedback controller would directly adjust the manipulated variable (T_W, the makeup cooling water supply). The kettle in Figure 11-6 has a cascade arrangement with the primary controlled variable still being the temperature of the reacting mass. An intermediate controlled variable for the secondary controller is the temperature of the circulating cooling water for the jacket of the kettle (T_{WI}). In this case, as is standard in cascade control arrangements, the manipulated variable for the primary is the set point for the secondary controller.

To better understand this use of cascade control, it will help to look at two potential disturbances for the example system. First, consider what happens when there is a change in inlet feed temperature (T_I) to the jacketed kettle. A change in this variable will change the basic temperature of the reacting mass. In general, the reacting mass will change in temperature by a first-order lag-type relationship; the time constant is the mass of the contents of the vessel times its heat capacity (capacitance) divided by the flow rate (F) of fluid into the vessel times its heat capacity (conductance). This time constant for the contents of the kettle probably will be quite large. As the reacting mass starts to change in temperature, it will be measured by the primary temperature transmitter (TT 1), which has its own dynamics, at least a first-order lag, and this first-order lag is interacting with the time constant of the kettle contents. The primary controller (TC 1) receives the error caused by the change in temperature, and corrective action is taken by adjusting the set point on the secondary controller (TC 2). Then, corrective action is taken by adjusting the manipulated variable, the incoming makeup cooling water flow (F_{WS}).

A second upset or disturbance happens if there is a change in the cooling water supply temperature (F_{WS}). When this occurs, there will be a rapid change in the temperature of the discharge of the cooling water circulating pump, and this change in temperature will be measured by TT 2 and sent to the secondary controller (TC 2). If there were no cascade control, this disturbance would not be sensed until the temperature change worked through the entire cooling water system and began to change the temperature of the reacting mass (T) inside the vessel. The advantages of the cascade arrangement for this type of upset are clearly evident. Figure 11-7 shows a general block diagram

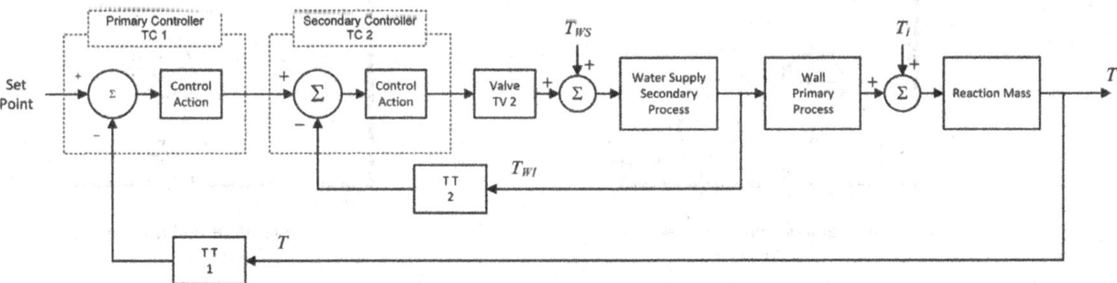

Figure 11-7. Block diagram for a jacketed kettle cascade control system.

layout for this type of cascade temperature control system with the two disturbances entering the system.

11.3 Implementing Cascade Control

A difficult question in implementing cascade control is how to find the most advantageous secondary controlled variable; specifically, the designer must determine how to identify and divide the process. Often, many choices of this intermediate point are available to the designer. The overall strategy should be to get as much of the process lag into the outer loop as possible while at the same time having as many of the disturbances as possible enter the inner control loop.

Figure 11-8 shows a general layout of a fired-charge heater that is used to increase the temperature of a fluid charge passing through a fired furnace. The feedback control of this arrangement is shown in Figure 11-8(a). Also shown in Figure 11-8 are three

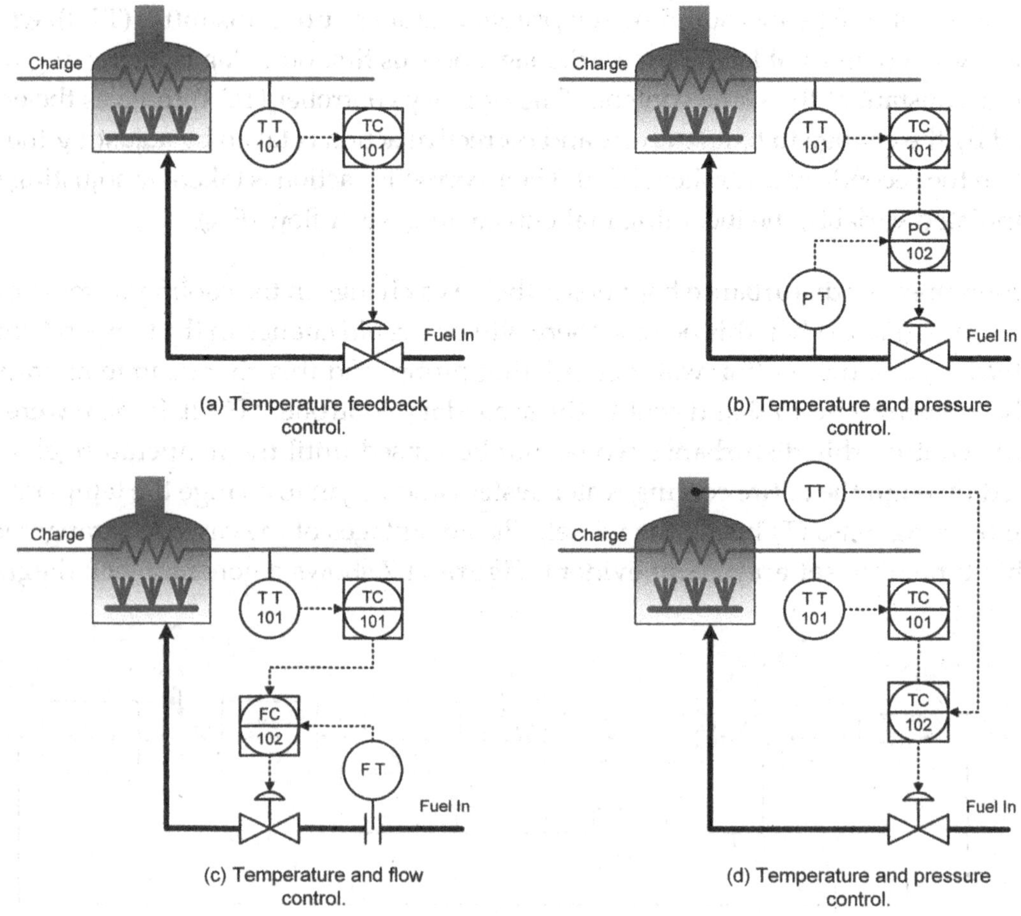

Figure 11-8. Alternate cascade control arrangements on a fired charge heater.

different methods for establishing a cascade control arrangement. The primary controlled variable is the same (temperature) for all three methods, but a different secondary controlled variable has been selected for each. The question is: Which cascade control arrangement is best?

To ascertain the best cascade control arrangement, the designer must determine the most likely disturbances to the system. It is helpful to make a list of these in order of decreasing importance. After this has been done, the designer may review the various cascade control options available and determine which best meets the overall strategy outlined earlier, that is, to have the secondary loop with the smallest time constant while at the same time receiving the bulk of the important disturbances.

Selecting the secondary control variable is so important in a cascade system that it is helpful to formalize some guidelines or rules of thumb to help guide the evaluation. The following discussion reflects what is commonly considered.

Make the secondary control loop include the input of the most severe disturbances; these disturbances that enter the secondary loop are the ones for which the cascade system will show the most improvement over conventional feedback control.

Make the secondary control loop's time constant as small as possible by including only minor lags of the complete control system. It is desirable, but not essential, that the inner control loop time constant be at least three times smaller than the outer time constant control loop.

Select a secondary variable with values that are useful and related to the values of the primary variable. During undisturbed operation, the relationship between the secondary variable and primary variable should be represented by a single line, and if it is a straight line, tuning the controllers will be simpler.

If it can be done while keeping the secondary control loop time constant relatively small, have the secondary control loop contain as many of the disturbance inputs as possible.

Choose a secondary control variable that will enable the primary controller to operate at the highest possible gain (lowest proportional band). This can be difficult to predict.

11.4 Selection of Cascade Controller Modes and Tuning

If both controllers in a cascade control system are three-mode controllers, there are a total of six tuning adjustments. It is doubtful that such a system could ever be tuned effectively. When selecting the modes to be included in both the primary and secondary controllers of a cascade arrangement, the designer should consider why a specific control mode should be added. For the secondary control loop, it is standard practice to include the proportional mode. There is less of a need to include the reset (integral) mode to eliminate offset because the set point for the inner controller will be reset continuously by the primary controller. Reset can be added to the secondary-loop controller if it is a liquid flow control loop because of the need to filter some of the transmission of high-frequency noise around the loop. An alternative is to operate the secondary loop at a low gain (high proportional band). This was discussed earlier in Chapter 10.

For the secondary loop, the controller should contain the proportional mode, and if the loop is sufficiently important to merit cascade control, it is probably true that reset should be included to eliminate any offset in the outer control loop. Rate or derivative control should be used in either loop only if the loop in question has a large amount of lag.

Tuning cascade controllers is the same as tuning all feedback controllers, but the practitioner must *work from the inside out.* The primary controller should be put on manual, that is, the loop should be broken and then the secondary control loop can be tuned using whatever tuning technique the practitioner finds most useful. After the secondary control loop is properly tuned, then the primary control loop may be tuned. When both control loops are tuned, the primary control loop will "see" the tuned secondary control loop as its final control element. This general procedure, when used in cascade systems, should provide stable system performance.

Note that many systems that involve a control valve are in fact cascade control systems. As discussed in Chapter 8, most control valves include a positioner, and a control valve positioner is a feedback control system.

11.5 Summary

Cascade control is only slightly more complex to implement than single-loop control, and when applied it greatly improves the performance of slow processes such as temperature that tend to demonstrate sluggish behavior. Cascade control improves a slow control loop by responding to disturbances of faster dynamics such as a flow or a pressure loop. The faster loop provides an early disturbance rejection, helping to maintain steady performance of the slower primary loop.

Exercises

11.1 Given the following heat exchanger, sketch a cascade arrangement for TC-FC (temperature control-flow control) in which a secondary flow loop has been provided for the steam.

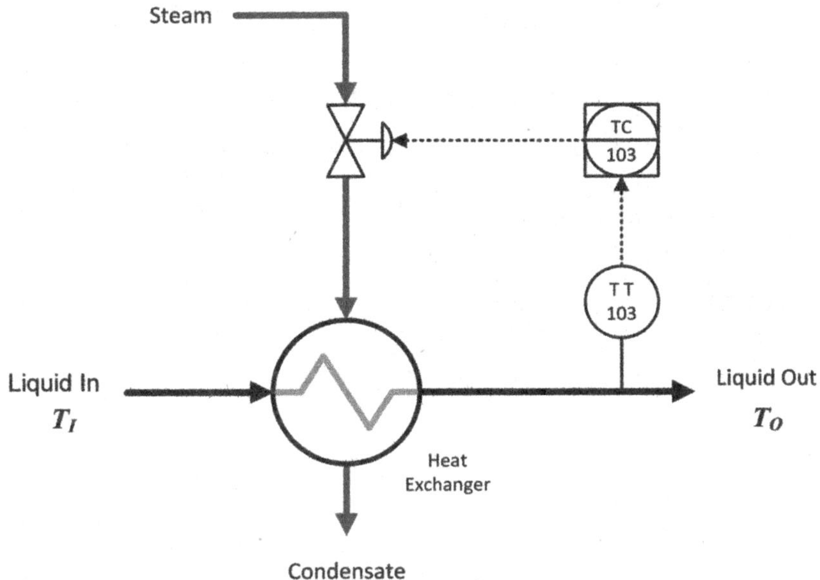

11.2 Redo Exercise 11-1 but provide TC-PC (temperature control-pressure control) with a secondary pressure control loop on the exchanger shell.

11.3 Compare the results of Exercises 11-1 and 11-2 and determine which gives a better initial response to disturbances in inlet process temperature (T_I).

12 Feedforward Control

Most control theory that was covered in Section 1 focused on feedback control. This chapter describes some of the inadequacies of feedback control and shows how feedforward control can be used when applicable. Feedforward control is based on the premise that the system designer has a clear understanding of the disturbances involved in the process, is able to quantify them, and then takes preemptive action to prevent the disturbances from affecting the process variable. Feedforward control could be useful in some cases even if the designer does not have a clear understanding of the disturbances. This will be illustrated using our home heating system example.

Learning Objectives

- Understand the fundamental concepts of feedforward control
- Define the structure of the control action equation contained in a feedforward controller
- Understand how feedforward control is used in practice, especially in combination with feedback control

12.1 Feedforward Control

Feedforward control is different in concept from feedback control. A manual implementation of feedforward control is illustrated in Figure 12-1.

In this control strategy, as a disturbance enters the process, the operator observes the disturbance and, based on the nature and quantity of the entering disturbance,

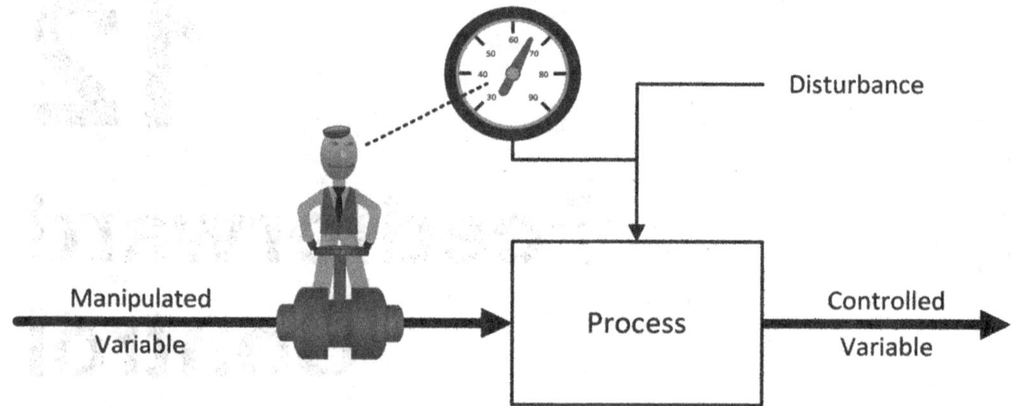

Figure 12-1. Manual feedforward control.

adjusts the manipulated variable in such a manner as to prevent any eventual change or variation in the controlled variable due to the observed disturbance. The conceptual improvement is obvious. Feedback control works to eliminate errors; feedforward control operates to prevent errors from occurring in the first place.

For example, in a home heating system (Figure 12-2), room temperature is maintained by controlling the amount of fuel to the furnace. The temperature transmitter (TT 100)

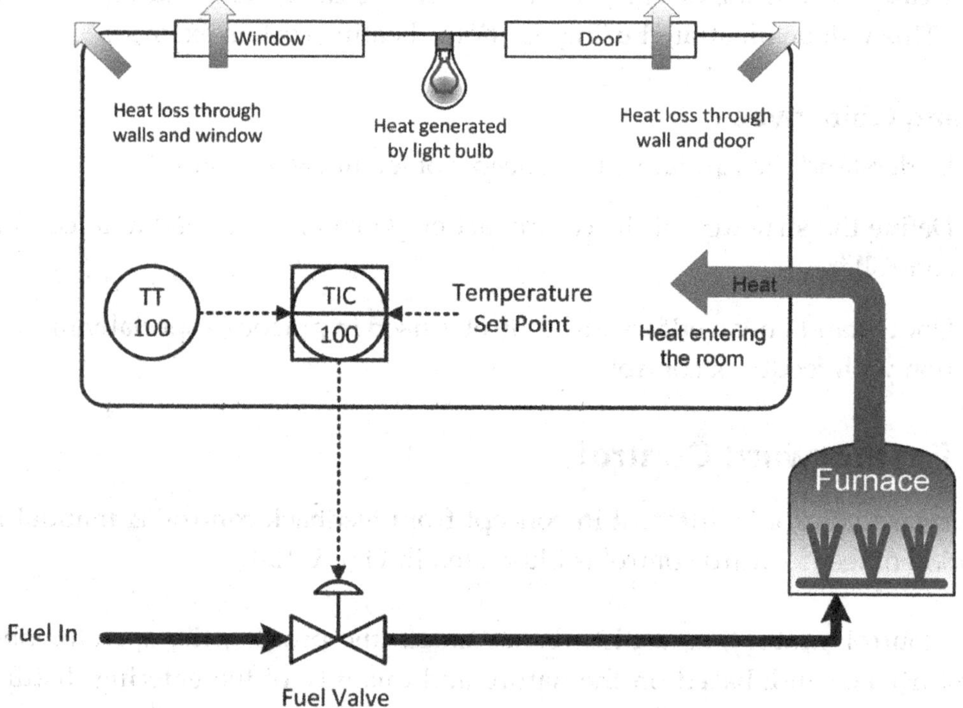

Figure 12-2. Control of home heating system.

reacts to disturbances in the room, and the controller calculates the required change to the fuel valve to maintain the desired room temperature. What happens if the outside air temperature suddenly drops? The temperature in the room will also drop as a result of increased thermal loss through the walls, window, and door. The temperature controller will adjust the fuel valve to increase the heat supplied to the room to compensate.

The addition of an outdoor temperature transmitter can provide an early warning to the controller and direct the temperature controller to increase the heat before the room starts to cool. In a similar manner, a temperature transmitter can be located close to the door. In this case, when the door is opened and warm air leaves the room, the second temperature transmitter can direct the temperature controller to increase the heat, compensating for the heat loss before the room starts cooling (see Figure 12-3).

Feedforward control, however, is not that simple. It escalates the technical understanding requirement to control the process. The operator or feedforward correction must know in advance what disturbances will be entering the process, how they might affect the process, and how to deal with them. This means knowing how the controlled variable must be adjusted to negate the disturbances. Therefore, ample provisions must be made to measure the possible disturbances. In addition, the system must

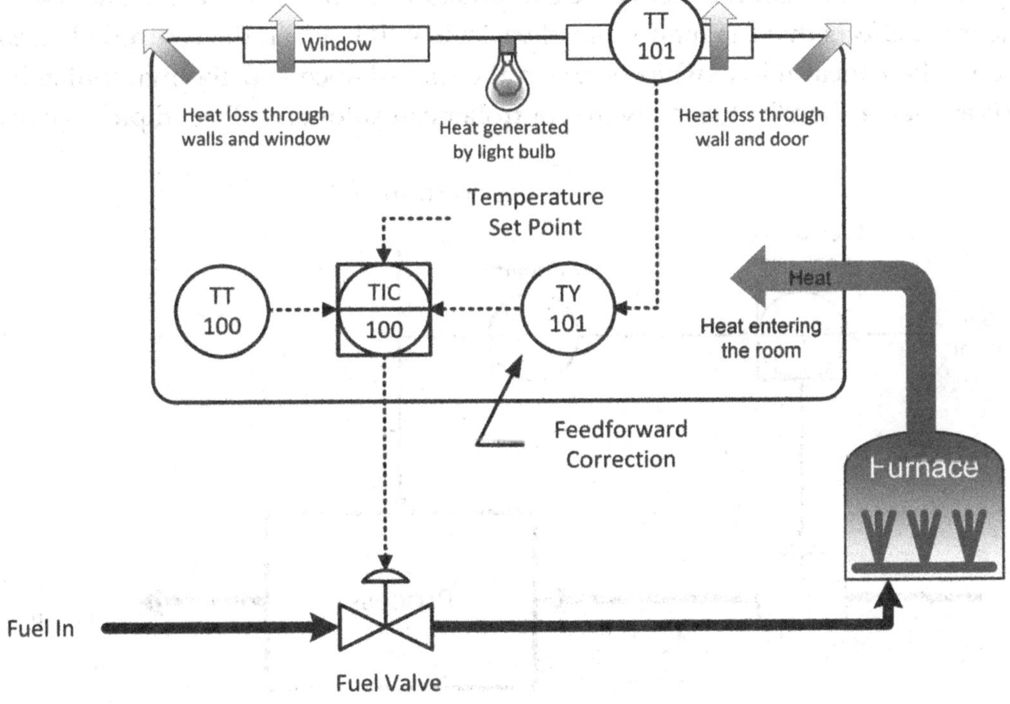

Figure 12-3. Home heating system with feedforward correction.

know specifically when and how to adjust the manipulated variable to compensate exactly for the effects of these disturbances. If the control system is designed with this explicit knowledge and ability to make the necessary adjustments to the controlled variable, the process will theoretically never deviate from its desired value or set point. If the system does not anticipate all the disturbances that might affect the process, then the controlled variable will deviate from its desired value and an uncorrected process variable may exist.

12.2 Steady-State Feedforward Control

Figure 12-4 shows the general conceptual structure of an automatic feedforward control strategy.

A disturbance is shown entering the process, and a sensor/transmitter measures the disturbance. Based on the measured values of the disturbance, the feedforward controller then calculates the needed values of the manipulated variables.

The feedforward controller must make sophisticated calculations. These calculations must reflect an awareness and technical understanding of the exact effects the disturbance will have on the controlled variable. With such understanding, the feedforward controller can then calculate the exact amount of manipulated adjustment required to compensate for the disturbance. These computations also imply specific understanding of the exact effects that the manipulated variable will have on the controlled variables. If the mathematical relationship between the disturbance and the manipulated variable is available, then the feedforward controller can automatically compute the needed

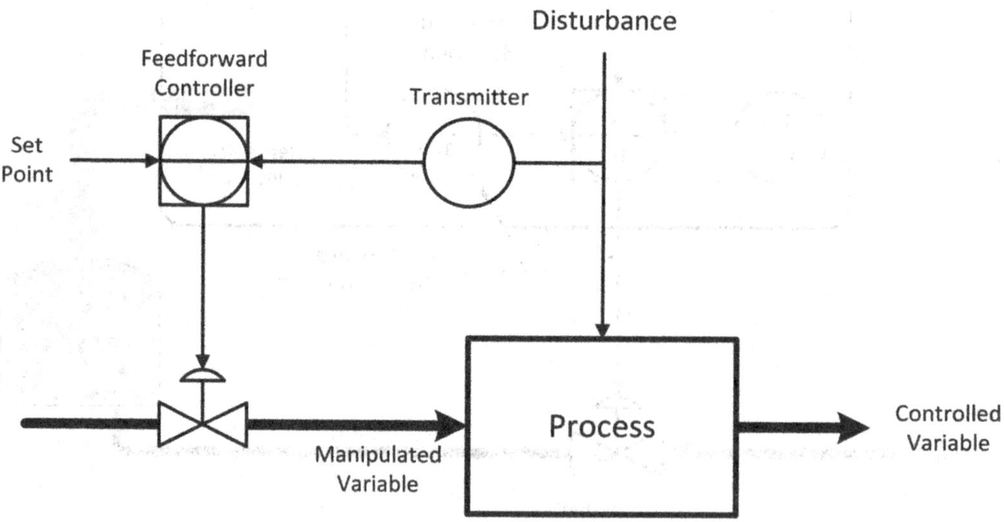

Figure 12-4. Feedforward control concept.

value of the manipulated variable to compensate for variation in the disturbance. The increase in the required theoretical understanding can be complex. Feedforward control, while conceptually more appealing, significantly escalates the technical and engineering requirements of the designer and practitioner. As a result, feedforward control usually is reserved for only a few of the most important loops within a plant. Although the number of applications is small, their importance is quite significant. Note that pure feedforward control is rarely encountered and is usually used in combination with feedback control as we saw in the previous example of the home heating system.

Two process conditions can make the overall effectiveness of feedback control quite unsatisfactory. One is the occurrence of large-magnitude disturbances, and the other is the occurrence of large amounts of lag within the process. Feedforward control can be used to deal with these large, anticipated disturbances.

A manipulated variable is chosen based on many of the same criteria typically used in feedback control. In feedforward control, however, a sensor is used to measure the disturbance as it enters the process, and the sensor transmits this information to a feedforward controller. The feedforward controller determines the needed change in the manipulated variable so that when the effect of the disturbance is combined with the effect of the change in the manipulated variable, there will be no change in the controlled variable. This perfect compensation is a difficult goal to achieve. It is, however, the objective for which feedforward control is structured. Just as in feedback control, it is necessary to provide the feedforward controller with a set point or desired value of the controlled variable.

There are some significant difficulties in feedforward control. The structure of a feedforward controller assumes that the disturbances are known in advance, that the disturbances have sensors associated with them, and that all the disturbances are measurable. Also, tremendous amount of theoretical knowledge is required to develop the feedforward controller's algorithm. In feedback control, relatively standard control algorithms, such as those used in a proportional-integral-derivative (PID) controller, are used, but each controller equation and/or algorithm must be specifically and uniquely designed for the particular control application involved. In effect, the feedforward control computations involve determining exactly how much change in manipulated variable is required for a specific disturbance. To make this computation accurately requires a considerable quantitative understanding of the process, its disturbances, and its operation.

Another significant aspect of feedforward-only control, as shown in Figure 12-4, merits consideration. In this case, there are no feedback phenomena whatsoever; if the controlled variable strays from its set point or desired value, the control system is uninformed and takes no corrective action to eliminate the deviation. This makes feedforward-only control somewhat impractical, and therefore it is rarely used in typical process control applications. The usual case is to combine feedforward control with feedback control.

Figure 12-5 shows an example application, a heat exchanger process controlled using feedforward. In this process, a liquid flows through the heat exchanger and is heated by condensing steam. The process variable for this example is the exit temperature of the liquid flowing through the exchanger. The manipulated variable is the steam flow to the heat exchanger. The steam flow is controlled by measuring the inlet temperature, and as the temperature decreases/increases, the steam inlet valve will increase/decrease the amount of steam to the heat exchanger. Essentially, feedforward control is used to calculate the desired flow to the heat exchanger. The measured disturbances to this process are assumed to be the inlet liquid temperature, and the flow rate of liquid through the exchanger is assumed to be constant.

The system can be greatly improved if in addition to measuring the inlet temperature, we also measure the inlet flow rate. An improved feedforward control strategy is shown in Figure 12-6.

Both the liquid intake temperature and its flow rate are measured, TT 102 and FT 102, respectively, and are transmitted to temperature controller TC 102. The temperature controller then calculates the required valve opening.

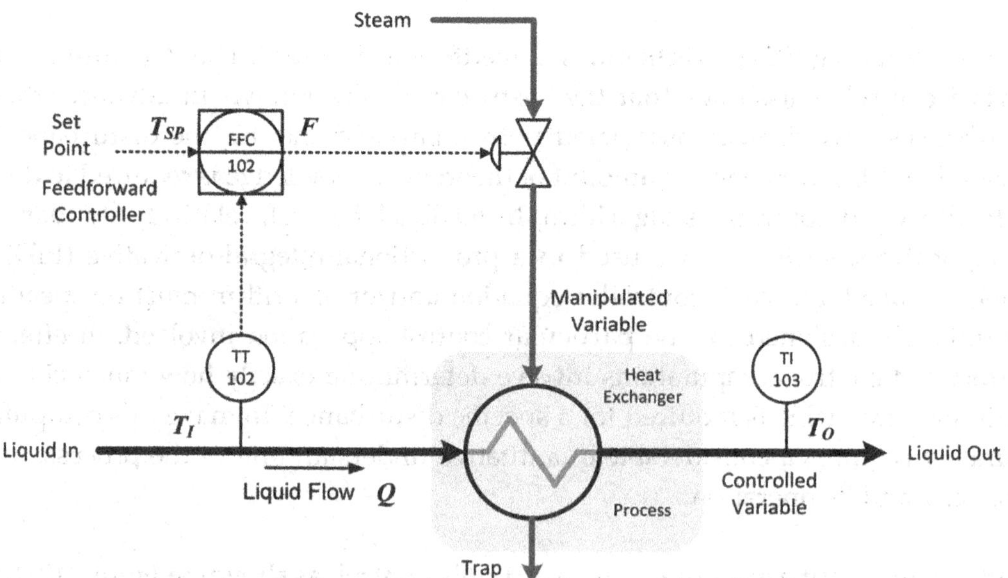

Figure 12-5. Feedforward control of a heat exchanger.

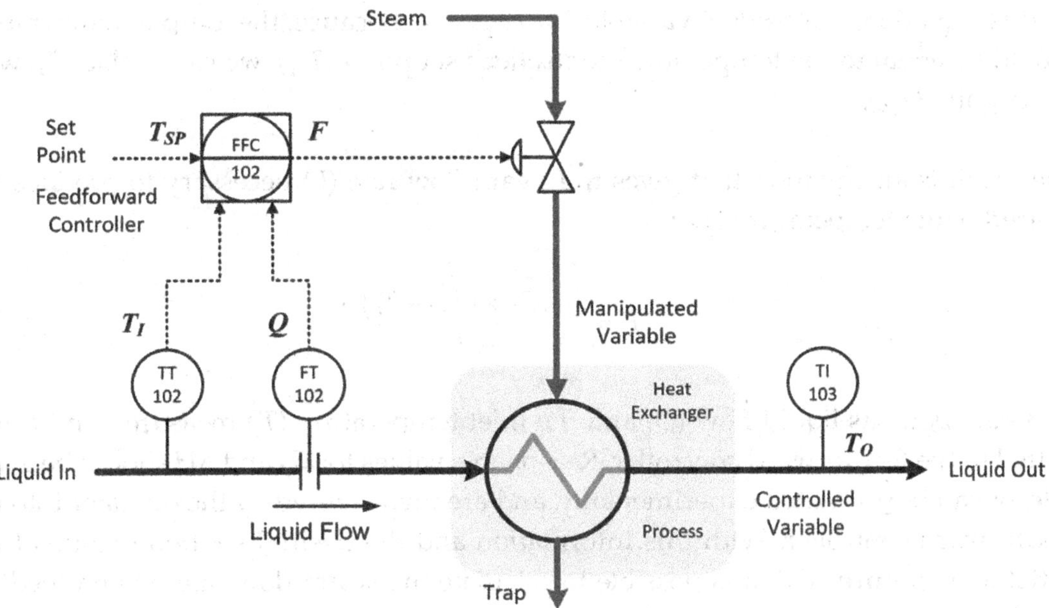

Figure 12-6. Feedforward control of a heat exchanger with both temperature and flow.

This calculation is based on a steady-state energy balance around the exchanger, meaning the amount of energy out of the system must equal the amount of energy into the system. Mathematically:

$$output(Btu/h) = input(Btu/h)$$
$$Q \times C_P \times (T_O - T_I) = F \times \Delta H \tag{12-1}$$

where

Q = liquid flow	in pounds/hour (lb/h)
C_P = liquid heat capacity	in Btu per degrees Fahrenheit/ pound (Btu/°F/lb)
T_I = liquid inlet temperature	in degrees Fahrenheit (°F)
T_O = desired outlet	in degrees Fahrenheit (°F)
F = temperature	in pounds/hour (lb/h)
ΔH = steam flow heat released by steam	in Btu/pound (Btu/lb)

Solving the equation for steam flow rate (F):

$$F = \frac{Q \times C_P \times (T_O - T_I)}{\Delta H} \tag{12-2}$$

In this equation, controlled variable T_O appears. Because the output temperature should be equal to the temperature controller's set point (T_{SP}), we can replace T_O with its set point (T_{SP}).

The result is an equation that gives the steam flow rate (F) necessary to produce the desired outlet temperature (T_O):

$$F = \frac{Q \times C_P \times (T_{SP} - T_I)}{\Delta H} \tag{12-3}$$

This equation has liquid flow (Q) and the inlet temperature (T_I) measured and transmitted to the feedforward controller. Reasonable values for C_P and ΔH should be available, or can be estimated experimentally, and are then entered by the operator into the feedforward controller. With this information and the desired set point value of the outlet temperature (T_{SP}), it is possible to calculate the steam flow rate (F) and feedforward control can be implemented as shown in Figure 12-6.

This example illustrates the complexity of feedforward control. Note the significant difference between the feedforward control and the feedback control where standard control equations usually are combinations of the three PID control modes. To design a feedforward control algorithm for a particular process control application requires significant understanding on the part of the designer. In addition, it means the controller hardware/software used in feedforward control will be unique in each installation. Also, the feedforward equation for a specific loop must be uniquely designed.

This example also shows that if there are other disturbances at work that are unforeseen, the control hardware will have no way of sensing the resultant change in the outlet temperature of the liquid and, as a result, no effective corrective action will take place.

We will see in Section 12.5 how feedforward control can be combined with feedback control, giving the system the advantage of both feedback and feedforward control.

12.3 Steady-State or Dynamic Feedforward Control

The previous example was a specific application of steady-state feedforward control, the energy balance was made around the heat exchanger, and it was a steady-state energy balance. It did not give any consideration to and quantitative recognition of the fact that process dynamics is associated with the exchanger. In practice, when the inlet conditions change, the resultant changes in the controlled variable will not occur instantaneously. Yet in the specific feedforward control equation, it was implicitly assumed that instantaneous correction was appropriate. When any condition changed in the feedforward control loop, there was an instantaneous (algebraic) calculation of a new steam flow.

For increased accuracy in feedforward control, it is desirable and often necessary to include the effects of process dynamics in the feedforward calculation of the manipulated variable. This is usually done using one of two strategies. In one case, it is possible to implement steady-state feedforward control and simply push the output signal through some type of dynamic compensator before that signal is used to adjust the manipulated variable itself. This approach to dealing with dynamic compensation is illustrated in Figure 12-7.

With dynamic compensation of this type, it is possible to manually adjust or tune the dynamic compensator so that it introduces proper dynamic corrections to the feedforward control action taken. In typical hardware applications, quite often the dynamic compensator is a simple lead-lag network or a simple second-order lag adjustment. In such cases, the operator tunes the ratio of the lag-lead time constants or tunes the ratio of the time constants in the second-order lag network. Such tuning is usually done after hardware installation.

The second and more accurate, but theoretically complicated, approach is to provide dynamic compensation through the general derivation of the feedforward control equation as a dynamic or unsteady-state relationship. This will be illustrated next.

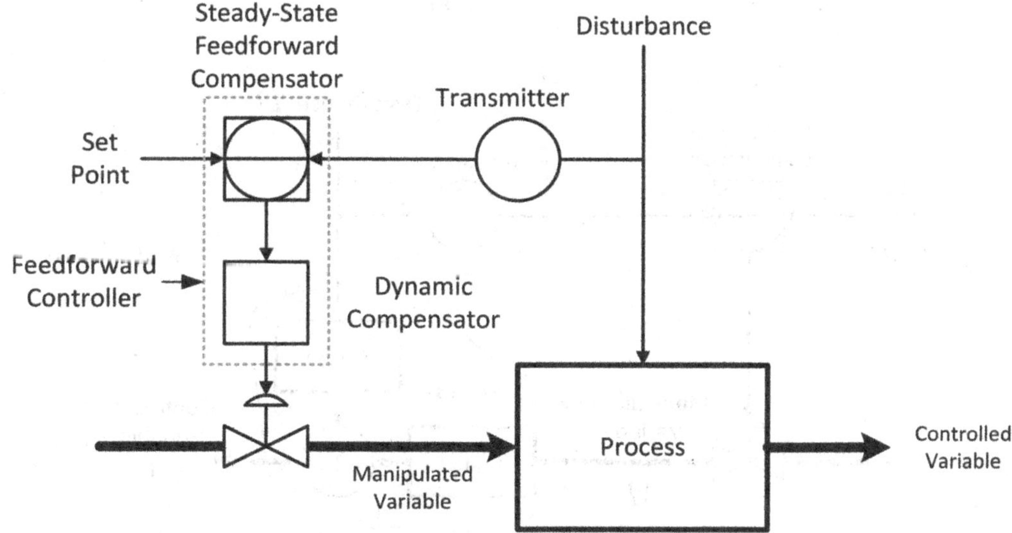

Figure 12-7. Feedforward control with dynamic compensation.

12.4 General Feedforward Control

It is possible to structure feedforward control in a more general sense than has been done up to this point. Figure 12-8 shows a process to be controlled in a feedforward manner. The controlled variable (*CV*) may be expressed in terms of two input forces working on the process, the disturbance (*D*) and the manipulated variable (*M*).

The symbols $G_1(s)$ and $G_2(s)$ represent the block diagram symbols for the dynamic relationship between the input to the block and the output from the block (*s*) being the Laplace differential operator (*d/dt*). The implication is that the overall relationship between input and output is dynamic and can be represented by differential equations.

The following relationship then follows:

$$CV = G_1(s) \times M + G_2(s) \times D \tag{12-4}$$

This may be solved for the manipulated variable (*M*):

$$M = \frac{CV - G_2(s) \times D}{G_1(s)} \tag{12-5}$$

But in every case of pure feedforward control, you do not actually measure and feed back the controlled variable; instead, the reference or desired value (*SP*) is taken and substituted for *CV* in the overall control equation. This gives:

$$M = \frac{SP - G_2(s) \times D}{G_1(s)} \tag{12-6}$$

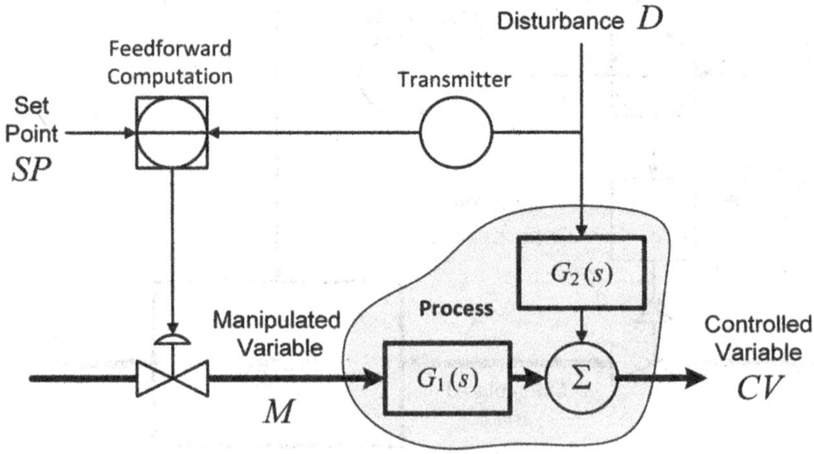

Figure 12-8. Generalized feedforward control block diagram.

This is the general feedforward control equation. The disturbance (*D*) is measured and the set point (*SP*) is provided. We must also provide in advance the specific relationships $G_1(s)$ and $G_2(s)$. A continuous calculation is then made of the needed value of the manipulated variable (*M*) that will maintain *SP* equal to *CV*. The significant factor, however, is that the designer of this control system must have enough understanding and knowledge of the process to make a specific quantitative determination of the mathematical relationship between *D* and *CV*, which is $G_2(s)$, and the mathematical relationship between the manipulated variable (*M*) and the controlled variable (*CV*), which is $G_1(s)$. The major increase in theoretical understanding required to implement feedforward control is evident.

One must understand that if there are two significant disturbances entering the process, each must have its own specific feedforward control computation and, by implication, there will be an additional control equation similar to Equation 11-6. The difficulties in handling multiple disturbances make feedforward control impractical to implement in many applications.

12.5 Combined Feedforward and Feedback Control

Feedforward-only control has some significant difficulties and disadvantages; the necessary computations are sophisticated and must take into account many effects. Sometimes they involve difficult mathematical relationships. Most models are based on empirical data and are not always correct but rather approximations. When inaccuracies exist in the model or computation, there is no corrective action. In addition, if disturbances other than the specific ones being measured for the feedforward controller enter the process, the automated feedforward control system does nothing, and errors build up.

As a result of these potential problems, feedforward-only control is rarely encountered; there is always some sort of feedback control added to complement the feedforward arrangement. The following examples will illustrate how this is done.

Figure 12-9 shows the heat exchanger from Figure 12-6, with feedback control added to the feedforward arrangement.

This configuration shows how a feedback control arrangement can supplement the feedforward control of the heat exchanger. In this particular example, the output of the feedback controller is used to adjust the set point of the feedforward controller.

The temperature controller (TC 103) retransmits the desired set point (T_{SP}) to the feedforward controller (FFC 102). The feedforward controller then makes the required

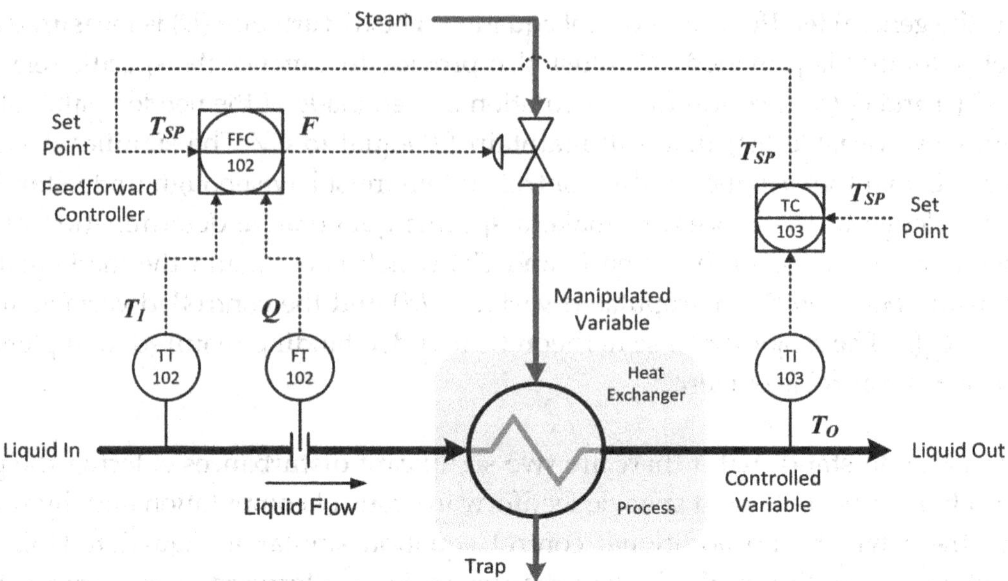

Figure 12-9. Feedforward control with temperature feedback.

adjustments to its output (F) based on the disturbances measured by the TT 102 (temperature) and FT 102 (flow) transmitters.

The system can be further improved by adding a second feedback controller to control the steam flow. This is illustrated in Figure 12-10. The addition of a feedback control loop is common when feedforward control is used. It stems from the general

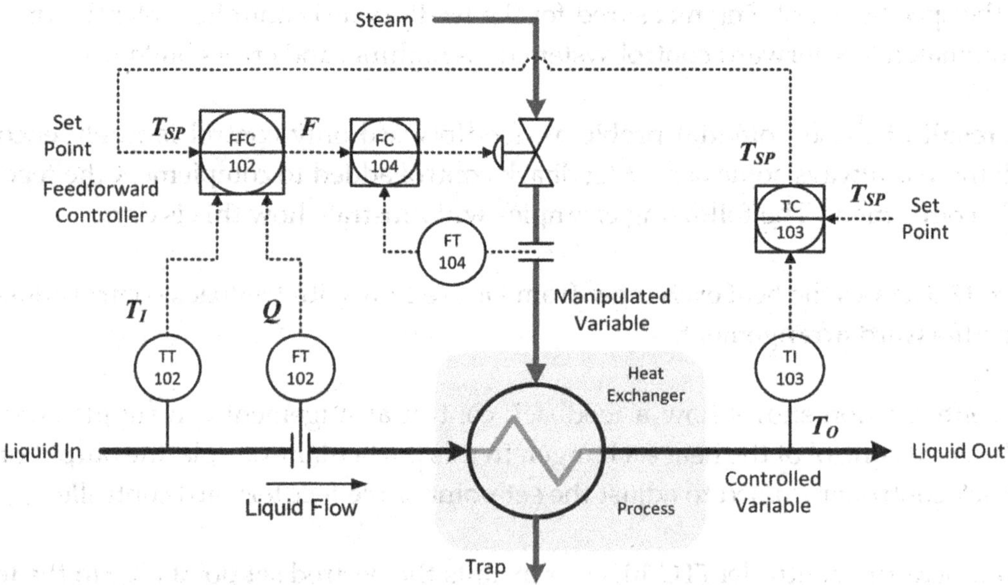

Figure 12-10. Feedforward control of flow and temperature feedback.

acceptance that if the variable is important enough to merit feedforward control, it is also important enough to ensure that the manipulated variable is maintained at its calculated value.

12.6 Summary

Feedforward control takes a preemptive approach to responding to disturbances. Disturbances are measured, and adjustments are made in the manipulated variable, preventing the controlled variable from deviating from its desired set point.

Exercises

12.1 Refer to the following figure. Design a steady-state feedforward control system for this unit.

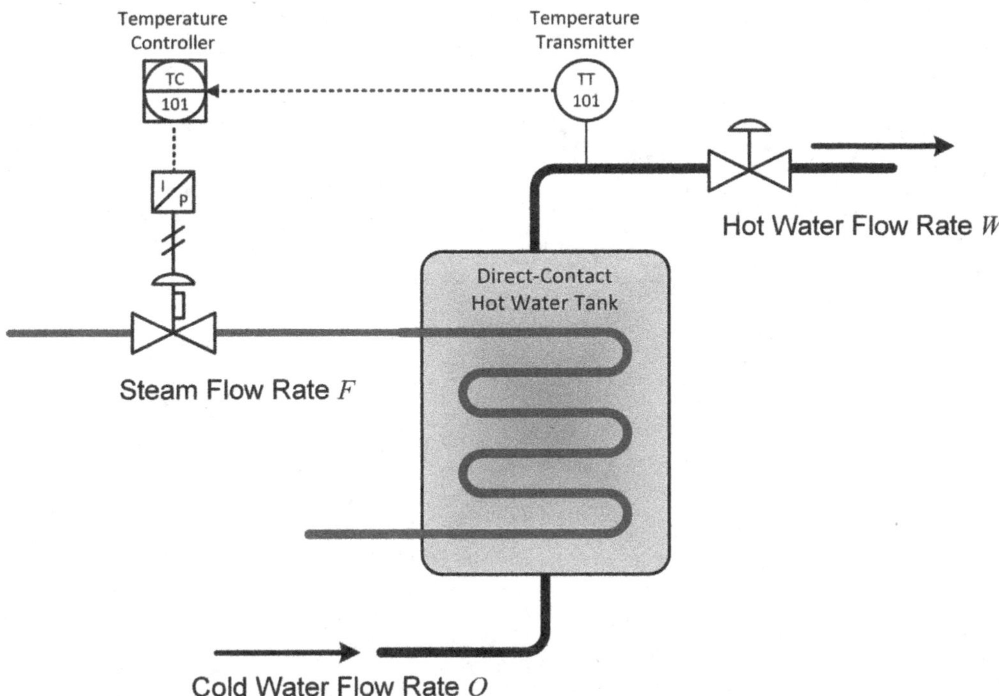

12.2 To Exercise 12-1, add a lag-lead dynamic compensation of the following form:

$$\frac{1+\tau 1s}{1+\tau 2s}$$

where the ratio of $\tau 1/\tau 2$ is adjustable (tunable).

12.3 Add cascade control of the manipulated steam flow to the drawing in Exercise 12-2.

13 Multivariable Control

In previous chapters, we viewed the control system as primarily having one or more independent control loops. In practice, there are processes in which different control loops interact with each other. Multivariable control is a method of dealing with multiple interacting control loops.

Learning Objectives

- Understand the concept of multivariable control
- Understand the problems associated with multivariable control
- Articulate possible multivariable control solutions

13.1 The Multivariable Control Concept

Consider, for example, a temperature control loop and a humidity control loop for a typical home. Decreasing the temperature will also decrease the relative humidity. Raising the humidity will raise the temperature. The required balance of comfort is difficult to resolve. In other words, one control loop affects the other.

The two control loops in Figure 13-1 illustrate the loop interaction. This represents what is called a 2×2 *process* (two controlled variables and two manipulated variables).

When there are two control loops on one unit, often the two control units will interact with one another. Specifically, when there is a change in the manipulated variable in Loop 1, not only will it produce a change in the controlled variable for Loop 1, but it will also produce a change in the controlled variable for Loop 2. Conversely, when

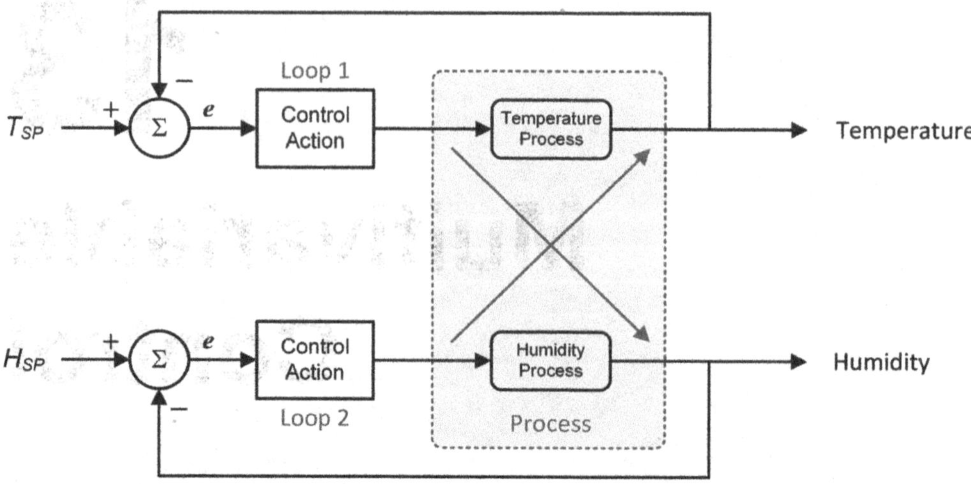

Figure 13-1. Conventional multivariable arrangement.

there is a change in the manipulated variable for Loop 2, it will produce a change in the controlled variable for Loop 2 but also in the controlled variable for Loop 1. This type of control interaction can produce significant control problems. When a change in the manipulated variable for Loop 1 occurs, there will be a change in the controlled variable for Loop 1. However, in addition, the change in the manipulated variable for Loop 1 will produce a change in the controlled variable for Loop 2. This, of course, will be sensed and fed around Loop 2, and will, in turn, produce a change in the manipulated variable for Loop 2. This not only produces the change in the controlled variable for Loop 2 but also produces a further change in the controlled variable for Loop 1. In effect, these loops start to chase one another in a figure-eight type of path as shown in Figure 13-2. To compensate for this, the control system must desensitize the loops, making it more complex and difficult to obtain quality feedback control.

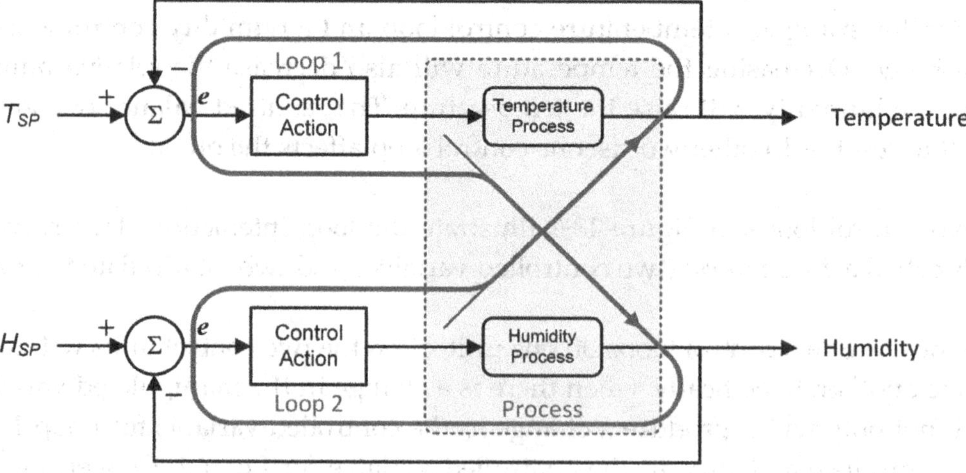

Figure 13-2. The effects of interaction in a 2 × 2 system.

13.2 Implementing Multivariable Control

When feedback control loops are interacting with one another, a control system is needed that will decouple the loops; to be specific, the interaction between the loops must be broken. The concept of breaking this interaction is referred to as *multivariable control*. The basic approach is to use conventional feedback control supplemented by a decoupler. This is illustrated in Figure 13-3.

In effect, a decoupler like the one shown is composed of two feedforward control elements. The overall operation is such that when a change in the controller output for Loop 1 occurs, it not only produces a change in the manipulated variable for Loop 1, but it also produces a change in the manipulated variable for Loop 2. Thus, when the interaction from Loop 1 to Loop 2 takes place, it will be compensated for by the change in the manipulated variable of Loop 2 by the decoupler. As a result, there will be no change in the controlled variable for Loop 2. Conversely, in Loop 2, when there is a change in the controller output, it produces a change in the manipulated variable for Loop 2, but it also produces a change in the manipulated variable for Loop 1. Thus, when the interaction from Loop 2 to Loop 1 takes place, it will be compensated for by the change in the manipulated variable in Loop 1, and there will be no change in the controlled variable in Loop 1.

The concept entailed by the decoupler as shown in Figure 13-3 is very appealing. The difficulty is in the design of the decoupler itself. The goal is to design the decoupler so that all interactions will be compensated for exactly. This is an admirable goal but often difficult to achieve in practice. For a 2 × 2 process, it is equivalent to designing two simultaneous feedforward controllers. The problem escalates rapidly, however, and if the system were a 4 × 4 process, the implementation of such a multivariable

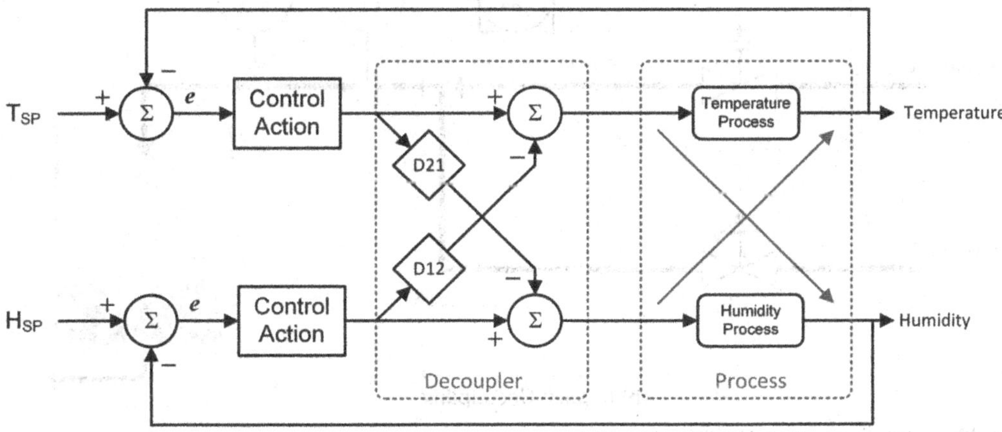

Figure 13-3. A decoupler for a 2 × 2 process.

control would require the implementation of 12 ($4^2 - 4$) simultaneous feedforward controllers. The increase in complexity is obvious.

Process interactions can be decoupled explicitly or implicitly. Often a decoupler such as the one shown in Figure 13-3 is implemented as a steady-state decoupler, that is, one consisting of algebraic terms only as shown in Figure 13-4(a). Explicit

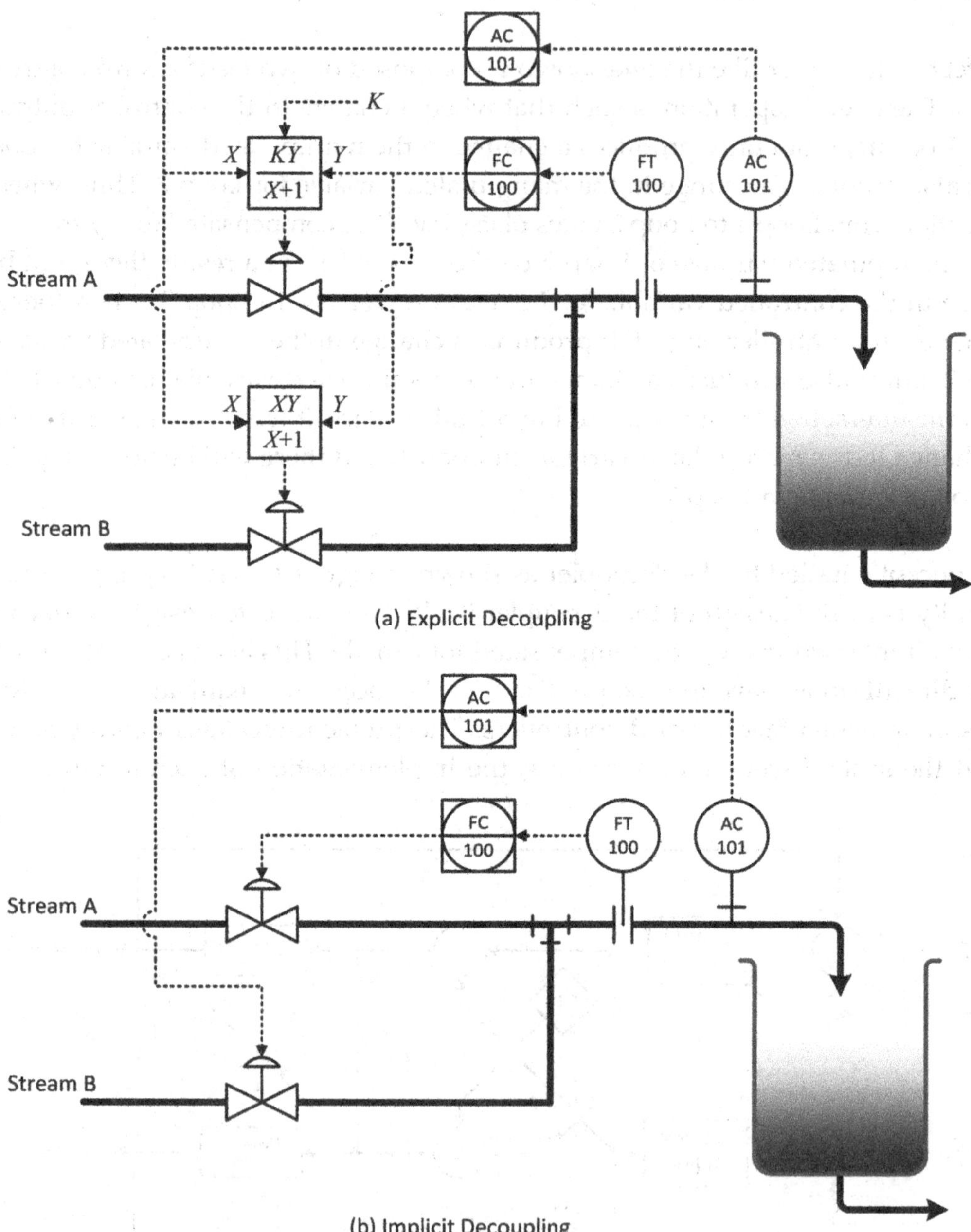

Figure 13-4. Implicit and explicit decoupling.

decouplers use a static or dynamic model of the process in order for each controller to influence other interacting controllers in such a way that any change in output reduces or eliminates the propagation of upsets to the other interacting controller. It is then possible to add dynamic compensation to the more important elements of the decoupler later—when the system is better understood—and gain further improvement.

Implicit decoupling involves rearranging and/or tuning the controller in ways that make the individual loops less interactive. The advantage is that no explicit decouplers are required. The recommended approach is to prioritize the control loops: those with more critical impact are tuned to respond faster to changes in the process, and lower priority controllers are detuned so that their actions have a lesser effect on the higher priority ones. Lambda tuning has proven to be successful in many applications where interactions are present.

Multivariable control is a challenging and difficult type of automatic control to implement. It requires significant design expertise and capability, as well as custom hardware and software. This typically implies the necessity of digital computation capabilities in the hardware. The cost of designing for multivariable control can be quite significant, but the improvement in control is equally significant. There are increasing numbers of commercial applications of multivariable control that have excellent economic justification; for example, multivariable control in heating, ventilation, and air-conditioning (HVAC) systems can bring significant savings in large buildings.

Multivariable control is a complex topic, and the presentation here is introductory at best. However, it does present the fundamental principle.

13.3 Summary

When two or more single-loop processes interact with each other, they should not be analyzed as single-loop systems but rather as a multivariable process. A multivariable process is said to have interaction when process input (manipulated) variables affect more than one process output (controlled) variable.

Exercises

13.1 Given the following 2×2 control system, draw the functional block diagrams for these two loops.

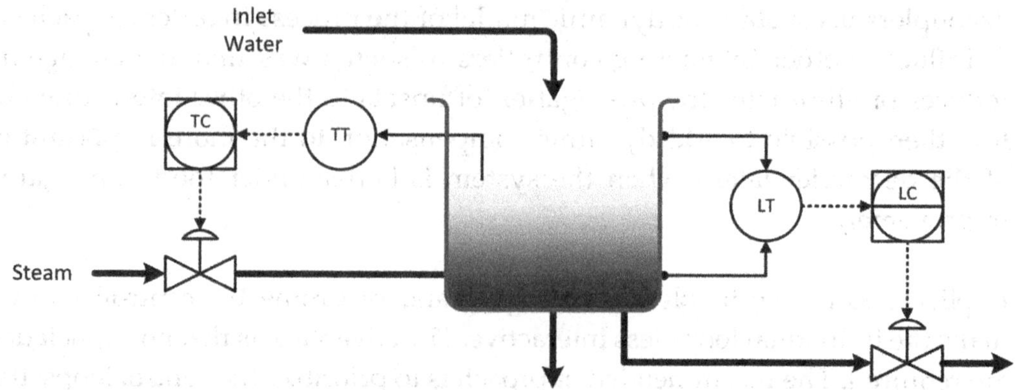

13.2 Sketch the functional outline of a decoupler for the system in Exercise 13.1.

14 Ratio Control

A commonly encountered type of multiple-loop feedback control system is *ratio control* (also called *fraction control*). Ratio control is used to ensure that two or more process variables are kept at a fixed ratio to each other, even though they are changing in value.

In industrial control systems, ratio control is used in maintaining the burner air/fuel ratio, in the mixing and blending of two liquids, in injecting modifiers such as pigments into resins before molding or extrusion, and in adjusting heat input in proportion to material flow.

Learning Objectives

- Recognize basic ratio control
- Understand the basic principles of ratio control
- Distinguish how and when to apply ratio control

14.1 The Ratio Control Concept

Ratio control is often confused with cascade control; however, the function of ratio control is quite different from that of cascade control. Ratio control resembles feedforward control but is, in effect, a special type of feedforward control. In ratio control, one loop adjusts a second control loop, and the second control loop is fractionally controlled by the first control loop in a feedforward manner, with their ratio maintained at a set fraction. Ratio control is often associated with process operations in which two or more streams must be mixed together continuously to maintain a steady composition in the resulting mixture, such as the air-to-fuel ratio (AFR) of an industrial combustion chamber.

In ratio control, we measure the flow of the first stream (referred to as *wild flow*) and then calculate (based on the required ratio) the value of the second stream (referred to as the *controlled flow*). A practical way to do this is to control one stream with a conventional flow controller and control the other stream with a ratio controller that maintains that stream flow at some preset ratio or fraction to the primary stream flow. Ratio control is then a form of feedforward control of the second steam. This is also referred to as a *wild feedforward loop*. You measure the disturbance of the first flow and calculate the ratio required for the second flow. This type of control is commonly referred to as a *multiple-loop feedback control system*.

This concept is best illustrated using the example of a home heating system. The furnace not only needs its fuel intake controlled to provide proper heating, but it also requires a proportional amount of air for efficient combustion, referred to as the *stoichiometric air-to-fuel ratio*.[1] Assuming the amount of fuel flow to the furnace is A and the amount of air required is B, the ratio of air to fuel can be expressed as $R = B / A$. Figure 14-1 shows that the fuel to the burner is measured using FT 102 and is then

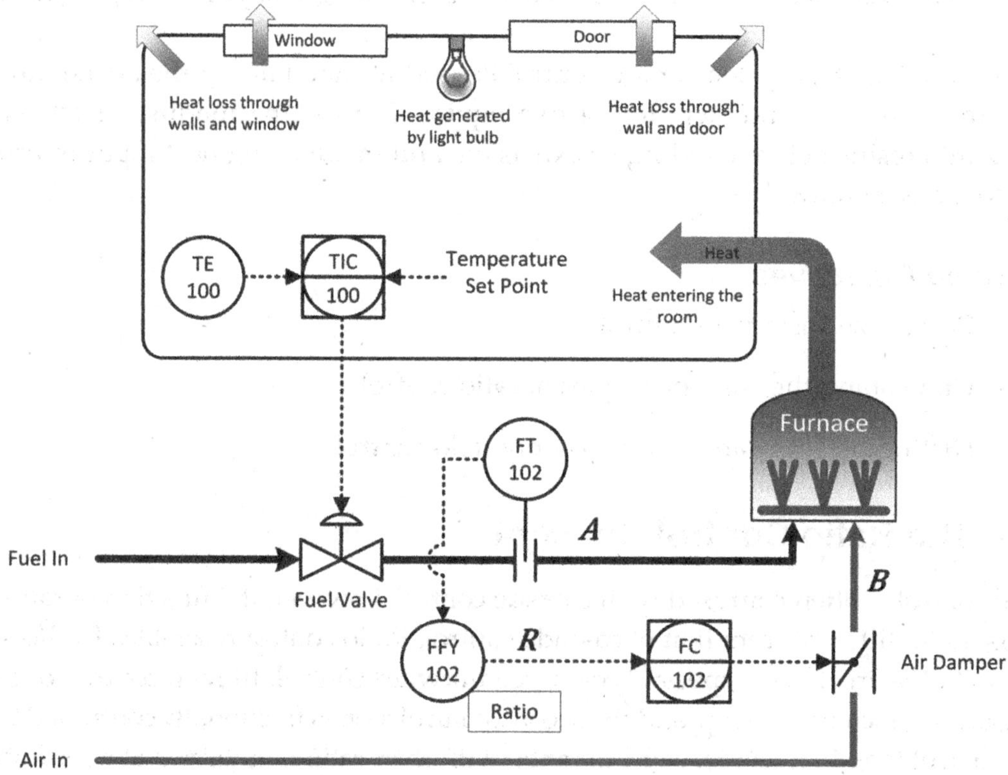

Figure 14-1. Home heating system showing ratio control of fuel and air.

1. For natural gas, the air-to-fuel ratio is approximately 10.

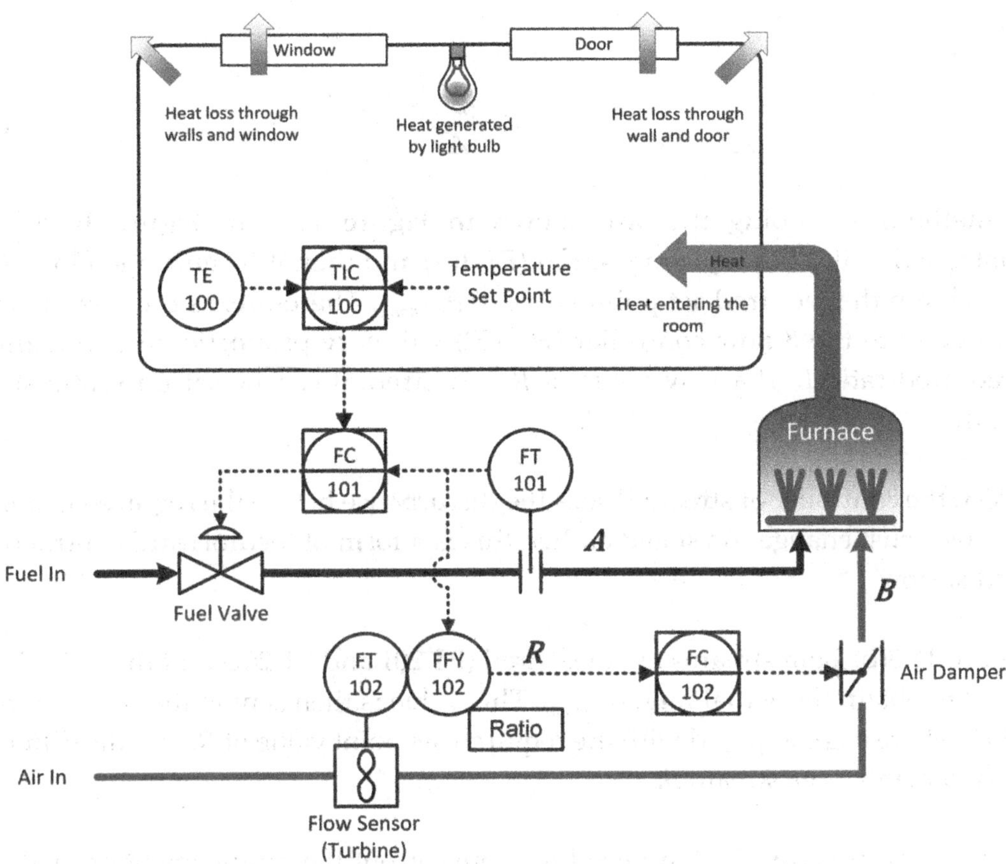

Figure 14-2. Home heating system showing conventional ratio control.

multiplied by FFY 102 to determine the required airflow rate *B* to the furnace. As a result, as flow A varies, the set point to flow rate *B* varies proportionally.

The designation FFY in the FFY 102 symbol denotes the flow rate (first F) and the ratio or fraction (second F); its function (Y) is defined outside the circle using the box symbol Ratio. Note that the airflow is entirely set by the variation of the fuel flow rate. Expressing this differently, the flow rate of the air (*B*) is set using feedforward control. In practice, a more conventional approach is used. Both the fuel flow and the airflow are controlled. Figure 14-2 shows that the fuel flow is controlled using a cascade arrangement, and the feedforward control of the air is further controlled using the fuel flow ratio as its set point.

14.2 Applying Ratio Control

One can get a better understanding of ratio control by considering a specific industrial examples. Assume two streams (*A* and *B*) that must be blended together in a proportion or ratio (R).

Then

$$R = \frac{B}{A} \tag{14-1}$$

Two methods for doing this are shown in Figure 14-3. In Figure 14-3(a), the uncontrolled (wild) flow A is measured (FT 101) and then it is multiplied by R (FFY 101) to obtain the required set point value of $B_{set\ point}$. Therefore, as the flow A varies, the set point to the B flow controller (FC 102) will vary proportionally to maintain the required ratio R. If a new value of R is desired, it can be set into ratio station FFY 101.

In effect, the controller of stream B sees the disturbance before the required change in the B flow must change. As stated earlier, this is a form of feedforward control of the second steam.

In Figure 14-3(b), both streams are measured (FT 201 and FT 202) and then divided by FFY 202 to obtain the actual ratio (R_{actual}). This value is then sent to the ratio controller (FC 202) where it is compared with the required set point value of R, and the difference is used to manipulate stream B.

Both methods in Figure 14-3 are used in industry, but the arrangement of Method 1, which is more linear than Method 2, is preferred for the following reason.

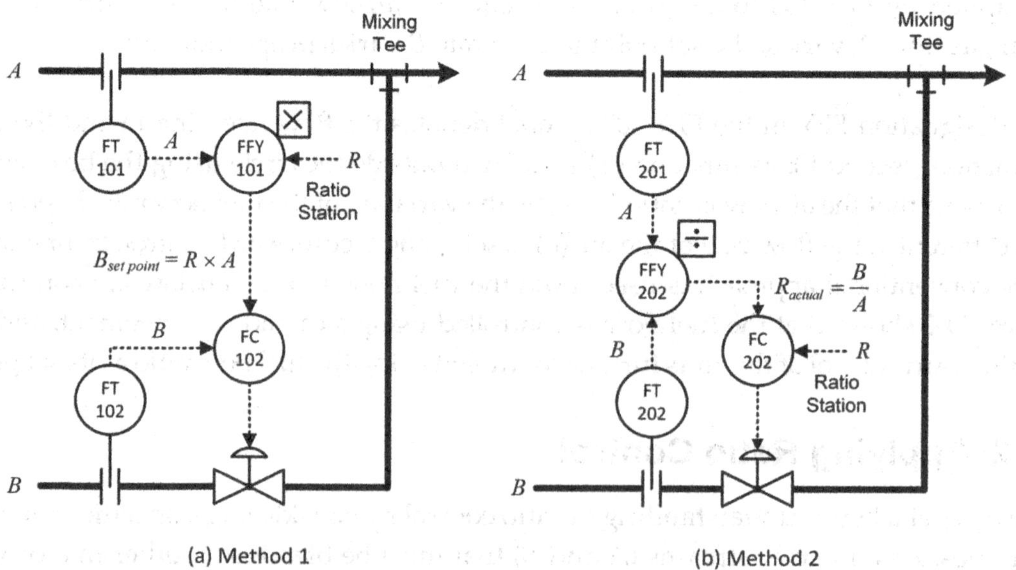

Figure 14-3. Two ratio control systems.

First consider Method 1, where ratio station FFY 101 calculates:

$$B_{set\ point} = R \times A$$

The gain of ratio station FFY 101 is the amount the output *changes* for a change in input. We can calculate this by differentiating $B_{set\ point}$ with respect to A.

then

$$\frac{dB_{set\ point}}{dA} = R$$

which is a constant.

In Method 2, R is calculated by FFY 202 as:

$$R_{actual} = \frac{B}{A}$$

Then the gain of ratio station FC 202 can be calculated by differentiating R_{actual} with respect to A:

then

$$\frac{dR_{actual}}{dA} = B\frac{d\left(A^{-1}\right)}{dA} = -\frac{B}{A^2}$$

As a result, when flow A changes in Method 2, the gain of the ratio station also changes and a nonlinearity is present, which makes the system less stable and more difficult to control.

14.3 Summary

Ratio control is primarily used to ensure that two or more process variables, such as material flows, are kept at a constant ratio even if they are changing in value.

Exercises

14.1 Diesel oil is injected into the suction of a fuel oil pump. (The impeller does the mixing to control viscosity in fuel oil.)

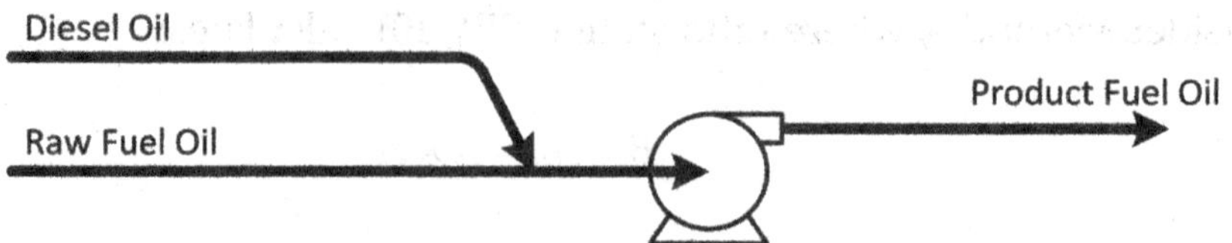

Design a ratio control system for this pump.

14.2 Install a viscosity meter as a sensor on the product fuel oil line of Exercise 14.1 and use it to adjust the ratio between the two streams.

15 Special Control Strategies

There are many important variations, extensions, and modifications of the standard control strategies. To illustrate the diversity of possible principles and concepts, this chapter will present several such implementations.

Learning Objectives

- Understand three-element boiler control
- Understand selective control
- Understand override control
- Recognize duplex and split-range control
- Recognize auto-selector or cutback control

15.1 Introduction

As we employ more advanced control strategies, the number of computational requirements increases. These additional computational functions are usually implemented by either specialty controllers or general-purpose controllers, with programmable functions primarily implemented by microprocessors. However, as we consider various control strategies, it is helpful to also consider the signal processing functions that are involved and the specific functions of the various computational operations.

The use of microprocessors extends this list virtually without limit. Complex calculations can be done without an increase in hardware. However, the increase in software development can be complex and expensive to implement and maintain. Previous chapters discussed the use of specialty controllers, such as in feedforward control (Chapter 12) and in multivariable control (Chapter 13). Special signal processing functions have also been introduced in ratio control (Chapter 14). Advanced process control functions are now standard in most programmable logic controllers (PLCs), programmable automation controllers (PACs), and distributed control systems (DCSs). Typical computational components used in process control include the following:

- **Addition/subtraction** – The output signal is the algebraic sum or difference of the input signals.
- **Multiplication/division** – The output signal is produced by multiplying or dividing the input signals.
- **Square root extraction** – The output signal is the square root of the input signal.
- **High/low selector** – The output signal is the highest/lowest of two or more input signals.
- **High/low limiter** – The output signal is the input signal limited to some preset high/low value.
- **Function generator** – The output signal is a function of the input signal.
- **Integrator** – The output signal is the time integral of the input signal. An integrator is also often called a *totalizer.*
- **Linear lag** – The output signal is the solution of a first-order differential equation in which the input signal is the forcing function. This is expressed mathematically as follows:

$$output = \left(\frac{K}{1+\tau s}\right) input \qquad (15\text{-}1)$$

where

K = gain
τ = time constant
s = Laplace transform operator d/dt

- **Lead-lag** – The output signal is the solution to the following differential equation. This is expressed mathematically as follows:

$$output = K\left(\frac{1+\tau_1 s}{1+\tau_2 s}\right) input \qquad (15\text{-}2)$$

where

K = gain

τ_1 = time constant 1

τ_2 = time constant 2

s = Laplace transform operator d/dt

A more complete list of such signal processing functions is included in ISA-5.1[1] Table 5.6. These signal processing functions are available both as analog hardware modules and as software in most DCS control packages.

One of the most frequently encountered special control strategies in industry is *three-element boiler control*. This control strategy combines feedback, feedforward, and cascade control to control the water level in a boiler steam drum.

15.2 Three-Element Boiler Control

Steam drums are used extensively throughout process industries and utilities to control the level of boiling water contained in boiler drums in process plants to provide a constant supply of steam. If the water level is too high, flooding of steam purification equipment can occur. If the level is too low, efficiency is reduced and pressure in the drum can build to dangerous levels. A drum-level control system tightly controls the water level no matter what disturbances are present, such as a change in level due to increase or decrease of steam demand or feedwater flow variations. A simplified piping and instrumentation diagram (P&ID) of a classic three-element strategy is shown in Figure 15-1.

Steam flow from the steam drum is measured, and this load on the system is compensated for by using feedforward control. Feedback control on the water level in the drum is handled in the conventional manner, and the manipulated flow into the steam drum (makeup water) is controlled through a cascade arrangement. This general arrangement is feedforward plus feedback plus cascade control, which explains the name *three-element feedwater regulator.* Figure 15-2 shows the block diagram of the strategy, clearly presenting the feedforward, cascade, and feedback control arrangements.

All three types of controls can be implemented using a single three-element regulator as shown in Figure 15-3. This is a classic example of this type of control, and there are literally thousands of these three-element regulators installed. In this arrangement, feedforward control, in effect, is used to provide compensation for significant variations in the major disturbance or the major load on the process. Arrangements such as this are often referred to as *load compensation.*

1. ISA-5.1-2009, *Instrumentation Symbols and Identification* (Research Triangle Park, NC: ISA [International Society of Automation]).

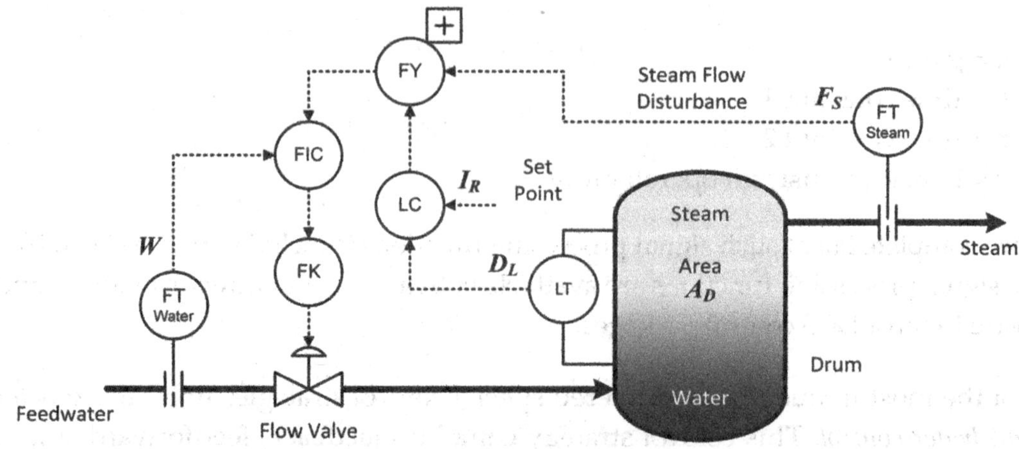

Figure 15-1. Three-element boiler control.

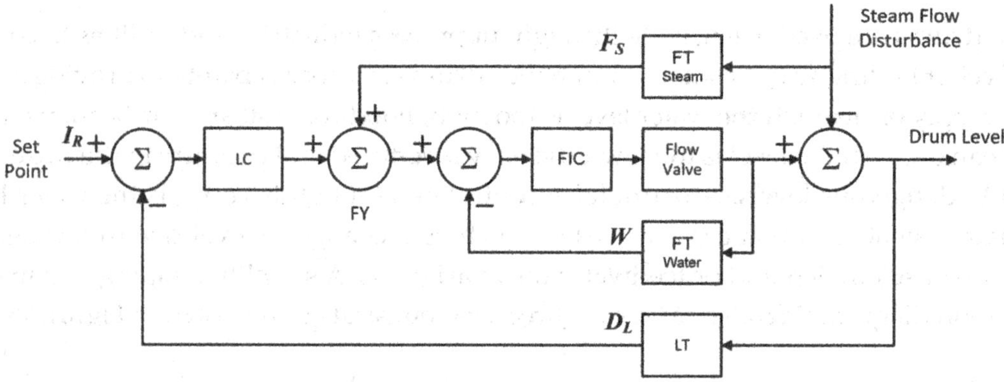

Figure 15-2. Three-element boiler control block diagram.

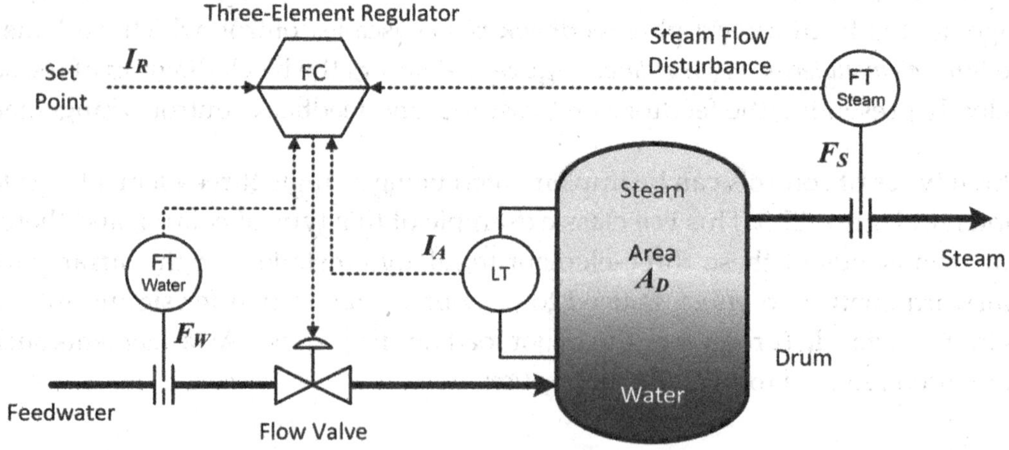

Figure 15-3. Boiler control using a three-element regulator.

15.3 Selective, Override, and Split-Range Control

There are many situations where more than one measured variable is used to control one manipulated variable. Also, some control configurations exist where two or more controllers control one manipulated variable. In both of these configurations, the selection of which measured value or which controller controls the manipulated variable is dictated by some prearranged criteria implemented in control logic.

In selective control applications, the lowest, highest, or median signal from among two or more signals is selected. In most applications, selective control is a form of multivariable control where the selectors dictate the online adjustment of control strategies as a function of changing operating conditions. The selectors enable the control strategies to be changed effectively and without disturbing the process. Override and selective control are usually implemented for safety and optimization considerations. These techniques often deal with multiple control objectives (controlled variables) and a single manipulated variable.

Selective Control

In discussing control loops, each controlled variable in a process must be paired with a specific manipulated variable at any given instant of time. However, because the number of controlled variables and the number of manipulated variables may not always be equal, a logical means of sharing variables among loops must be devised. It may be necessary to switch a controller from one controlled variable to another, or it may be necessary to switch a controller from one manipulated variable to another. Depending on what needs to be selected, this can be done with selective devices that have two or more inputs and produce a single output. The use of such selective devices is a way to achieve constraint control on flows and other operating conditions.

Selecting Measured Variables

A simple low-selector control system is illustrated in Figure 15-4. Characteristic of selective control is that there are multiple inputs to the selector. In this example, temperature is measured at two different locations in the room, but only the lower of the two is directed to the controller. In this example, a second temperature transmitter is placed in the room close to the door. When the door opens and warm air escapes, the temperature close to the door will drop and will be lower than the temperature sensed by the central temperature transmitter. The low selector will select the lower of the two temperatures (TE 2), thereby directing the temperature controller (TIC) to increase fuel to the furnace to compensate for the drop in temperature. This simple strategy will reduce the time constant of the room in returning to a comfortable temperature. Effectively, the dead time of the room has been reduced because the controller reacts to

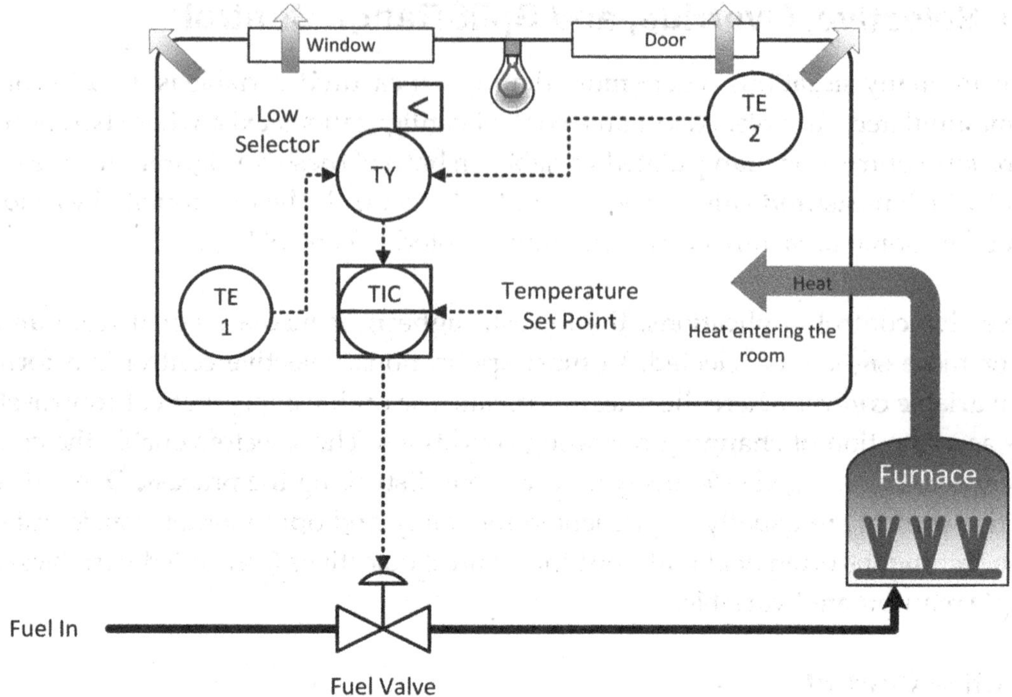

Figure 15-4. Simple low-selector temperature control system.

the temperature at the door (TE 2) immediately instead of having to wait for the room temperature sensor (TE 1) to measure the change in temperature.

This type of selective control is also often referred to as *auto-selector control,* and when the selector relay is a low selector, it is also referred to as *cutback control.*

Selecting Manipulated Variables

In selective control, we use similar types of selecting devices and logic when limits must be placed on one or more controlled variables with a single manipulated variable. In such cases, the controller output is the point at which a selection is made. Refer to Figure 15-5 for a strategy in which pressure is controlled in a pipeline by a throttle valve on a pump discharge whenever the pump motor current is below its rated limit. If the pump motor current should increase to its limit, the current controller (IP) in the pump motor takes over manipulation of the control valve from the pressure controller (PC) and closes the valve further. As a result, pipeline pressure is controlled below the set point on the PC. (This can also be viewed as override control.)

The reset feedback structure in both controllers is provided to eliminate reset (integral action) windup. Without such feedback, the controller action is delayed until the controller reset action unwinds, and the override protection might take effect too late.

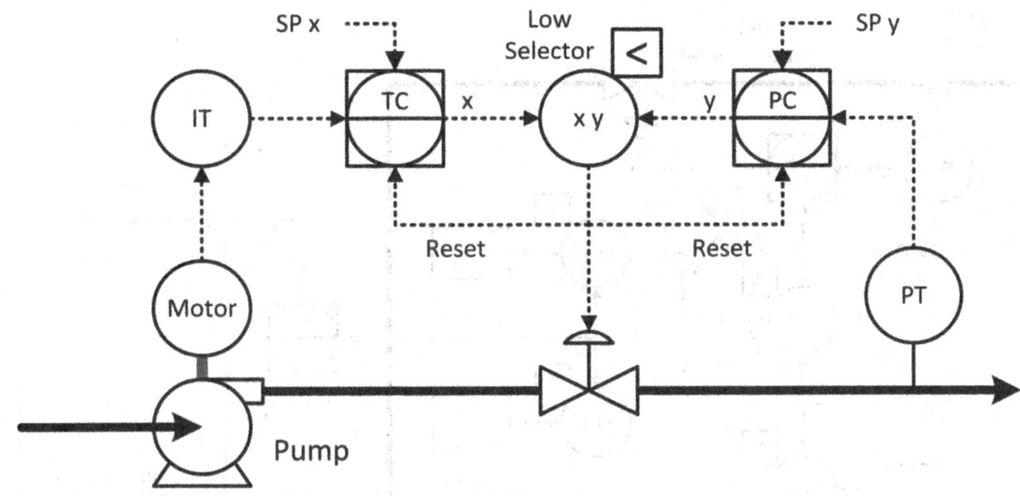

Figure 15-5. Selective control of pressure in a pipeline.

Figure 15-6 illustrates a more complex control strategy that employs many computational functions; the temperature of a process heater is controlled by selecting one of two fuels, the first being of limited availability and less expensive (fuel *A*) and the second being more expensive (fuel *B*). The objective is to use fuel *A* to the limit of its availability before starting to use fuel *B*.

Assume that primary fuel *A* is burned up to the high limit of its availability and is then supplemented with fuel *B*. Fuel *A* has a high limit that is an input to a low selector on the set point of *A* (FY-C). This is shown as a manual input of the high limit, but it could also be provided automatically.

Any difference between the output of the temperature controller and the flow of fuel *A* is converted to a set point signal for fuel *B*. In reality, the temperature controller output represents the total heat input to the system from the two combined fuels. The set point F_A of the flow controller FC-A for fuel *A* is calculated as:

$$F_A = m\left(1 + \frac{B}{A}\right) \qquad (15\text{-}3)$$

where

F_A = fuel *A* set point (FC-A)
m = temperature controller output (TIC)
A = maximum heat input by fuel *A*
B = maximum heat input by fuel *B*

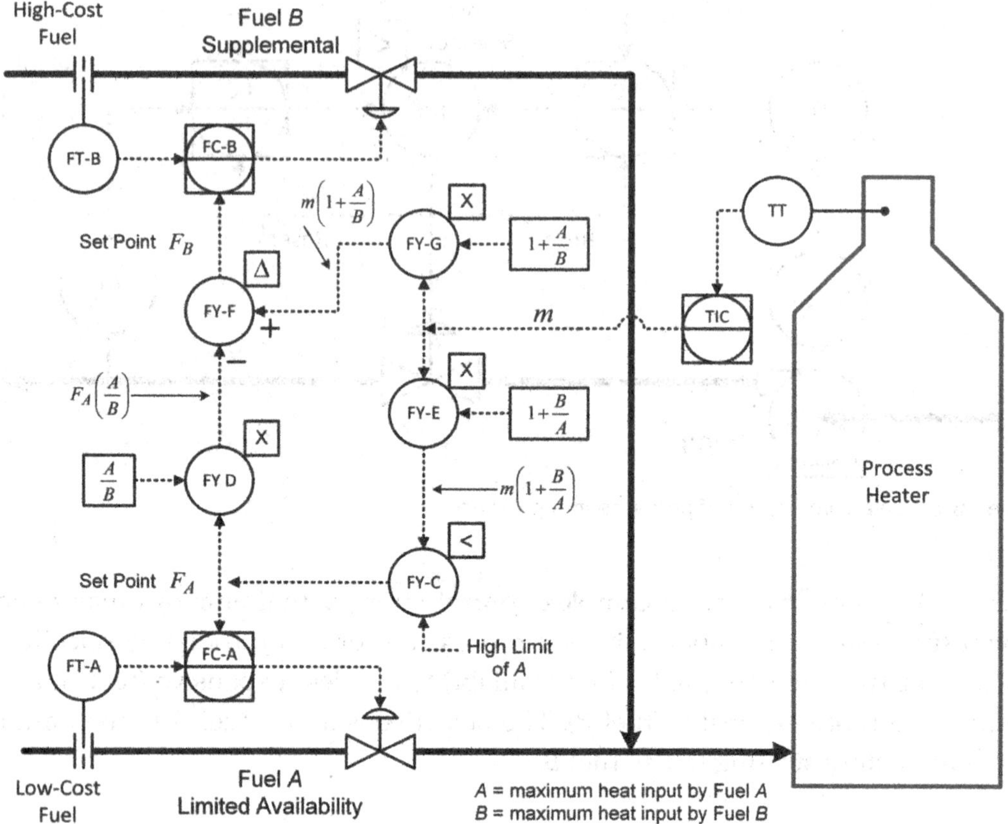

Figure 15-6. Selective control of two manipulated variables and one controlled variable.

The ratio B/A represents the full-scale heat input that can be generated by fuel B to that of fuel A. The temperature controller output (m) represents the total heat input requirement of the system, which has to be met by the two fuels. Through this correction, the controller output (m) and the fuel A set point (F_A) will be on the same scale.

These computations are needed to ensure a smooth transition between the two modes of operation. The fuel B set point (F_B) is calculated by converting the output (m) of the temperature controller (TIC) and the equivalent flow of fuel A into a set-point signal for the supplemental fuel flow controller (FIC-B). The set point for fuel B is then calculated as:

$$F_B = m\left(1+\frac{A}{B}\right) - F_A\left(\frac{A}{B}\right)$$

then

$$F_B = m\left(1+\frac{A}{B}\right) - m\left(1+\frac{B}{A}\right)\left(\frac{A}{B}\right) \tag{15-4}$$

As long as the requirement for fuel *A* is at or below the high limit set by FY-C, the two inputs to the subtracting function (FY-F) will cancel one another, meaning that fuel *B* is not being used and the *B* set point will be zero:

$$m\left(1+\frac{A}{B}\right)-m\left(1+\frac{B}{A}\right)\left(\frac{A}{B}\right)=0$$

When the high limit of fuel *A* is exceeded, F_A becomes constant and F_B starts to increase.

Any difference between the output of the temperature controller and the flow of fuel *A* is converted to a set point signal for fuel *B*.

Override Control

Override control is often considered as a special type of selective control. The override control concept is a strategy by which a process variable is kept within a confined limit, usually for protective purposes. (*Interlock control* is an alternate protective method, but it is usually oriented toward equipment malfunction, and it acts to shut down the process. Override control is less drastic.) Override control maintains the process in operation but within and under safe confined conditions.

Surge Tank Example

To illustrate override control, consider the flow control from the surge tank shown in Figure 15-7. A hot saturated liquid enters a process surge tank, then it is pumped into the process. Normally, the surge tank operates at a normal safe level as shown, but if the level gets too low, the liquid will not have enough net positive suction head (NPSH) and the pump will start to cavitate. Override control provides the required protection to the pump. The strategy for providing this protection is shown in Figure 15-8.

In the override control application shown in Figure 15-8, the tank level is now controlled. The variable-speed pump will, of course, pump more liquid as the energy input to it increases. It follows that the flow controller must, therefore, be a reverse-acting controller (output increases as input decreases), and the level controller must be a direct-acting (output increases as input increases) controller. The output of each controller is connected to a low-level selector relay, and its output goes to the pump.

Under normal operating conditions, the level set point is above the normal surge tank level, and the level controller will attempt to speed up the pump. Normally, the output of the flow controller will be less, and the low-level selector relay will select the flow controller output to manipulate pump speed. If the flow of hot, saturated liquid

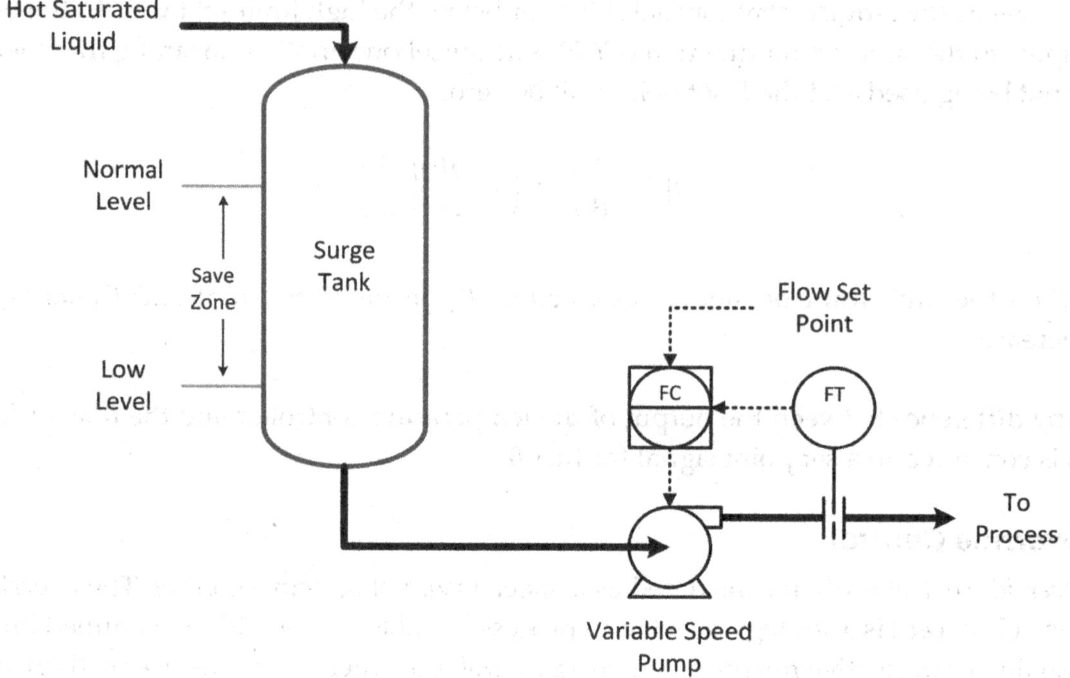

Figure 15-7. A system needing override control.

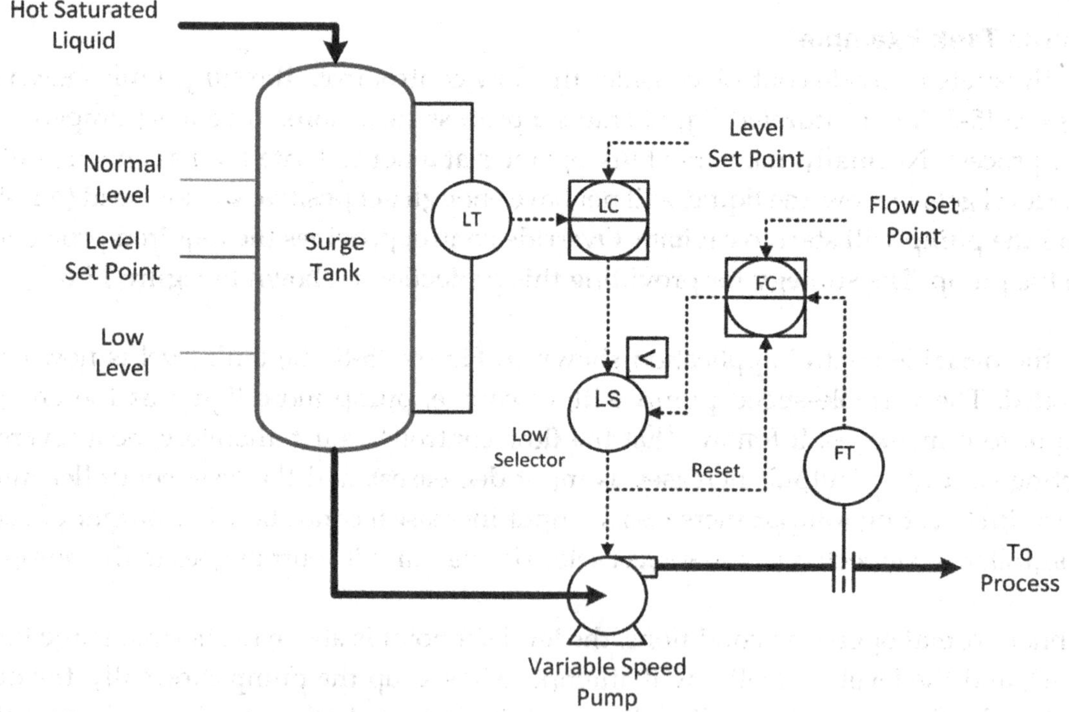

Figure 15-8. Override control system.

slows down and the level drops, the level controller will try to slow down the pump by reducing its output. When the output of the level controller drops below the output of the flow controller, the low-level selector relay will select the output of the level controller to control the pump. Now the level controller *overrides* the flow controller. Override control is also often called *constraint control* for this reason.

In override control, it is essential that any controller that has integral (reset) action must also have reset windup protection. This is shown in Figure 15-8 for the flow controller. The reset feedback structure prevents the integral action from saturating the controller when a changeover occurs. Reset windup in an override control system is a simple and straightforward way of preventing windup. Without such protection, the controller action is delayed until the controller reset action unwinds, and the override protection will most likely be too late. In the example shown, the flow controller will not wind up because of the reset feedback signal.

Distillation Column Example

As another example of override control, consider the distillation column differential pressure control shown in Figure 15-9. Temperature control is usually used to provide manipulation of steam flow to the reboiler. But if the column experiences downcomer flooding, it will greatly increase column differential pressure, and the steam flow rate must be reduced. The strategy shown in Figure 15-9 enables the temperature controller to set the reboiler steam flow rate as long as the column differential pressure is below its set point. This set point is the maximum desired differential pressure. In the example shown, both the temperature controller and the differential pressure controller are assumed to have the reset mode, both can be overridden, and thus both need reset windup protection. In this example, neither the temperature controller (TC) nor the differential pressure controller (PDC) will wind up because of the reset feedback signal. As long as one of the controllers is in control, the other controller is prevented from winding up.

Split-Range Control

In a split-range control strategy, a process parameter (e.g., temperature) is adjusted by more than one manipulated variable. A single controller's output is connected to more than one final control element.

A common application for split-range control is when a temperature control loop must apply both heat and cooling to a process. This is normally accomplished by using two final control elements, one for heating and one for cooling. In industry, this is most often implemented using control valves equipped with positioners. Each valve operates through one-half of the controller's output range, applying the maximum

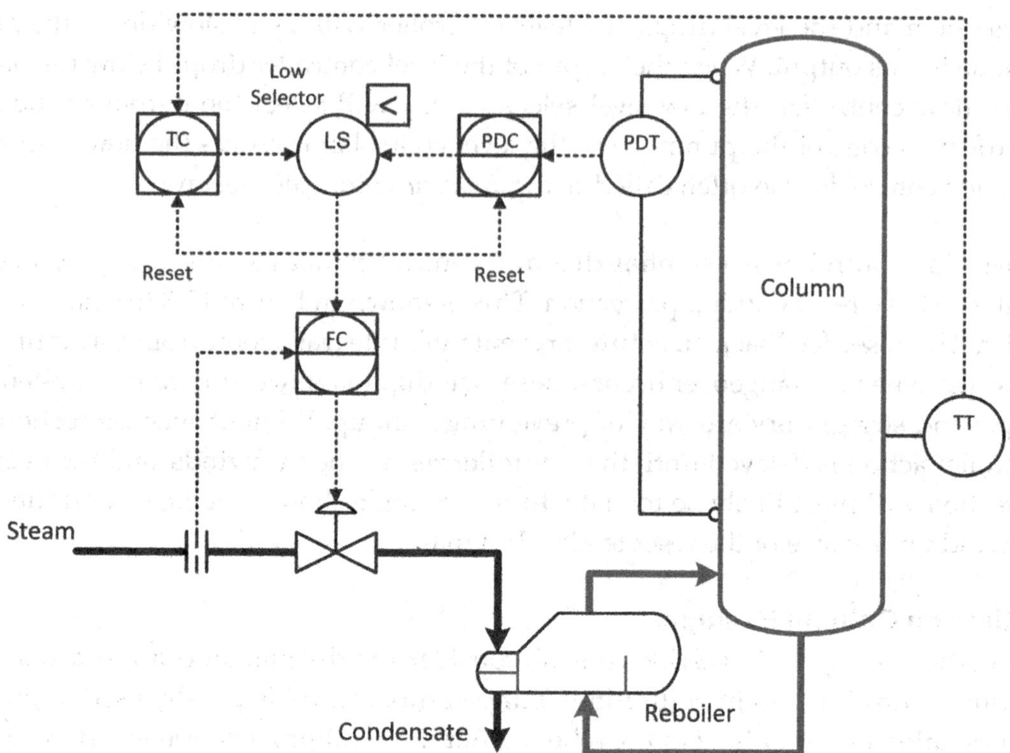

Figure 15-9. Column differential pressure override control.

heating or cooling at the extremes of the controller's output range. At the midpoint of the range, neither heating nor cooling is applied (see Figure 15-10).

Positioners are normally used, rather than relying on the valve-spring to realize the required ranges. Positioners will provide a more accurate and precise range splitting.

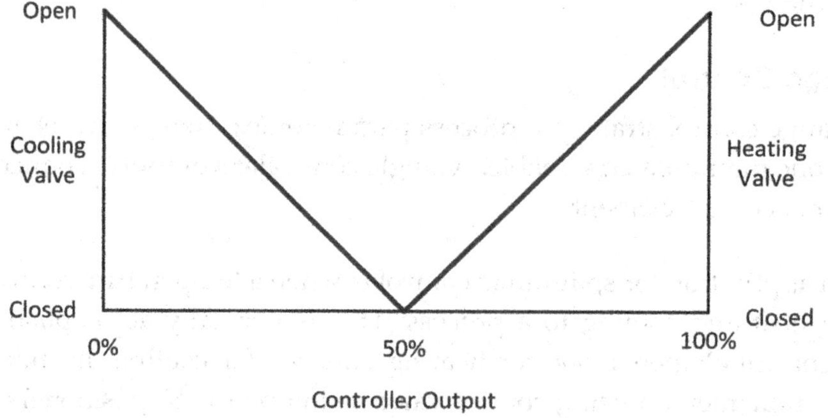

Figure 15-10. Valve positions versus controller output.

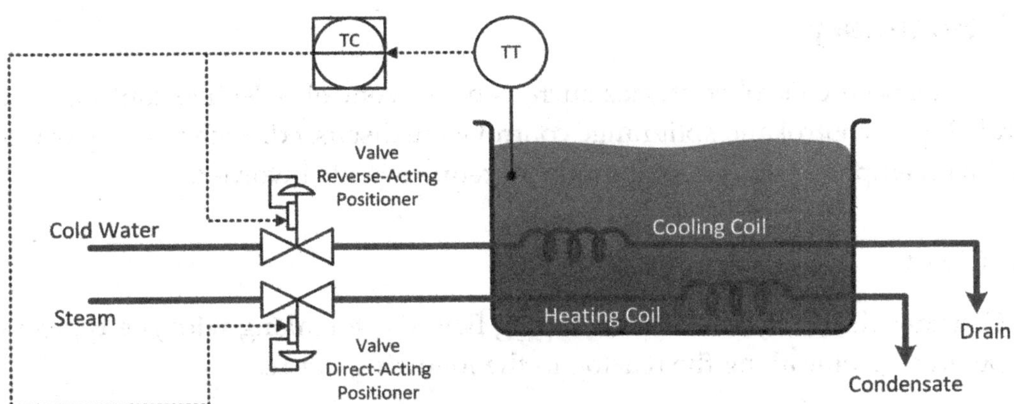

Figure 15-11. Split-range control of bath temperature.

The controller in split-range control has one input and two outputs. A typical example of the application of split-range control is shown in Figure 15-11. Assume that the temperature controller (TC) is a proportional-only controller; its output is fed to both control valves. Usually, the cold water valve signal varies from 0% to 100%, as is typical in split-range control; the valve action would be set so that the valve travels from full open to full close as input to the positioner changes from 0% to 50%. In addition, the steam valve would go from fully closed to fully open as the signal changes from 50% to 100% (see Figure 15-10). The system is designed so that the controller output produces a 50% signal when the measurement and set point agree, and, at this point, both valves are closed. If measured temperature rises above or falls below the set point, either water or steam will be circulated in proportion to the difference between the set point and the measured value of temperature.

Usually, for safety reasons, the cooling valve would operate from 0% to 50% and the steam valve would operate from 50% to 100%. This way, if a failure occurs and the controller output goes to zero, the system will then go to full cooling.

One problem with this setup is that the system may oscillate continuously between cooling and heating when the controller output is around 50%. To avoid this, a deadband between the two ranges can be added; an example of this would be setting a range of 0% to 49% for the cooling valve and a range of 51% to 100% for the heating valve. Alternatively, a small overlap can also be introduced so that at 50% both valves are slightly open; an example of this would be setting a range of 0% to 52% for the cooling valve and a range of 48% to 100% for the heating valve.

Split-range control is conceptually similar to selective control, as was shown in Figure 15-5, which also used a single controlled variable to constrain or choose between two manipulated variables.

15.4 Summary

Several advanced control strategies such as boiler control, selective control, override control, duplex control and split-range control were discussed. Each strategy was articulated with emphasis on the computational requirements involved.

Exercises

15.1 Consider the catalytic exothermic plug flow reactor, along with the typical temperature profile along the reactor, in the following figure.

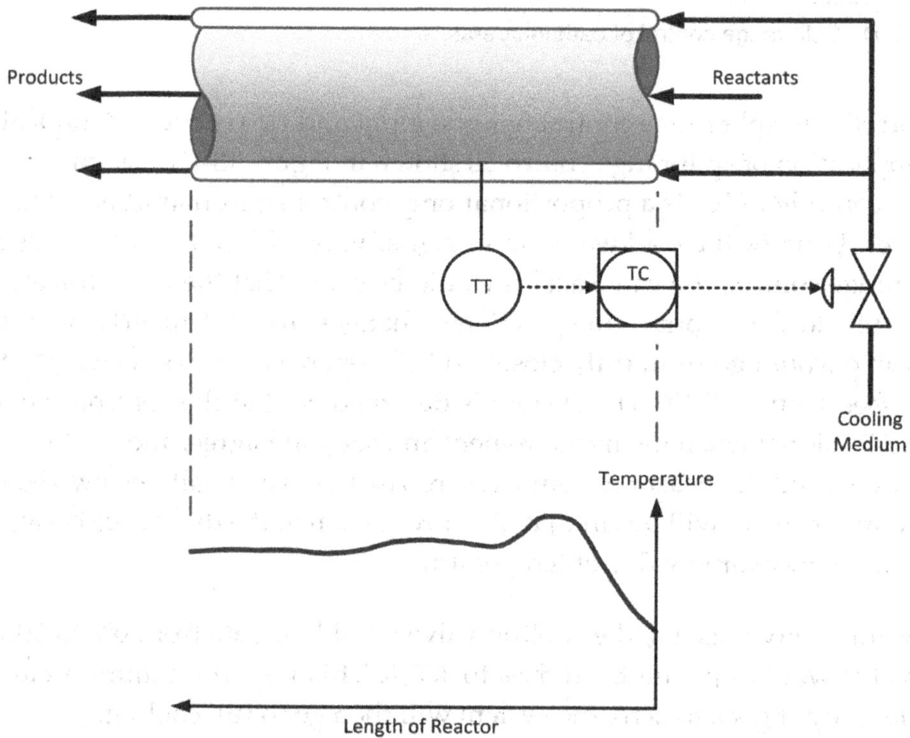

As conditions change and as the catalyst ages, the hot spot moves. Design a selective control scheme so that the measured variable moves as the hot spot moves. Use multiple temperature measurements with a high selector.

15.2 To operate a pumping station on a pipeline efficiently, it is best to operate the control valve wide open on all stations but one. This station becomes the throttling unit. It is necessary to cut back on the control valve position if the:

- Suction pressure drops too low
- Motor load rises too high
- Discharge pressure rises too high

Sketch an auto-select/cutback control system to handle this.

16

Statistical Process Control

Statistical process control (SPC) refers to using statistical methods to measure process quality, consistency, and variations in an effort to produce a higher quality product and reduce product rejects. SPC is a primary tool to improve product quality by reducing process variation. It is most often used when a large number of items, such as prepared food and pharmaceutical products, are produced.

The focal point of SPC is its ability to provide a continuous quality-improvement tool kit. SPC can be used as the primary tool for monitoring, control, and diagnostics.

Learning Objectives

- State the basic approach of statistical process control (SPC) and the tools used therein
- Understand the purpose of SPC
- Interpret control charts and variability
- Explain the meaning of the following terms:
    - Sample and population
    - Mean or average
    - Variability and range
    - Standard deviation

16.1 Introduction

SPC, in addition to other control strategies, is often used to monitor the consistency of processes used to manufacture a product as it is designed and specified. Its aim is to maintain and keep processes under control despite variability in the materials, methods, machines, personnel, and environment used in the production of the product. No matter how good or bad the design, variability is inevitable. SPC helps to ensure that the product is being manufactured as designed and specified. SPC will not improve a poorly designed product's reliability or quality, but it can be used to assist and maintain a consistency in the manufacture of a product.

16.2 Statistics (Mean and Standard Deviation)

Statistics is the knowledge and science of variation, and for that reason it is an excellent tool to monitor and control variations.

Statistics requires collecting data. The problem is that after the data is collected, what do we do with it? Statistical techniques help us visualize massive amounts of data and, with proper analysis, make the data useful in maintaining product quality. To fully understand a process and its variability, and the effect of changes in process variables, rules of thumb and experienced people are valuable. However, significant decisions should be based on properly collected and analyzed data.

We start by reviewing some fundamental concepts and terminology. In dealing with statistics, we distinguish between the *sample* data and the *population* that the sample data represents. We use the sample data to draw conclusions about the population. In dealing with SPC and improving quality, it is more natural to speak of processes than populations.

- **Sample** – Refers to a collection of data that has actually occurred; a set of existing measurements that have been recorded in some way, usually electronically. The term *population* is used in statistics to refer to all possible samples that could have been collected. In manufacturing and in SPC, we usually draw conclusions based on the sample rather than the population.
- **Distribution** – A summary of the sample (the data points collected) that is usually presented in the form of a histogram.
- **Range** – The difference between the largest and smallest values of a sample.
- **Mean** – The mean, or average, is a measure of location of the distribution. This is one of several statistical parameters that describe the sample. The

mean is the sum of all the values in a sample, divided by the number of data points. The mean of a whole population is typically denoted by the Greek letter μ (pronounced "mu"), whereas the mean of a sample is usually denoted by $\bar{x}$ (x bar).

Mathematically:

$$\bar{x} = \frac{1}{n}\sum_{i=1}^{n} x_i \tag{16-1}$$

where

$\bar{x}$ = mean (or average of sample)

n = number of samples

x_i = i^{th} sample in the data

In addition to the mean, information about the distribution of values in the sample is required to quantify the sample, that is, the spread, or variation of the data points in the sample.

- **Variance** – The variance is a measure of the spread of a sample from its mean, abbreviated s^2.

Mathematically:

$$s^2 = \frac{1}{n-1}\sum_{i=1}^{n} (x_i - \bar{x})^2 \tag{16-2}$$

where

s^2 = variance

$\bar{x}$ = mean (or average)

n = number of samples

x_i = i^{th} sample

Note that in most engineering sample sets the value for n is large, so using a divisor of n, which you might come across in some texts, rather than $n - 1$ as shown here, makes little difference. Technically, using n rather than $n - 1$ calculates the variance of a population, and $n - 1$ the variance of a sample. This adjustment corrects for an inherent bias in the sample mean.

- **Standard deviation** – Usually signified with the Greek letter σ (pronounced "sigma") and calculated as the square root of the variance (s^2).

Mathematically:

$$\sigma = \sqrt{s^2} \tag{16-3}$$

where

s^2 = variance

σ = standard deviation

The *range* is often used to estimate the standard deviation. But the standard deviation is a more practical measure of spread, and it is easier to visualize on a histogram. It also has the advantage of being in the same units as the variable.

Histogram

A histogram is a graphical representation of a sample that is distributed over a specific range. It shows how the data points are dispersed (spread). Histograms provide visual representations of large amounts of data that are not easy to recognize in a tabular or spreadsheet form. Histograms are used to identify how the output of a process meets design expectations.

To better understand the application of the statistical terms, consider the following sample data: daily average temperature measurements were collected from a process with a set point of 70°F over a 30-day period, as shown in Table 16-1.

There are 30 measurements, but it is difficult to draw any specific conclusions about the data without further analysis. We start our analysis by graphing the data over the 30-day period. See Figure 16-1.

Using the statistical terms described, we can draw the following conclusions based on the sample.

Table 16-1. Daily measurements of average temperature (°F) over a 30-day period.

71.00	67.00	70.00	73.50	69.50	69.50
70.00	68.00	71.00	74.50	71.00	66.50
70.50	68.50	70.50	70.00	68.00	71.50
69.50	69.00	72.50	72.50	70.00	72.00
65.50	68.50	72.00	69.00	70.50	73.50

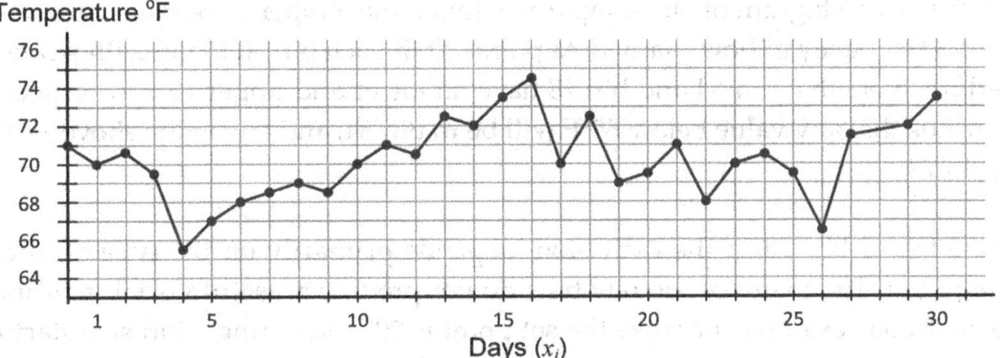

Figure 16-1. Daily temperature over a 30-day period.

The average temperature of the sample can be calculated by using Equation 16-1:

$$\overline{x} = \frac{1}{30}\sum_{i=1}^{30} x_i = 70.17°F$$

Where x_i represents each of the 30 samples in Table 16.1.

This is 70.17 – 70.00 = 0.17°F on average above its required set point of 70°F.

The measurements vary widely from day to day with a range of variability of:

$$x_{max} - x_{min} = 73.10 - 65.00 = 8.10°F$$

However, the standard deviation (using Equation 16-2) is:

$$\sigma = \sqrt{\frac{1}{30-1}\sum_{i=1}^{30}(x_i - 70.17)^2} = 2.09°F$$

A histogram can be constructed to provide a visual indication of the parameters and provides more usable information.

The first step in constructing a histogram is to decide how the process should be measured and what data should be collected that gives meaningful information. The data must be variable data, or data that is measured on a continuous scale, such as volume, size, weight, time, or, as in our example, temperature.

To construct a histogram of our sample, we divide the temperature range into *bins* and count the frequency of how many data points fit in each bin. Table 16-2 illustrates this procedure. Note that bin 64 and bin 78 have no lower and upper values, respectively. This means that any value below 65°F will be in bin 64, and any value above 77°F will be in bin 78.

The number of bins, and their division, depends primarily on the type of information required. Too many or too few bins do not provide a useful visual indication of the data. In our example, because the set point is 70°F, selecting a bin size starting at (70 ± 1)°F seems appropriate. Figure 16-2 shows the constructed histogram.

The histogram provides a more informative visual summary of the data. It is easy to see that the average temperature is around 70°F as required, with few readings below and above. The frequency *distribution* tells us a lot about the probability of the temperature going out of control.

If this sample is representative of the overall population of days (randomly drawn), and if the sample is large enough, we could conclude that the probability of finding days when the temperature is over 74°F or under 65°F is very low.

Table 16-2. Daily average temperature over a 30-day period divided into bins.

Temperature Range in °F	**–65**	**65–67**	**67–69**	**69–71**	**71–73**	**73–75**	**75–77**	**77–**
Bin	**64**	**66**	**68**	**70**	**72**	**74**	**76**	**78**
Data Point in Each Bin		65.50	68.00	69.00	71.50	73.50		
		66.50	68.00	69.50	72.00	73.50		
		67.00	68.50	69.50	72.00	74.50		
			68.50	69.50	72.50			
			69.00	70.00	72.50			
				70.00				
				70.00				
				70.00				
				70.50				
				70.50				
				70.50				
				71.00				
				71.00				
				71.00				
Frequency	0	3	5	14	5	3	0	0

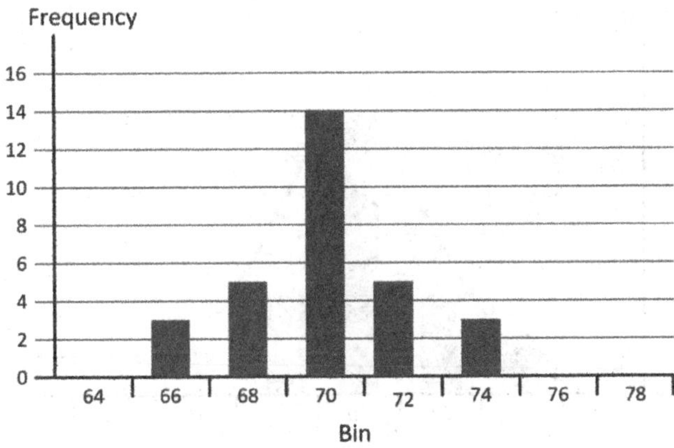

Figure 16-2. Histogram of temperature over a 30-day period.

Normal Distribution

The shape of this distribution, which is common both in nature and industrial settings, is called a *normal distribution.* A normal distribution resembles a bell-shaped curve. A theoretical normal distribution is shown in Figure 16-3. A normal distribution is characterized by being symmetrical about the mean, meaning that 50% of all data points are below the mean and 50% are above the mean. Also, 68% of all data points are within 1 standard deviation of the mean, 95% of all data points are within 2 standard deviations of the mean, and 99.7% of all data points are within 3 standard deviations of the mean.

Understanding normal distribution and its relationship to standard deviation helps in understanding normal expected variations in a process. Statistically, this means that we can expect that any data point value is:

- Likely to be within 1 standard deviation (68 out of 100 samples)
- Very likely to be within 2 standard deviations (95 out of 100 samples)
- Almost certainly within 3 standard deviations (99.7 out of 100 samples)

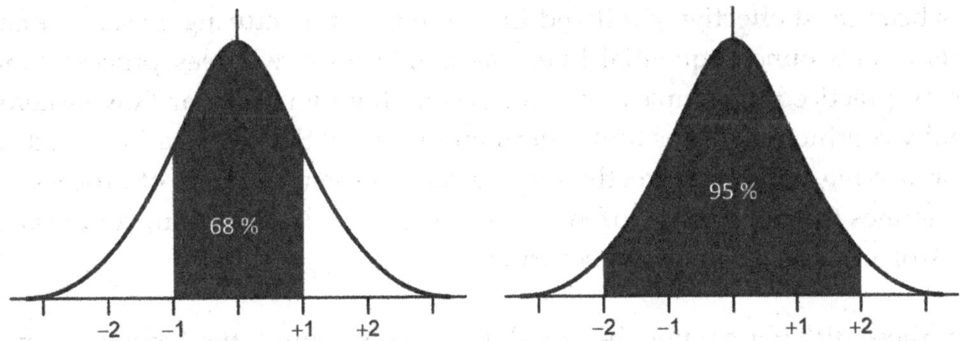

Figure 16-3. Theoretical normal distribution.

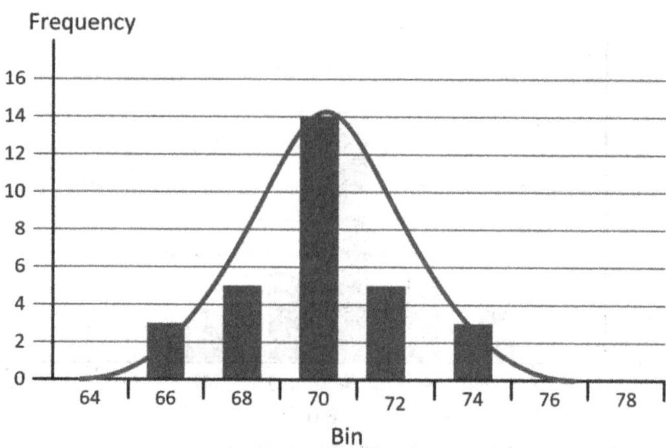

Figure 16-4. Histogram of temperature over a 30-day period compared to normal distribution.

The previous histogram based on our sample is shown again in Figure 16-4 overlaid with a theoretical normal distribution curve based on the sample's calculated mean and standard deviation.

All processes and products have variability; each bottle of ketchup may not have the same amount of ketchup in the bottle. The variability that exists when a process is running well is referred to as *short-term* or *inherent variability*, and it is usually measured by its standard deviation.

Most process measurements have a target value (set point) that is to be maintained. Too much ketchup in the bottle may become too expensive for the manufacturer, but too little may lead to customer complaints, or worse, to legal fines imposed by regulators.

The primary purpose of SPC is to provide an indication of when the process moves away from its target value. A secondary purpose is to give an indication of when sample variability has increased. In both cases, corrective action is required.

SPC has been most effectively utilized in discrete manufacturing. Discrete manufacturing is built around sequential linear assembly lines, whereas process control is most often practiced in complex, interconnected, highly nonlinear flow systems that are usually continuous. Terms quite common in typical SPC applications, such as *zero defects* or *no scrap*, do not fit directly into the conventional concepts of process control. This illustrates the significant differences between the use of SPC in continuous process control and in discrete manufacturing.

Some of these differences must be noted. In process control, the focus is on material and energy balances and their systematic adjustment. The process is said to be *in control* when sensors that measure these material and energy balances are at or near their target values or set points. In SPC, a process is considered *in control* when sampled process measurements are normally distributed and fall within a window defined as a set number of standard deviations about a mean, which means variability is reduced to within an acceptable range. In some environments, the term *in control* is used even if the mean is significantly different from the target or set point for that particular measurement.

Standard SPC assumes there is no autocorrelation in the variables being controlled. Conventional control loops essentially ensure that autocorrelation will occur. The basic focus of SPC is on the monitoring and study of the process in a systematic and statistically meaningful way. In the short term, this focus is on identifying and measuring process problems that might otherwise go undetected. In the long term, the focus is on using these measurements to identify potential problems and employ improvements. The net result is that, in the short term, we accept variations that are statistically normal and correct those that exceed what is statistically outside the norm.

Conventional process control assumes that all process relationships are deterministic in nature. SPC, on the other hand, assumes nature is inherently random and cannot be modeled simply by deterministic cause and effect. Only probabilities and distributions can describe various things. Conventional process control focuses on process dynamics, whereas the focus of SPC is static. Another contrast is that while process control is reactive, SPC is proactive.

16.3 The Tools of SPC

The tools of SPC are control charts and graphs, which are usually computer generated and monitored by operators, but reviewed by a quality control team before any specific action is taken. Note that the operator/engineer is very much in the loop in SPC.

Control Chart

The most common quality control chart is the Shewhart chart.[1] To construct such a chart, periodic measurements of a process variable are made and recorded as shown in Figure 16-5.

Note that the upper control limit (UCL), the lower control limit (LCL), and the expected mean ($\bar{x}$) are all determined by historical data collected. If a measurement falls outside

1. The Shewhart chart is named after Walter A. Shewhart (1891–1967), an American physicist, engineer, and statistician who is known as the father of statistical quality control.

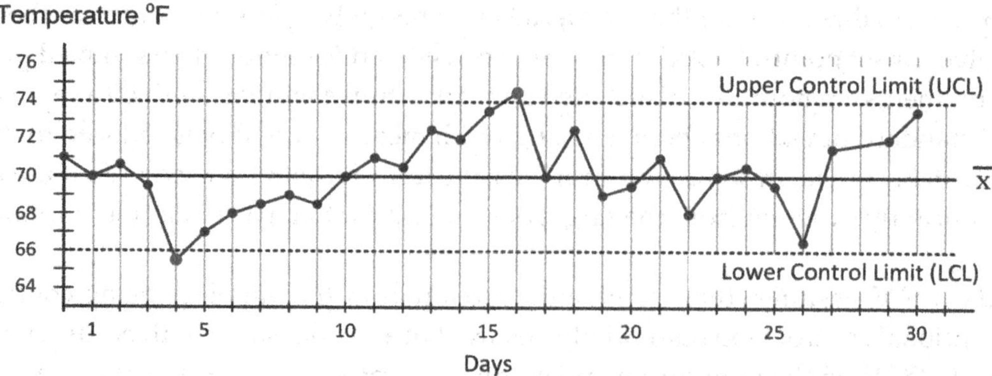

Figure 16-5. Typical quality control chart.

the control limits or if successive data points (for example, 7 data points) fall on only one side of the mean, this might be an indication for attention. In SPC terminology, the process is out of control until the cause of the abnormal event has been identified and corrected. Note that SPC control charts are not intended to take the place of conventional process control loops. Instead, they complement, enhance, and optimize conventional process control. Just as there can be conventional process control that is advanced beyond simple proportional-integral-derivative (PID) feedback control, SPC also has a range of more advanced tools.

Note that in this example, the UCL and LCL were set at two standard deviations. The decision of where to place the UCL and LCL depends on the product specification.

Pareto Analysis

When applying SPC to a process, an important early step is to identify the most serious problem to address. Pareto analysis[2] is a tool that is helpful for separating, in an organized and quantitative manner, the many minor problems from the (hopefully few) critical ones.

Pareto analysis is a decision-making tool used for the selection of a limited number of tasks that produce a significant overall effect. It uses the Pareto principle (also known as the *80/20 rule*). The idea is that by doing 20% of the work, you can generate 80% of the benefit of doing the entire job. In other words, a vast majority of problems (80%) are produced by a few key causes (20%). This technique is also referred to as *the vital few and the trivial many*. Figure 16-6 shows a simplified Pareto analysis. In this case study, it is clear that the temperature sensor requires special attention.

---

2. Pareto analysis is named after Vilfredo Pareto (1848–1923), an Italian engineer, sociologist, economist, political scientist, and philosopher. He made several important contributions to economics, particularly in the study of income distribution and in the analysis of individuals' choices.

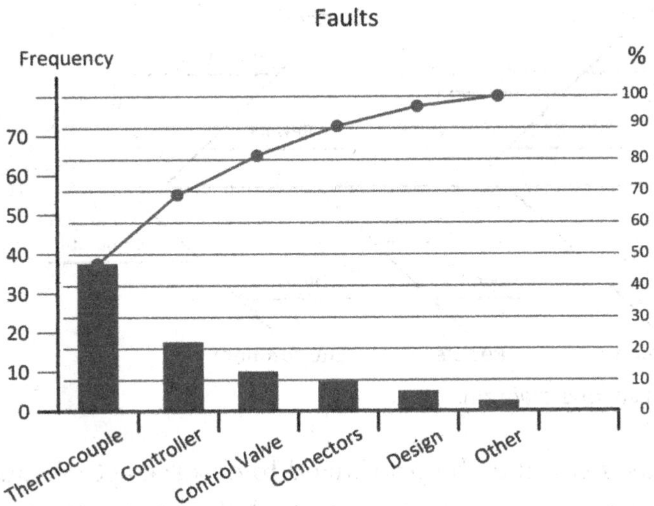

Figure 16-6. A typical Pareto analysis.

A Pareto chart displays contributions to problems using a bar graph. As a quality-improvement tool, Pareto analysis:

- Defines categories of defects that cause a particular defect
- Counts the frequency (number) of occurrence of each defect
- Displays graphically as a bar chart, sorted in descending order the frequency of defects
- Uses a second *y*-axis to show the cumulative percentage of defects

Ishikawa or Fishbone Diagram

After the determination of the categories of defects and the frequency of occurrence has been made, all possible reasons for the problem are listed, and then these possible reasons are entered onto an Ishikawa[3] or *fishbone* diagram. A typical fishbone diagram is illustrated in Figure 16-7. Note especially the headings for the various possible causes:

- Materials
- Methods
- Machine
- People
- Environment

---

3. The Ishikawa diagram, also known as a fishbone diagram, is named after Dr. Kaoru Ishikawa (1915–1989), a Japanese mechanical engineer and professor in the field of quality management.

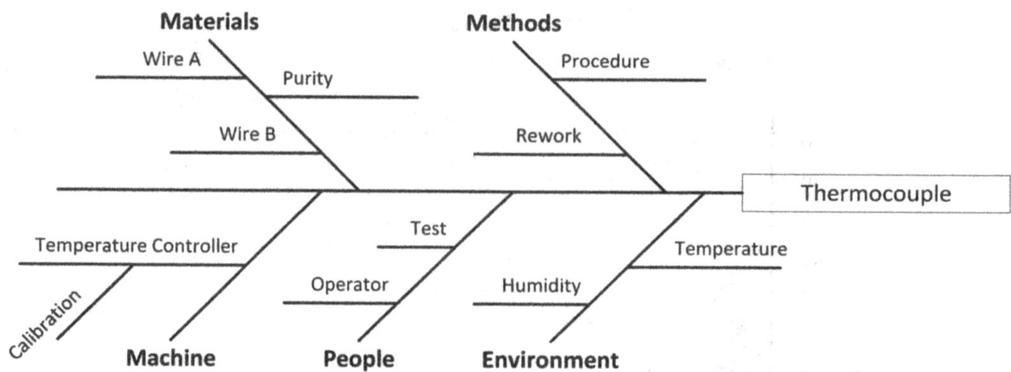

Figure 16-7. Typical fishbone diagram.

These possible causes can then be prioritized to observe actual cause-and-effect relationships. As cause-and-effect data are collected, they can be entered onto SPC charts to document relationships and to communicate the problem. Over several months, you will gain a much more detailed understanding of the process and its operation, which should lead to a much higher quality level.

Note that SPC tends to give discipline and method to what good engineers and good management should be doing all along.

16.4 Statistical Process Optimization

A method that is currently starting to receive more attention is statistical process optimization (SPO). This method suggests a knowledge-based, self-learning, online system. Data is collected in the form of controlled variables, process variables, and process results, such as laboratory test data and/or cost data. This data is then used to build a statistical process model, such as a multivariable, nonlinear regression model. The model can contain artificial intelligence and expert systems (see the next section) to improve (i.e., help the model to *learn*) through new process data. Then new values for controller set points can be selected based on the consequences from using them.

16.5 Summary

SPC stands for statistical process control. SPC *does not* refer to a particular technique, algorithm, or procedure, but it is an *optimization strategy* concerned with *continuous process improvements.* SPC uses a collection of (statistical) tools for data and process analysis making inferences about process behavior. Decision-making based on SPC is considered a key component of maintaining and improving quality control.

Exercises

16.1 Raw material used in a chemical process contains some impurities. To ensure that the impurities are below the requirement specified, three samples are tested at regular intervals. When the process is running satisfactorily, the mean percentage of impurities was found to be 16.7 with a standard deviation of 3.4.

Set up a control chart for means and for standard deviations for the following set of samples.

Sample 1	Sample 2	Sample 3	Mean	Standard Deviation
16.9	19.3	20.2		
14.2	19.1	5.2		
9.3	12.2	8.1		
24.2	25.6	22.0		
22.7	19.3	23.1		

16.2 In the production of widgets, four samples are taken and measured at regular time intervals. The following table shows the measurement of the four samples at eight sequential time intervals. The target measurement value is 9 mm.

Sample	Time interval							
	1	2	3	4	5	6	7	8
1	9.0	10.4	8.2	7.9	8.2	8.4	7.4	7.6
2	8.1	9.0	9.2	7.7	9.0	8.1	8.0	8.5
3	8.7	7.9	7.9	9.7	7.4	8.4	8.9	8.1
4	7.5	7.2	7.7	9.3	8.6	8.7	9.8	8.8

A. Calculate the mean sample range and the standard deviation of the process.

B. Using the result of A., draw a control chart for means showing 95% upper and lower limits, and plot the eight samples.

C. Briefly discuss your observations of the patterns of the chart.

17 Fuzzy Logic Control

Fuzzy logic control is a relatively new concept in process control and is particularly useful when systems are difficult to model mathematically.

Learning Objectives

- Understand the basic concepts of fuzzy logic control
- Explain the meaning of the following terms:
    - Rule-based
    - Fuzzification, defuzzification
    - Membership function
- Explain and delineate the membership functions

17.1 Origin of Fuzzy Logic

The concept of fuzzy logic was originally conceived by Lotfi Zadeh, a professor at the University of California at Berkley, and presented not as a control methodology but as a way of processing data by enabling partial set membership rather than crisp set membership or nonmembership. This means that adjectives such as *hot* or *cold* are not exact (or not crisp), but rather they depend on one's perception. This approach to set theory was not applied to control systems until the 1970s. Professor Zadeh reasoned that people do not require precise, numerical information (input), and yet they are highly capable of adaptive control functions (output). A perfect example of this is driving a car, as we saw in Chapter 1. If feedback controllers could be programmed to accept noisy, imprecise input, they could be much more effective and perhaps easier to implement.

17.2 What Is Fuzzy Logic?

Fuzzy logic is a problem-solving control method that lends itself to implementation in systems ranging from simple, small, embedded microcontrollers such as room temperature controllers, to large, distributed networks of data acquisition and control systems. It can best be implemented in software or in a combination of both software and hardware. Fuzzy logic provides a simple way to arrive at a definitive conclusion based on vague, ambiguous, imprecise, noisy, or missing input information. In its simplest approach, fuzzy logic mimics how a person would make decisions.

Fuzzy logic offers several unique features that can make it a good choice for many control applications.

- Fuzzy logic is inherently robust, particularly in a noisy environment, because it does not require precise inputs and can be programmed to fail safely if a feedback sensor quits or is destroyed.
- Fuzzy logic is not limited to a single feedback input and control output. This makes fuzzy logic control a candidate in multivariable control applications. Any sensor data that provides some indication of a system's actions and reactions can be incorporated.
- Fuzzy logic can also be used to control nonlinear systems that are difficult or impossible to model mathematically.

17.3 Operating Rules

Fuzzy logic is based on rules such as IF x AND y THEN z. This approach to solving control problems mimics rational thinking and does not require an attempt to mathematically model the system to be automated. The fuzzy logic approach is empirically based, relying on experience rather than a detailed technical understanding of the system. For example, instead of dealing with temperature control in terms of set point and proportional-integral-derivative (PID) control, fuzzy logic deals with linguistic rules, taking the form of "IF THEN" statements. Each antecedent (condition) and each consequent (result) is determined by membership functions that are often geometrically represented by triangles or other shapes. The fuzzy logic controller deals with temperature as follows:

Antecedent	Consequent
IF (process is too cool) AND (process is getting colder)	THEN (add heat to the process)
IF (process is too hot) AND (process is heating rapidly)	THEN (cool the process quickly)

These terms are imprecise yet very descriptive of what must be done. Consider what you do at home if the temperature is too cold: you turn up the thermostat. Fuzzy logic mimics this approach as illustrated in the earlier statements.

In the process industry, a fuzzy logic controller that performs similar to a proportional-integral (PI) controller has successfully been implemented using four fuzzy logic rules. For a reverse-acting controller, these rules are shown in the following chart.[1]

Rule 1	IF the error is negative AND the change in error is negative, THEN the change in output is positive.
Rule 2	IF the error is negative AND the change in error is positive, THEN the change in output is zero.
Rule 3	IF the error is positive AND the change in error is negative, THEN the change in output is zero.
Rule 4	IF the error is positive AND the change in error is positive, THEN the change in output is negative.

The implementation of these rules is the part of the controller referred to as the *fuzzy logic engine.*

Fuzzy logic is considered a multi-valued logic that enables you to define intermediate values, such as *warm,* between the binary extremes of hot/cold; true/false; or 1/0 as is done in binary logic. The relation of the quality *warm* to temperature can then be described by what is called a *membership function.* Membership functions can take on values between 0 and 1. This causes numerical values such as 70°F to be associated with linguistic statements like *fairly cool* or *slightly cool.* The set of fuzzy values between 0 and 1 that are associated with various real-world numerical values constitute what we call a *fuzzy set.*

Fuzzy logic does require numerical parameters to operate correctly. In process control, the error (*set point – measured value*) and its derivative (rate of change of error) are

1. Nicholas P. Sands and Ian Verhappen, eds., *A Guide to the Automation Body of Knowledge,* 3rd ed. (Research Triangle, NC: ISA [International Society of Automation], 2018).

typically used, but exact values of these numbers are usually not significant unless extremely responsive performance is required. For example, a simple temperature control system could use a single temperature feedback sensor whose data is subtracted from the required set point to compute the *error* and then time-differentiate this error to yield the rate of change or slope of the error.

To apply fuzzy logic to typical process control applications, the following steps are essential:

1. Explicitly define the control objectives and criteria, and any possible (probable) system failure modes.
2. Determine the input and output relationships and decide on a minimum number of variables for input to the fuzzy logic engine (typically error and its derivative, or the derivative of the process variable). Although it is possible to use a single, instantaneous error parameter without knowing its rate of change, this reduces the system's ability to minimize overshoot for a step input. The more inputs that are used, the more complicated the rules become, and the better the processes behave as required.
3. State the control application in a series of IF x AND y THEN z statements that define the desired system output response for given system input values. The number and complexity of rules depends on the number of input parameters that are to be processed and the number of fuzzy variables associated with each parameter.
4. Create fuzzy logic membership functions that define the meaning (values) of input/output terms used in the rules.

17.4 Linguistic Variables

Fuzzy logic uses the concept of linguistic or *fuzzy* variables rather than numbers. Fuzzy variables are words or objects. The input(s) to the fuzzy logic engine is a noun such as temperature, flow, pressure, or velocity. Because the error is the difference between the measured value and the set point, it can be thought of in the same way. The fuzzy variables themselves are then adjectives that modify the variable (e.g., large positive error, small positive error, zero error, small negative error, and large negative error). As a minimum, one could simply have positive, zero, and negative variables for each of the parameters. Additional ranges, such as very large and very small, could also be added to extend the responsiveness to exceptional or very nonlinear conditions, but they are not necessarily required in a basic process control application.

Table 17-1. Typical 3 × 3 rule matrix.

Rate of Change of Error	(error = set point – measured value)					
	Rule	Negative	Rule	Zero	Rule	Positive
Negative	1	No Change	2	Heat	3	Heat
Zero	4	Cool	5	No Change	6	Heat
Positive	7	Cool	8	Cool	9	Heat

Linguistic variables are best summarized in what is called a *rule matrix*. The simplest rule matrix consists of 3 columns and 3 rows (3 × 3), giving a total of 9 possibilities as illustrated in Table 17-1. The columns represent negative error, zero error, and positive error inputs.

The rows represent the negative, zero, and positive rates of change of error input. This rule matrix has two input conditions, error and rate of change of error, and one output response at the intersection of each row and column. There are nine possible logical products (AND) output response conclusions, but only three possible outcomes. In this example, the output response can be heat, cool, or no change for a typical temperature control process.

Rule matrices usually have an odd number of rows and columns, to accommodate a zero region where no action is required. This is not a requirement and is not always possible, but it is highly desirable to avoid continuous dithering of the output. It is also possible to have a different number of rows than columns. This may occur when numerous degrees of inputs are needed. The maximum number of possible rules is simply the logical product of the number of rows and columns. Definitions of all these rules may not be necessary because some input conditions may never occur in practical situations. The primary objective of this construct is to ensure all inputs are considered to keep the system under control.

Linguistic rules describing the control system consist of two parts: an antecedent block (between the IF and THEN) and a consequent block (following THEN). Depending on the system, it may not be necessary to evaluate every possible input combination (for 5 × 5 and larger matrices) because some may rarely or never occur. By performing this type of evaluation (usually done by an experienced operator), fewer rules can be evaluated, thus simplifying the processing logic and perhaps even improving the fuzzy logic system performance.

After transferring the conclusions from the nine rules to the matrix, there is a noticeable symmetry to the matrix. This suggests (but does not guarantee) a reasonably well-behaved (linear) system. This implementation may prove to be too simplistic for some control problems; however, it does illustrate the process. Additional degrees of error and error-dot may be included if the desired system response calls for it. This will increase the rule base size and complexity but may also increase the quality of the control.

17.5 Defuzzification

Defuzzification is the last step of a fuzzy controller. This step calculates the actual controller output value that is a "crisp" value, meaning it is not fuzzy. This is done by the fuzzy controller's defuzzifier.

Much has been written about the different techniques of defuzzification, and there are many different ways defuzzification can be done. The following are some of the best-known methods:

- Center of sums method (COS)
- Center of gravity (COG)/centroid of area (COA) method
- Center of area/bisector of area method (BOA)
- Weighted average method
- Maxima methods
- First of maxima method (FOM)
- Last of maxima method (LOM)
- Mean of maxima method (MOM)

Different defuzzification methods give slightly different results, and different methods might be more relevant depending on the application. Another consideration is the computational complexity required by the different techniques.

17.6 Single-Input Control Example

To better understand how a fuzzy logic controller actually works, we will design a simple fuzzy temperature controller. The system will consist of a room, a gas furnace, and a fuzzy controller with a temperature sensor; see Figure 17-1. The gas flow to the furnace is regulated by a valve, and the system is essentially balanced at 70°F with a valve opening of 50%.

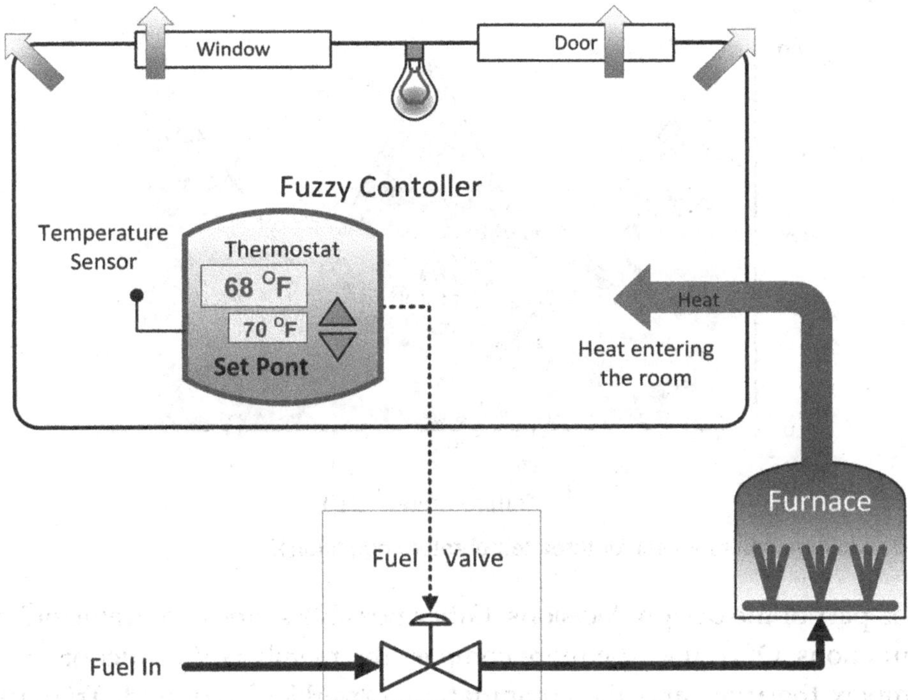

Figure 17-1. Fuzzy logic control of a room.

The fuzzy controller will sample the temperature and then, based on a set of fuzzy rules, will increase or decrease the valve opening as required to maintain the room temperature at 70°F despite disturbances. A simplified block diagram is illustrated in Figure 17-2.

For simplicity, we will define only three temperature conditions: *warm*, *comfort*, and *cold*. Each condition is represented on the graph in Figure 17-3 and takes the shape of a triangle. The shaded area is a membership function, and the shape of each membership

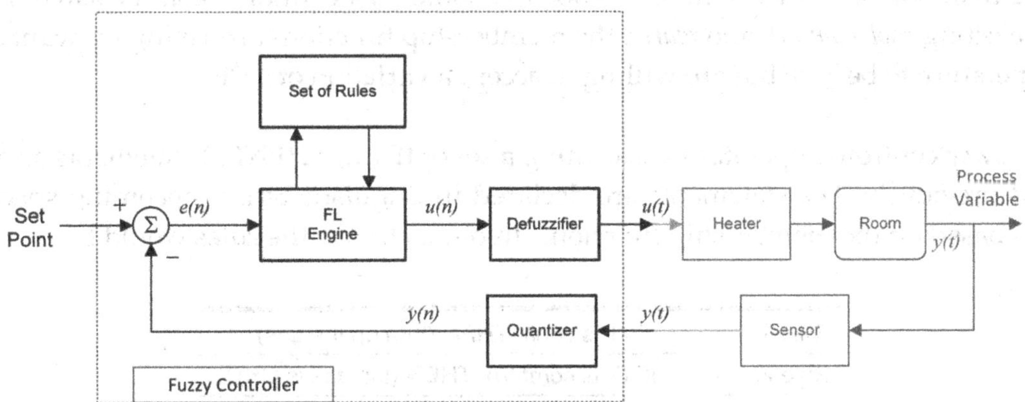

Figure 17-2. Block diagram of a typical fuzzy controller.

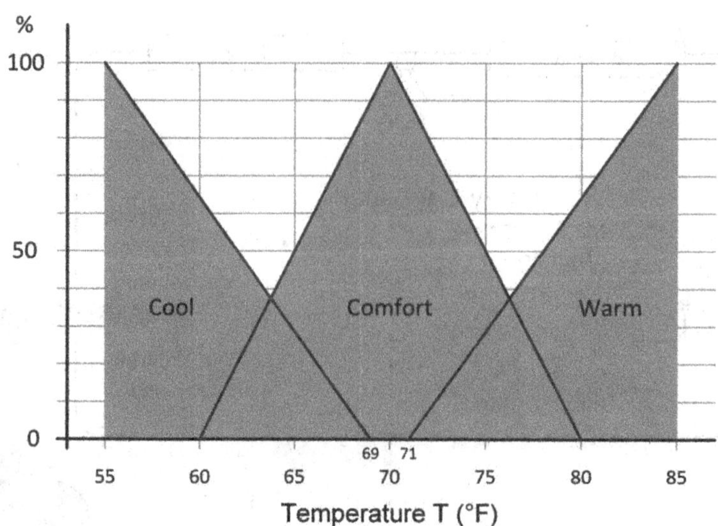

Figure 17-3. Input membership set for three temperature conditions.

function is part of the design decisions. Other possible shapes are trapezoid or exponential functions. Of course, the more complex the membership function shapes, the more complex the rules and the computational overhead required. Triangular and trapezoidal shapes are primarily used and tend to work well.

The membership functions each represent a range of temperatures that are all members of the same grouping, and the degree of relevance as a function of temperature. For example, referring to Figure 17-3, we can say that *comfort* is anything between 60°F and 80°F and 100% true at 70°F. At the same time, any temperature below 69°F is considered cool; however, using the term *cool* is truer at 55°F than it is at 60°F. The terms *comfort* and *cool* are fuzzy sets because they express a range of values rather than a specific temperature value. Notice also that the fuzzy sets overlap between 60°F and 69°F and between 71°F and 80°F. This means that a specific temperature can be a member of more than one set; 65°F is a little bit cool and somewhat comforting at the same time. By selecting *cool, comfort,* and *warm,* the membership functions are saying we want our temperature to be 70°F but are willing to accept a variation of ±1°F.

The fuzzy controller operates by executing a set of IF (…) THEN(…) statements as outlined previously. The statements are declared in linguistic terms, becoming specific rules based on the membership functions. In our example, the rules would be:

Rule 1:	IF (T is *cool*), THEN (turn up the gas)
Rule 2:	IF (T is *comfort*), THEN (the gas is OK)
Rule 3:	IF (T is *warm*), THEN (turn down the gas)

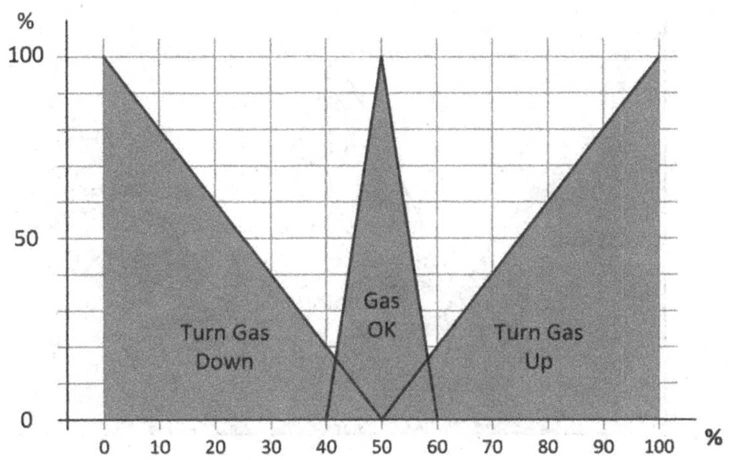

Figure 17-4. Output set for three kinds of action.

These statements clearly define what the system must do to maintain the temperature at a comfortable level. What remains to be defined are the terms Turn Gas Up and Turn Gas Down. Figure 17-4 shows this in yet another graph of fuzzy sets. Here we see that the triangle for Turn Gas Down corresponds to turning the gas valve toward off, anywhere from 0% to 50%. Gas OK corresponds to not changing the gas valve or at most moving it ±10% in either direction, and Turn Gas Up corresponds to turning the gas valve toward fully open anywhere from 50% to 100%. With these two graphs and the set of rules, the fuzzy logic controller can now compute the desired controller output.

Let us assume that the door is opened, and cold air enters the room such that the temperature sensor now reads 62°F.

The fuzzy controller would determine that 62°F is 50% *cool* and 20% *comfort.* Figure 17-5 shows this superimposed on our input membership functions. This means that Rule 1 applies at 50%, and Rule 2 applies at 20%. Rule 3 does not apply in this case because 62°F is outside its membership. The fact that Rule 3 does not apply makes sense because the temperature is far below its desired level of 70°F and must increase, not decrease.

The next step is to calculate the actual controller output value that is a crisp value, meaning it is not fuzzy. This is done by the fuzzy controller's defuzzifier.

Rule 1 says: If the temperature is *cool,* then turn up the gas. Because Rule 1 is 50% true, we set the maximum value of Turn Gas Up at 50%. Similarly, Rule 2 is 20% true, so we adjust the Gas OK triangle to 20% maximum value. Technically, the output of the

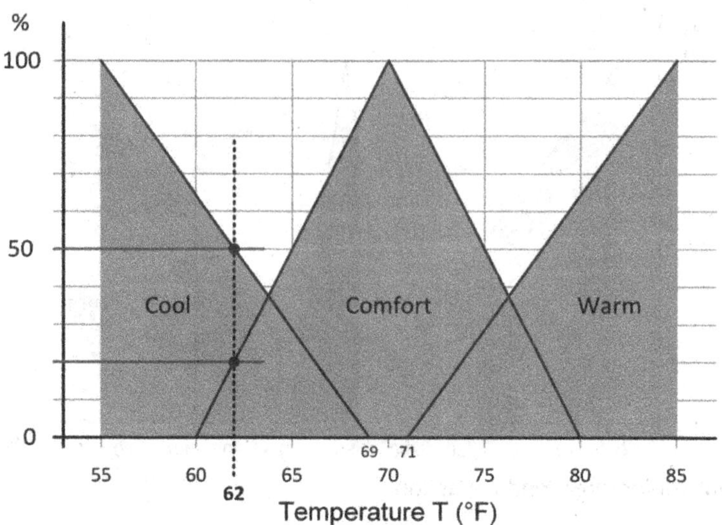

Figure 17-5. Input membership set showing the membership contribution at 62°F.

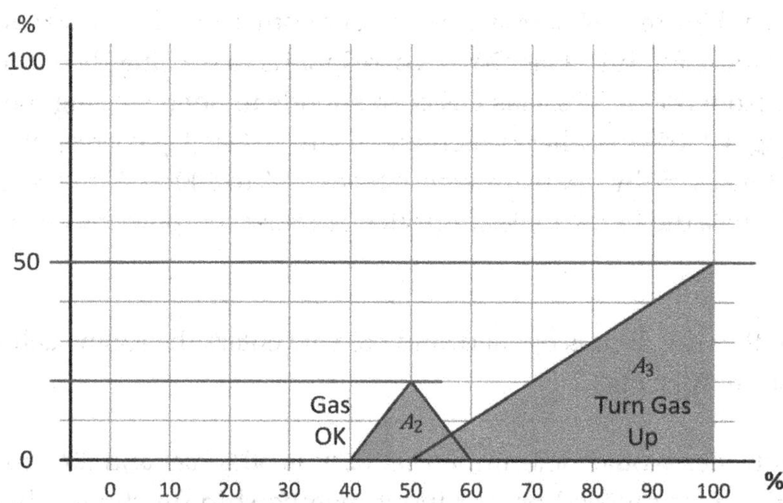

Figure 17-6. Output weighted by input conditions.

fuzzy logic controller is the union of these two sets; see Figure 17-6. However, the valve needs a specific crisp command from the controller, so we must "defuzzify" the output set to obtain a crisp value. This is done by finding some "best" compromise between the rules that apply.

Choosing a defuzzification method can be difficult because different methods will give slightly different results and might be more applicable depending on the application. Another consideration is the computational complexity required by the different techniques.

The COS method is selected for our example because it is mathematically less complex, gives an acceptable crisp value, and is also one of the most commonly used defuzzification techniques used. In our example, we find the point on the horizontal axis that is the average of the two triangles' center of gravity (centroids), and then apply Equation 17-1:

$$x^* = \frac{\sum_{i=1}^{k} A_i \times \bar{x}_i}{\sum_{i=1}^{k} A_i} \tag{17-1}$$

where

x^* = defuzzified value
A_i = area of each contributing triangle
x_i = i^{th} centroid of each triangle

Calculate Areas (A_i)

$$A_2 = \frac{1}{2} \times (0.6 - 0.4) \times 0.2$$
$$= 0.5 \times 0.2 \times 0.2 = 20\%$$

$$A_3 = \frac{1}{2} \times (1 - 0.5) \times 0.5$$
$$= 0.5 \times 0.5 \times 0.5 = 12.5\%$$

Calculate Centroids (x_i)

$$x_2 = 0.5$$

$$x_3 = \frac{0.5 + 1 + 1}{3}$$
$$= \frac{2.5}{3} = 0.883$$

Calculating x^*

$$x^* = \frac{0.02 \times 0.5 + 0.125 \times 0.833}{0.02 + 0.125}$$
$$= \frac{0.01 + 0.104}{0.145} = 78.7\%$$

The final controller output is then 78.7%, meaning the gas valve will open, increasing the furnace heat output. In this case, the gas is turned up 28.7% from 50% (see Figure 17-7). This makes sense because 62°F borders on *cool.*

An interesting exercise is to see what will happen next. As the gas is increased to 78.7%, the room will start to warm up from 62°F to, say, 65°F. Figure 17-8 shows how the controller output is reduced from 78.7% to 70% as would be expected. It is left as an exercise for the reader to do the actual calculations.

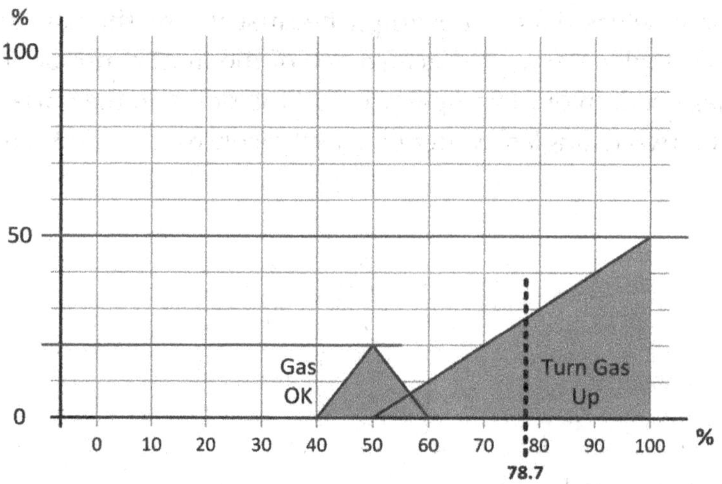

Figure 17-7. Output weighted by input conditions, showing defuzzified output to be 78.7%.

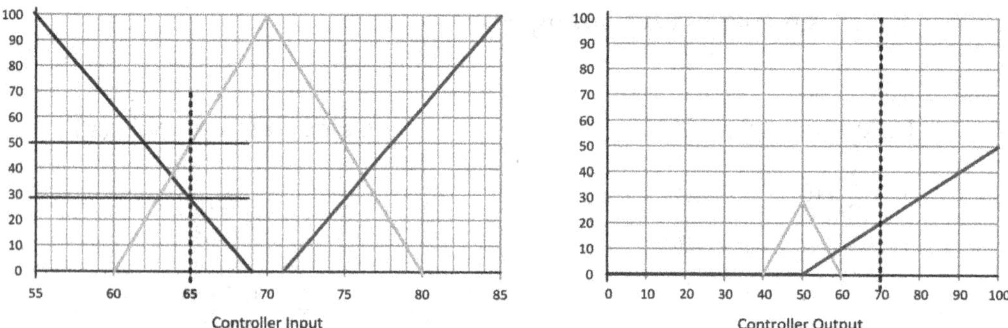

Figure 17-8. Controller output at $T = 65°F$.

17.7 Two-Input Control System Example

The previous example clearly demonstrates the effectiveness of fuzzy control. However, this type of control will most likely produce an initial overshoot in our temperature application and some decaying oscillation because of the inertia of the furnace. We can reduce this by adding a second input to the fuzzy controller. The second input is similar to differential control. In this application, the controller will also know whether the temperature is increasing or decreasing. The controller will then adjust its output accordingly. This means if the temperature is below the set point and decreasing, the controller will provide a more aggressive output. When the temperature is increasing, it will provide a less aggressive output.

Using the same example, we add "rate of temperature change" to the controller. This information could be derived from differentiating the measured temperature, or it could be measured separately. Now the controller knows not only what the temperature (T) is, but also how fast it is increasing or decreasing. This rate of change in temperature can be expressed as $\Delta T/dt$.

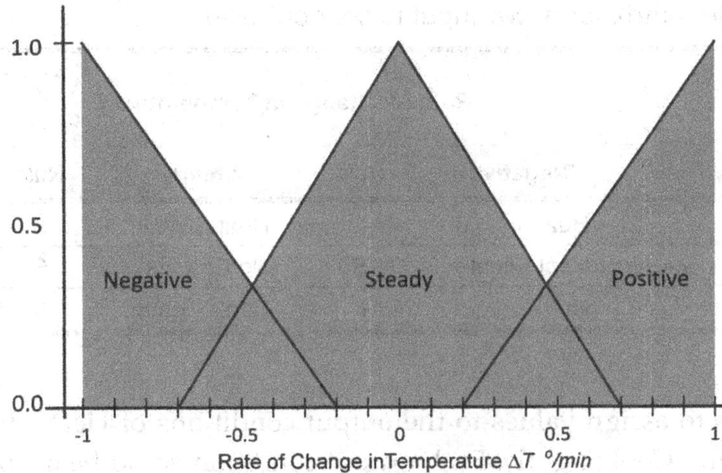

Figure 17-9. Input membership showing the rate of change.

We will define the membership functions of $\Delta T/dt$ in three simple categories of slope: Positive, Steady, and Negative (see Figure 17-9). Also, the rate of change must be explicitly specified. In this example, we define the temperature rate of change as 0.20°F/min to 1°F/min (see Figure 17-9). The categories are again represented as triangles, where the height represents the extent to which that term is true, going from a probability of 0 to a probability of 1, or from 0% to 100%.

The Rule Matrix

We can now deal with the important step of establishing the rules of the controller. There are now two times three, or nine possible sets of inputs for T and $\Delta T/dt$. This is shown in Table 17-2. For each set of fuzzy inputs, a fuzzy output is specified. Notice that these rules are just common sense but can sometimes be difficult to assign.

Based on Table 17-2, we can establish the following nine rules.

Rule 1:	IF (*T* is *cool* and *negative*), THEN (Heat)
Rule 2:	IF (*T* is *cool* and *steady*), THEN (Heat Slightly)
Rule 3:	IF (*T* is *cool* and *positive*), THEN (No Change)
Rule 4:	IF (*T* is *medium* and *negative*), THEN (Heat Slightly)
Rule 5:	IF (*T* is *medium* and *steady*), THEN (No Change)
Rule 6:	IF (*T* is *medium* and *positive*), THEN (Cool Slightly)
Rule 7:	IF (*T* is *warm* and *negative*), THEN (No Change)
Rule 8:	IF (*T* is *warm* and *steady*), THEN (Cool Slightly)
Rule 9:	IF (*T* is *warm* and *positive*), THEN (Cool)

Table 17-2. Rule matrix for a two-input fuzzy controller.

Temperature T	Rate of Change in Temperature $\left(\frac{\Delta T}{dt}\right)$					
	Rule	**Negative**	Rule	**Steady**	Rule	**Positive**
Cool	1	Heat	2	Heat Slightly	3	No Change
Medium	4	Heat Slightly	5	No Change	6	Cool Slightly
Warm	7	NC	8	Cool Slightly	9	Cool

The next step is to assign values to the output conditions of Heat, Heat Slightly, No Change, Cool, and Cool Slightly. Each output condition could be a fuzzy set, as they were in our previous example. But this time we will use discrete values.

Each position in the table will now correspond to a specific rule and will contain a number from 0% to 100%, which correspond to the percent increase or decrease (see Table 17-3).

For example, if the temperature is *cool* and *negative* (Rule 1), then turn the gas way up (+30%). If the temperature is *cool* and *steady*, then turn the gas up just a little (+15%). Table 17-3 shows all values based on the previous rules. The choice of how much to increase or decrease the gas can be a difficult decision and is part of "tuning" the controller. In this example. we used very conservative numbers. We can now use an example to see how it works.

Assume that the temperature is 62°F (same as in our previous example) and that it is rising somewhat rapidly at 0.6°F/min. From graph (a) in Figure 17-10, we see that 62°F corresponds to 50% *cool* and 20% *comfort*, and from graph (b), the rate of change ($\Delta T/dt$), we see that 0.6°F/min corresponds to 25% *steady* and 50% *positive*.

Table 17-3. Rule matrix for a two-input fuzzy controller.

Temperature (T)	Rate of Change in Temperature $\left(\frac{\Delta T}{dt}\right)$					
	Rule	**Negative**	Rule	**Steady**	Rule	**Positive**
Cool	1	+0.3	2	+0.15	3	0.0
Medium	4	+0.15	5	0.0	6	−0.15
Warm	7	0.0	8	−0.15	9	−0.3

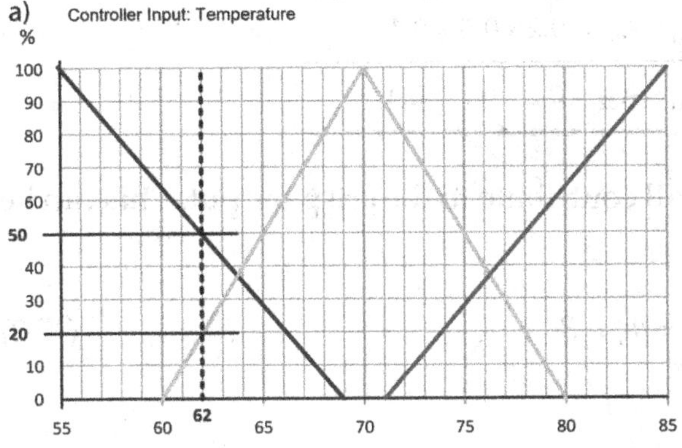

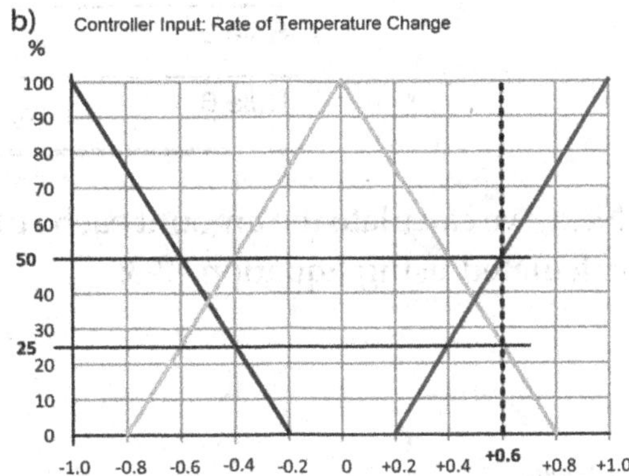

Figure 17-10. Inputs of a two-input fuzzy controller.

Looking at the list of rules, we determine that only four of the nine rules actually apply:

Rule 2:	IF (*T* is *cool* and *steady*), THEN (+15%)
Rule 3:	IF (*T* is *cool* and *rising*), THEN (No Change)
Rule 5:	IF (*T* is *medium* and *steady*), THEN (No Change)
Rule 6:	IF (*T* is *medium* and *positive*), THEN (–15%)

Compatibility

Because all the variables are fuzzy, each rule will contribute—to a different degree—to the final crisp output value. The first step toward arriving at a single crisp output value is to compute the *compatibility* for each rule. The probabilistic compatibility is the product of two or more probabilities (rules) that occur at the same time. Compatibility can be expressed mathematically using Equation 17-2:

$$compatability\ (w_{i1}) = A_{i1} \times A_{i2} \tag{17-2}$$

where

w_i = compatibility (influence of each ith rule)
A_{i1} = probability of the ith rule of first condition (T)
A_{i2} = probability of the ith rule of second condition ($\Delta T/dt$)

Applying this equation to each of the applicable rules of this example, we get:

Rule 2	$w_2 = A_{21} \times A_{22} = 0.5 \times 0.25 = 0.125$
Rule 3	$w_3 = A_{31} \times A_{32} = 0.5 \times 0.5 = 0.25$

Rule 5	$w_5 = A_{51} \times A_{52} = 0.2 \times 0.5 = 0.1$
Rule 6	$w_6 = A_{61} \times A_{62} = 0.2 \times 0.5 = 0.1$

Next, we calculate the amount each rule will contribute to the crisp output. This can be calculated using Equation 17-3:

$$y_i = w_i \times B_i \tag{17-3}$$

where

y_i = output of rule i
w_i = compatibility of rule i
B_i = appropriate value of output table (Table 17-3)

Using Equation 17-3 to calculate the output for each applicable rule gives us the following:

Rule 2	$y_2 = w_2 \times B_2 = 0.125 \times 0.15 = 0.01875$
Rule 3	$y_3 = w_3 \times B_3 = 0.25 \times 0 = 0$
Rule 5	$y_5 = w_5 \times B_5 = 0.1 \times 0 = 0$
Rule 6	$y_6 = w_6 \times B_6 = 0.1 \times -0.15 = -0.03375$

These four outputs of the four rules now must be defuzzified (combined) to form one single crisp output to command the heater. A simple average would not work because some conditions are evoked more strongly than others. One approach is to use a *weighted mean*, as specified in the following equations:

$$w_{wm} = \frac{w_1 y_1 + w_2 y_2 + \ldots}{w_1 + w_2 + \ldots} \tag{17-4}$$

where

w_{wm} = final value of controller output
w_i = compatibility of rule i
y_i = output of rule i

Applying Equation 17-4, we get:

$$
\begin{aligned}
w_{wm} &= \frac{w_2y_2 + w_3y_3 + w_5y_5 + w_6y_6}{w_2 + w_3 + w_5 + w_6} \\
&= \frac{0.125 \times 0.01875 + 0.25 \times 0 + 0.1 \times 0 + 0.1 \times -0.3375}{0.125 + 0.25 + 0.1 + 0.1} \\
&= \frac{0.00234 + 0 + +0 + (-0.3375)}{0.575} \\
&= \frac{-0.335}{0.575} = -0.583 \text{ or } 58.3\%
\end{aligned}
$$

Therefore, the final controller output must be reduced by (50 – 58)% = 8.3% even though the temperature in the room is somewhat cool (64°F). The reason, of course, is that the temperature is rising fast.

Fuzzy logic controllers will be used in an increasing number of products and systems. On the consumer side, they are already being used in automobile cruise control, air conditioners, and microwave ovens, to name a few. For industrial systems, they are already used in process control for temperature, humidity, and flow control.

Fuzzy logic is also used in combination with traditional PID controllers, where the task of the fuzzy logic is to adapt the PID parameters to changing conditions.

Fuzzy logic is also well suited to multivariable control situations, where mathematical models are difficult, if not impossible, to derive.

Specifying the right inputs and the correct set of rules and determining the nature and ranges of the fuzzy variables can be time-consuming.

17.8 Summary

- Fuzzy logic provides a linguistic, nontraditional approach to a control problem. This method focuses on what the system should do rather than trying to understand how it works. Control problems are solved using linguistic descriptions and memberships rather than modeling the system mathematically.
- Fuzzy logic control uses an imprecise but very descriptive language to deal with input data more like a human operator. It is forgiving of input data and often works when first implemented with little or no tuning.

Section 3

Advanced Topics

This section (Chapters 18 through 22) introduces advanced topics to encourage the reader/student toward further studies in process control.

The section starts by continuing and formalizing the discussion we started in Chapter 6, "Process Characteristics." Chapter 18, "Process Modeling and Simulation," provides an overview of process modeling and explains the science and mathematics behind the models. It also introduces the idea of simulating process control systems based on the models derived.

Chapter 19 discusses controlling dead time and the benefits of model-based control. By creating an accurate model of the process, a better controller for the process can be designed and processes with long dead time and/or interacting variables can be controlled more effectively.

To some extent, all real applications have nonlinear characteristics. The control system must either compensate for the nonlinearities or adapt to them. Chapter 20 discusses both nonlinear and adaptive control.

Chapter 21 provides a historical view of distributed control and control system architecture and how it has evolved. It also looks at the structure of large control systems and provides insight into their advantages and disadvantages.

Chapter 22 introduces digital systems, specifically digital signal processing, computer implementation of process models, and controller design. The chapter examines how computer computations, models, and simulations are implemented in a typical controller and/or computer.

18

Process Modeling and Simulation

One of the most difficult aspects of designing an automated system is translating its connectional physical implementation into a set of equations that accurately describes the process.

Deriving models can be a complex topic, and in keeping with this book's philosophy, we will outline only some of the fundamental concepts.

Learning Objectives

- Understand the concept of developing a model based on the laws of science
- Develop the required differential equations and transfer functions
- Distinguish between static and dynamic models
- Distinguish between an experimental and a mathematical model
- Develop an appreciation of system simulation based on derived models

18.1 Introduction

Usually, it is impossible to describe a system completely. In the design process, every effort must be made to derive a set of equations that describes the physical system approximately and adequately, within its intended use and context.

We discussed the idea of modeling in Chapter 5, "Dynamics of Systems," and Chapter 6, "Process Characteristics." In Chapter 5, we used experimental measurements to derive a first-order plus dead time (FOPDT) model of a process; and in Chapter 6, we derived models of integral, self-regulating (first-order lag), and second-order lag processes from first principles. We also showed in Chapter 6 how the time-domain and frequency-domain points of view are related by the Laplace transform.

After we have derived a set of equations, we compare the data generated by the equations with data collected from the physical system by measurements, if possible. If the two sets of data agree or are sufficiently close within their intended use and context, we then gain confidence that the set of equations will lead to a good description of the real-world system. The equations can then be used to make predictions about the dynamic and static, long-term behavior of the system under different scenarios. It is important to keep in mind that the set of equations stays valid only as long as the two sets of data are close, and the intended use and context have not changed. If a prediction from the equations leads to some conclusions that are not close to the real-world behavior, then we should modify and correct the underlying equations. Generating good equations is not an easy exercise.

The set of equations, once derived, is referred to as a *model* of the system.

Dynamic versus Static

Process models can be classified as *dynamic* or *static* (meaning steady-state). In dynamic models, most variables are a function of time, and these models are used to predict how systems behave when subjected to changes in set points, load changes, or disturbances. Static models are often derived from their dynamic counterpart by setting the derivative terms to zero. which is referred to as the *final value property of the Laplace transform* (see Appendix C).

Static models describe the systems once a steady state has been reached and are most useful for design calculations,

Experimental versus Theoretical

Experimental models, as the name implies, are derived from tests performed on an actual process. These models tend to be simpler and easier to manipulate, and they are usually first- or second-order approximations of the theoretical model (see Chapter 5). Experimental models are an alternative to physics-based models and use input/output data to develop empirical dynamic and/or static models of systems.

Theoretical models are derived from the fundamental laws of physics and chemistry. They tend to be complex but provide a better understanding of the process (see Chapter 6) and are often required in the design stages. The development of chemical models is essentially a complete course in chemical engineering, with practical experience as well, and will not be covered here.

In applying the laws of science, we apply the principles of conservation; these include conservation of mass, momentum, energy, and species.

18.2 How to Build a Model

Usually, building a model is an engineering task. However, it is important for the automation professional and technologist to understand the underlying principles and assumptions that were made, and perhaps how to make modifications to an existing model when the actual physical system undergoes minor changes. It is important to remember that any model is at best an approximation of the real world, especially if the model or parts of it are derived by empirical means.

Models of processes consist of ordinary differential and partial differential equations, plus related algebraic equations. The basic steps in building a model of a process can be summarized as follows:

1. Identify the intended purpose and context of the model, and how accurately it must describe the system it is to emulate.
2. Draw a piping and instrumentation diagram (P&ID) and label all the process variables.
3. Develop a detailed block diagram of the system, and describe each block using a derived differential equation and/or its transfer function. Break down the system as much as possible so that each block can be described by a clear transfer function. For each component of the system (process, measurement, actuator, etc.), we may also need to draw its block diagram, which represents the subsystem. Then we need to determine the *input* and *output* signals for each component block.
4. Clearly state all assumptions on which the model will be based. These assumptions should describe the relationships among the quantities to be studied.
5. Write the appropriate conservation equations (mass, component, energy, species). Introduce equilibrium relations and other algebraic equations (from thermodynamics, transport phenomena, chemical kinetics, equipment geometry, etc.).
6. Simplify the model, if possible, by arranging the equations so that the dependent variables (outputs) appear on the left side and the independent variables (inputs) appear on the right side. This form of the model equation is convenient for computer simulation and subsequent analysis.
7. Classify inputs as disturbance variables or as manipulated variables, such as set point changes.

Models of processes—mechanical, chemical, or otherwise—are based on the fundamental laws of science and conservation.

Electrical Systems	Ohm's law, Kirchhoff's voltage, and current laws
Hydraulic and Pneumatic Systems	Conservation of mass
Mechanical Systems	Newton's law
Thermal Systems	Conservation of energy

18.3 Differential Equations of Physical Systems

The mathematical description of a process depends on the properties of its basic elements (namely resistance, capacitance, inertia, and dead time—see Chapter 5), not on the type of system (thermal, liquid flow, gas flow, mechanical, or electrical). Two different systems may have the same process characteristics or behavior. That is, they may have the same equations but different variables. This means we can extend our knowledge of one system to other systems that have the same process characteristics. This becomes very useful for process modeling and simulation.

Before we outline the different type of processes, we will first describe the basic differential equations that describe the dynamics of systems. The differential equations describing the dynamic performance of systems are obtained by utilizing the laws of physics and the concepts of:

- Variable *through* element
- Variable *across* element
- Integrated *across* element
- Integrated *through* element

Table 18-1 identifies the variables for the different types of systems. Students who are familiar with electrical systems and Ohm's law may find it easier to compare the different systems to their electrical counterpart.

Table 18-2 shows the fundamental equations that relate the variables of the different types of systems and their respective components.

Table 18-3 shows typical applications of the differential equations.

To better understand the modeling process, we will examine two cases, one based on an electromechanical system and the other on the two-tank system we discussed in Chapter 6.

Table 18-1. Through and across variables.

System	Variable Through		Variable Across		Integrated Across		Integrated Through	
Electrical	Current	i	Voltage	v	Charge	q	Flux linkage	λ
Translational Mechanical	Force	F	Velocity	v	Momentum	P	Displacement	y
Rotational Mechanical	Torque	T	Angular velocity	ω	Angular momentum	H	Angular displacement	θ
Fluid	Volumetric rate	Q	Pressure	P	Volume	V	Pressure momentum	γ
Thermal	Flow rate	Q	Temperature	T	Heat energy	H	N/A	

18.4 Case Study: Direct Current Motor

A typical separately excited field direct current (DC) motor can be modeled using the electrical and mechanical circuits shown in Figure 18-1.

DC Motor Model

Based on the electrical schematic and the mechanical free body diagram, we can write the following two differential equations that describe the electrical and mechanical model of the DC motor, respectively.

Electrical	Mechanical
$v_a = iR_a + L_a \frac{di}{dt} + K_e\omega$	$T = K_T i = J\frac{d\omega}{dt} + K_d\omega$

where

v_a = armature voltage (input)	volts (V)
i = armature current	amperes (A)
R_a = armature resistance	ohms (Ω)
L_a = armature inductance	henries (H)
K_e = back emf constant	volts per radians per second (v / rad/s)
K_T = torque constant	ounce inch per ampere (oz · in/A)
J = total inertia of armature and load	ounce inch second square (oz · in · s^2)
ω = motor shaft speed	radians per second (rad/s)
K_d = viscous damping constant	ounce inch per radiant seconds (oz · in / rad/s)

Table 18-2. Equations of different system components.

Type of System	Physical Element	Symbol	Equation	Energy (*E*) Power (*P*)
Electrical	Resistance	R	$v = iR$ $i = \frac{1}{R}v$	$P = vi$ $= \frac{1}{R}v^2$ $= i^2R$
	Capacitance	C	$i = C\frac{dv}{dt}$ $v = \frac{1}{C}\int i\,dt$	$E = \frac{1}{2}Cv^2$
	Inductance	L	$v = L\frac{di}{dt}$	$E = \frac{1}{2}Li^2$
Translational, Mechanical	Mass	m	$F = m\frac{dv}{dt}$	$E = \frac{1}{2}mv^2$
	Damper	f	$T = fv$	$P = fv^2$
	Spring	K, F	$v = \frac{1}{K}\frac{dF}{dt}$	$E = \frac{1}{2}\frac{F^2}{K}$
Rotational, Mechanical	Mass	J	$T = J\frac{d\omega}{dt}$	$E = \frac{1}{2}J\omega^2$
	Damper	f	$T = f\omega$	$P = \omega^2$
	Spring	K	$w = \frac{1}{K}\frac{dT}{dt}$	$E = \frac{1}{2}\frac{T^2}{K}$
Fluid	Resistance	R_f	$Q = \frac{1}{R_f P_f}$	$P = \frac{1}{R_f}P^2$
	Capacitance	C_f	$Q = C_f\frac{dP_f}{dt}$	$E = \frac{1}{2}C_f P_f^2$
	Inertia	I	$P_f = I\frac{dQ}{dt}$	$E = IQ^2$
Thermal	Resistance	R_{TH}	$q = \frac{1}{R_{TH}}T$	$E = C_{TH}T$
	Capacitance	C_{TH}	$q = C_{TH}\frac{dT}{dt}$	$E = C_{TH}T$

Table 18-3. Application of the fundamental laws of conservation.

System	Diagram	Input	Output	Physical Principle	Differential Equation
Electrical, Parallel	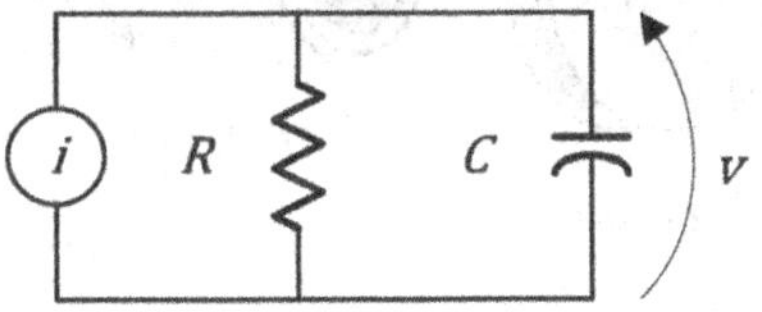	Current i	Voltage v	$\Sigma i = 0$	$i = C\frac{dv}{dt} + \frac{v}{R}$
Electrical, Series	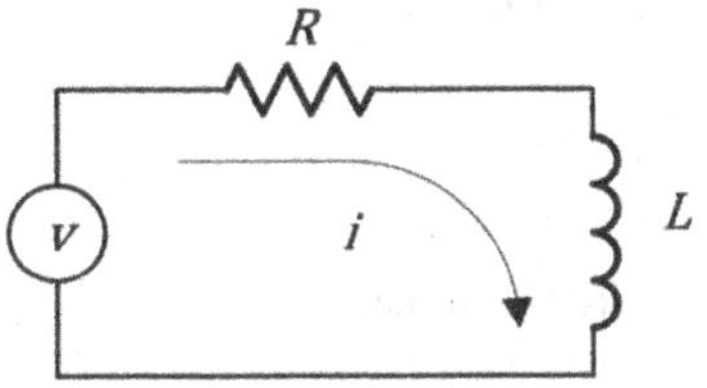	Voltage v	Current i	$\Sigma v = 0$	$v = L\frac{di}{dt} + R \times i$
Translational, Mechanical	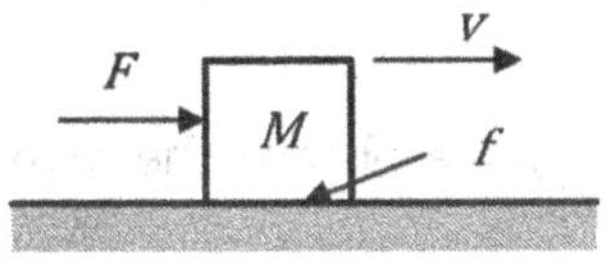	Force F	Velocity v	$\Sigma F = 0$	$F = M\frac{dv}{dt} + f \times v$
Rotational, Mechanical	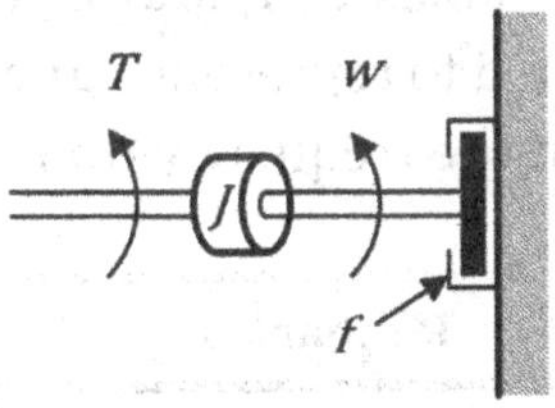	Torque T	Angular velocity ω	$\Sigma T = 0$	$T = J\frac{d\omega}{dt} + f \times \omega$
Fluid	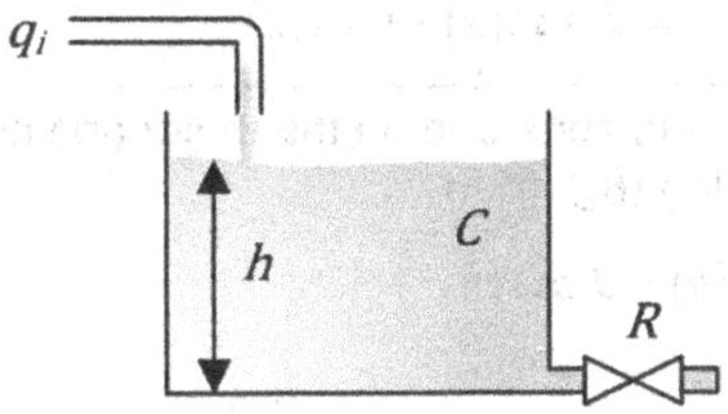	Inflow q_i	Head h	$\Sigma q = 0$	$q_i = C\frac{dh}{dt} + \frac{h}{R}$
Thermal	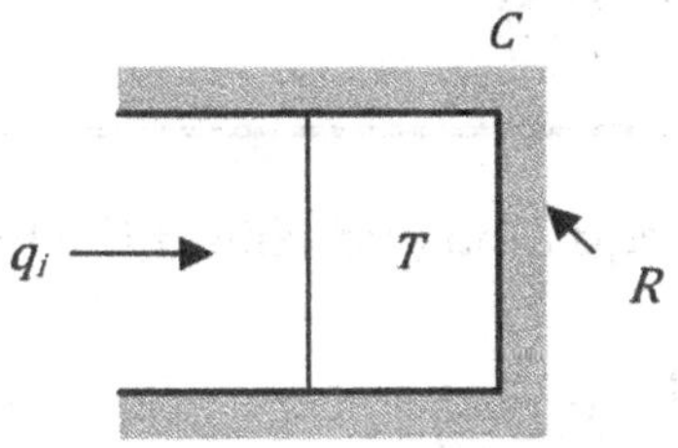	Heat q_i	Temperature T	$\Sigma q = 0$	$q_i = C\frac{dT}{dt} + \frac{T}{R}$

From the equations, we can see that the armature voltage (v_a) is the input and the speed (ω) is the output. The armature current depends on the armature voltage, and the speed depends on the armature current.

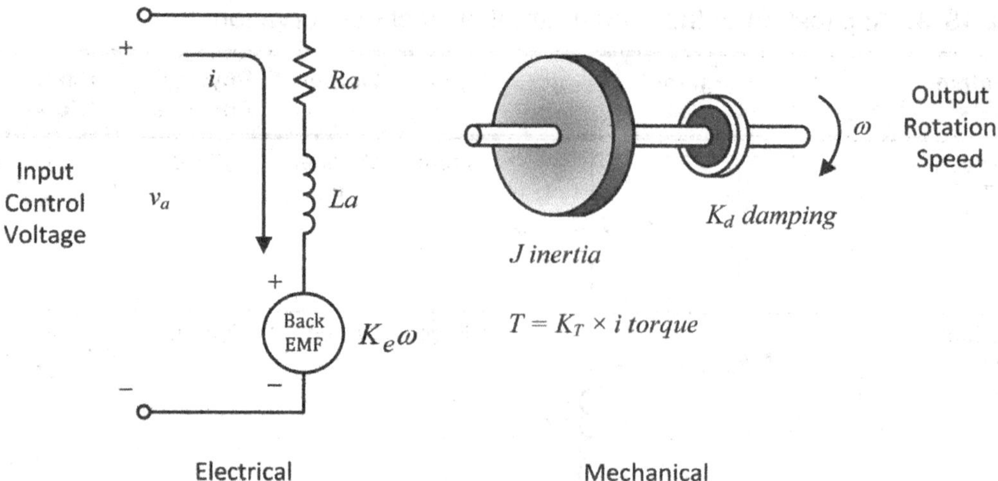

Figure 18-1. Schematic diagram of a DC motor.

Using the Laplace transform and solving for $\frac{\Omega(s)}{V_a(s)}$, we obtain the armature-voltage transfer function. Note that the time-dependent variables v_a, i_a and ω_a are now $V_a(s)$, $I_a(s)$, and $\Omega(s)$. Remember that lowercase letters are used to represent signals in the time domain and uppercase letters are used for signals in the frequency domain.

Electrical	Mechanical
$V_a(s) = R_a I(s) + L_a\ s\ I(s) + K_e\Omega(s)$	$K_T\ I(s) = J\ s\ \Omega(s) + K_d\Omega(s)$
Neglecting the inductance (L_a), then $V_a(s) = R_a I(s) + K_e\Omega(s)$	Assuming zero load on the motor and neglecting damping (K_d), then $K_T\ I(s) = J\ s\Omega(s)$ or $I(s) = \frac{J\ s\ \Omega(s)}{K_T}$

Substitute the torque equation into the voltage equation and collect the terms.

$$V_a(s) = R_a \frac{Js\Omega(s)}{K_T} + K_e\Omega(s)$$

$$= \frac{R_a\ J s\Omega(s) + K_T K_e\Omega(s)}{K_T}$$

$$= \Omega(s)\left(\frac{R_a\ J\ s + K_T\ K_e}{K_T}\right)$$

$$\frac{V_a(s)}{\Omega(s)} = \left(\frac{R_a\ J\ s + K_T\ K_e}{K_T}\right)$$

$$= \frac{K_T}{R_a\ J\ s + K_T\ K_e}$$

$$= \frac{\dfrac{K_T}{K_T\ K_e}}{\dfrac{R_a\ J}{K_T\ K_e}s + 1}$$

$$\frac{\Omega(s)}{V_a(s)} = \frac{\dfrac{1}{K_e}}{\dfrac{R_a\ J}{K_T\ K_e}s + 1}$$

This is a standard first-order system with a time constant (τ) of

$$\frac{R_a\ J}{K_T\ K_e}.$$

Note that we clearly stated all assumptions in developing our transfer function. Of course, depending on the motor, we may choose not to neglect the inductance and damping ratio (L_a and K_d) in our model derivation. If we include these parameters, the transfer function will then be:

$$\frac{\Omega(s)}{V_a(s)} = \frac{\dfrac{K_T}{L_a J}}{s^2 + \left(\dfrac{R_a}{L_a} + \dfrac{K_d}{J}\right)s + \left(\dfrac{R_a\ K_d + K_T K_e}{L_a J}\right)}$$

giving us a second-order model. This derivation is provided next.

DC Motor Transfer Function Derivation

Electrical	Mechanical
$V_a = iR_a + L_a \dfrac{di}{dt} + K_e\omega$	$K_T i - J\dfrac{d\omega}{dt} + K_d\omega$

Calculate the Laplace transformation function of each of the equations:

$V_a(s) = R_a I(s) + L_a\ s\ I(s) + K_e\,\Omega(s)$	Solve the torque equation for $I(s)$: $K_T\ I(s) = J\ s\Omega(s) + K_d\,\Omega(s)$ $I(s) = \dfrac{J\ s\Omega(s) + K_d\,\Omega(s)}{K_T}$

Substitute the torque equation into the voltage equation and collect terms:

$$
\begin{aligned}
V_a(s) &= R_a I(s) + L_a\ s\ I(s) + K_e\Omega(s) \\
&= R_a \frac{J\ s\Omega(s) + K_d\Omega(s)}{K_T} + L_a\ s \frac{J\ s\ \Omega(s) + K_d\Omega(s)}{K_T} + K_e\Omega(s) \\
&= \frac{R_a\ J\ s\ \Omega(s) + R_a K_d \Omega(s) + L_a J\ s^2\ \Omega(s) + L_a\ K_d\ s\ \Omega(s) + K_T K_e \Omega(s)}{K_T} \\
&= \Omega(s)\left(\frac{R_a\ J\ s + R_a K_d + L_a J\ s^2 + L_a\ K_d\ s + K_T K_e}{K_T}\right) \\
&= \Omega(s)\left(\frac{(L_a J)\ s^2 + (R_a\ J + L_a\ K_d)s + (R_a K_d + K_T K_e)}{K_T}\right)
\end{aligned}
$$

Solve for the transfer function:

$$V_a(s) = \Omega(s)\left(\frac{(L_a J)\ s^2 + (R_a\ J + L_a\ K_d)s + (R_a K_d + K_T K_e)}{K_T}\right)$$

$$\frac{V_a(s)}{\Omega(s)} = \frac{(L_a J)\ s^2 + (R_a\ J + L_a\ K_d)s + (R_a K_d + K_T K_e)}{K_T}$$

$$\frac{\Omega(s)}{V_a(s)} = \frac{K_T}{(L_a J)\ s^2 + (R_a\ J + L_a\ K_d)s + (R_a K_d + K_T K_e)}$$

$$\frac{\Omega(s)}{V_a(s)} = \frac{\dfrac{K_T}{L_a J}}{s^2 + \left(\dfrac{R_a}{L_a} + \dfrac{K_d}{j}\right)s + \left(\dfrac{R_a\ K_d + K_T K_e}{L_a J}\right)}$$

DC Motor Simulation

Consider the following data from a manufacturer:

$R_a = 0.73\ \Omega$
$L_a = 100\ \mu\text{H}$
$K_e = 11.3$ V/1000 rpm
$K_T = 15.29$ oz · in/A
$J = 0.017$ oz · in · s²
$K_d = 1.65$ oz · in/1000 rpm (motor only)

Note that the transfer function speed (ω) is expressed in rad/s, and the manufacturer's data is in V/1000 rpm. An appropriate conversion is required.

$$K_e = \frac{11.3\ \text{V}}{1000\ \text{rpm}} \times \frac{1\ \text{rev}}{2\pi\ \text{rad}} \times \frac{60\ \text{s}}{\text{min}} = 0.1079 \frac{\text{V}}{\text{rad/s}}$$

Substituting these values into the equations, we get:

First-Order Model	Second-Order Model
$\frac{\Omega(s)}{V_a(s)} = \frac{\frac{1}{K_e}}{\frac{R_a J}{K_T K_e} s + 1}$	$\frac{\Omega(s)}{V_a(s)} = \frac{\frac{K_T}{L_a J}}{s^2 + \left(\frac{R_a}{L_a} + \frac{K_d}{J}\right) s + \left(\frac{R_a K_d + K_T K_e}{L_a J}\right)}$
$\frac{\Omega(s)}{V_a(s)} = \frac{9.268}{0.645 s + 1}$	$\frac{\Omega(s)}{V_a(s)} = \frac{8.99 \times 10^6}{s^2 + 7300 s + 9.77 \times 10^5}$

Both equations describe a dynamic model of the motor and can be used to determine the static model of the motor or gain by using the Laplace transforms final value theorem.[1]

The motor's DC gain is the relationship between the input voltage (v_a) and the output speed (ω) of the motor. It is the final value of the output after the system has settled down as t approaches infinity.

DC gain is important to know for the feedback element. The feedback element measures the speed in rad/s and must produce a voltage proportional to the motor's required response.

Therefore, the tachometer must convert radians per second to revolutions per minute using the following transfer function:

$$\frac{1\ \text{rad}}{\text{s}} = \frac{1\ \text{rad}}{\text{s}} \times \frac{1\ \text{rev}}{2\pi\ \text{rad}} \times \frac{60\ \text{s}}{\text{min}} = \frac{60}{6.28}\ \text{rpm}$$

First-Order Model	Second-Order Model
$\frac{\Omega(s)}{V_a(s)} = \frac{9.268}{0.645s + 1}$	$\frac{\Omega(s)}{V_a(s)} = \frac{8.99 \times 10^6}{s^2 + 7300s + 9.77 \times 10^5}$
$\frac{\omega}{V_a} = \frac{9.268}{1} = 9.268 \frac{\text{rad/s}}{\text{V}}$	$\frac{\omega}{V_a} = \frac{8.99 \times 10^6}{9.77 \times 10^5} = 9.20\ \text{rad/V}$
$\frac{\omega}{V_a} = 9.268 \frac{\text{rad/s}}{\text{V}} \times \frac{1\ \text{rev}}{2\pi\ \text{rad}} \times \frac{60\ \text{s}}{1\ \text{min}}$	$\frac{\omega}{V_a} = 9.20 \frac{\text{rad/s}}{\text{V}} \times \frac{1\ \text{rev}}{2\pi\ \text{rad}} \times \frac{60\ \text{s}}{1\ \text{min}}$
$\frac{\omega}{V_a} = 88.5\ \text{rpm/V}$	$\frac{\omega}{V_a} = 87.9\ \text{rpm/V}$

1. See Appendix C.

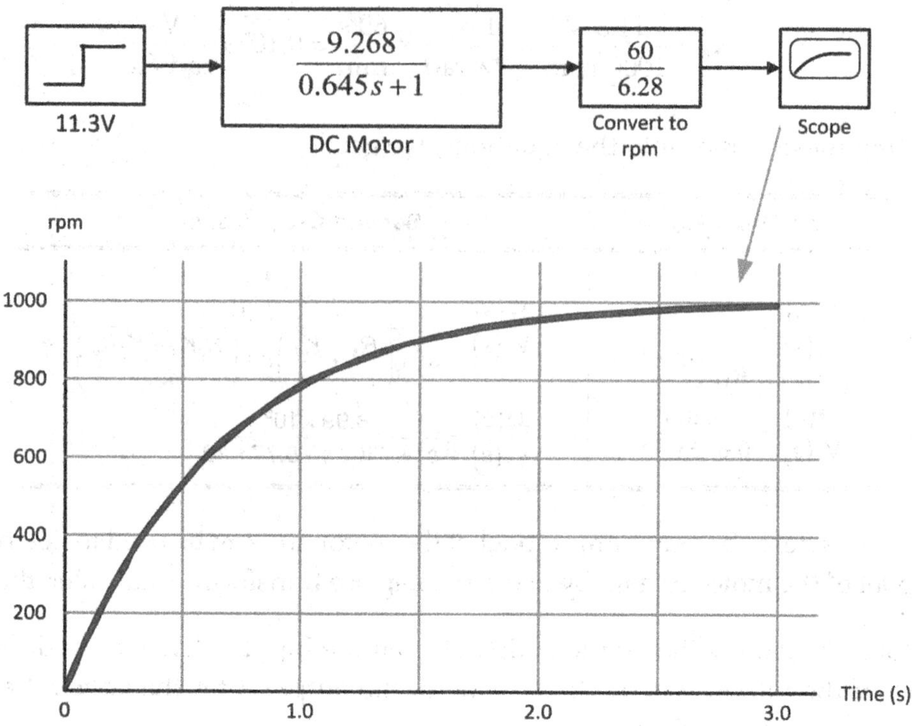

Figure 18-2. Simulink simulation.

This means that for each input voltage, the output shaft will increase in rotation by 88.5 rpm based on the first-order model.

Then at 11.3 V input, we have $11.3 \times 88.5 = 1000$ rpm for the first-order model, and $11.3 \times 87.9 = 993$ rpm for the second-order model. Both closely agree with the original specification ($K_e = 11.30$ V/1000 rpm), which does not consider the motor's damping.

Using this information, we could simulate the open-loop response of the motor. Simulation software is available from different software providers. Simulink from MathWorks enables users to simulate systems by simply drawing their block diagrams, as shown in Figure 18-2.

18.5 Case Study: Two-Tank Level Control (Revisited)

In the chemical processing industry, a two-tank liquid level control system is often used. The setup is shown in Figure 18-3.

Two-Tank Level Control Model

The setup consists of two tanks: Tank 1 receives liquid via a control valve and passes the liquid to Tank 2 through a connecting pipe at the bottom of the tank, and the liquid

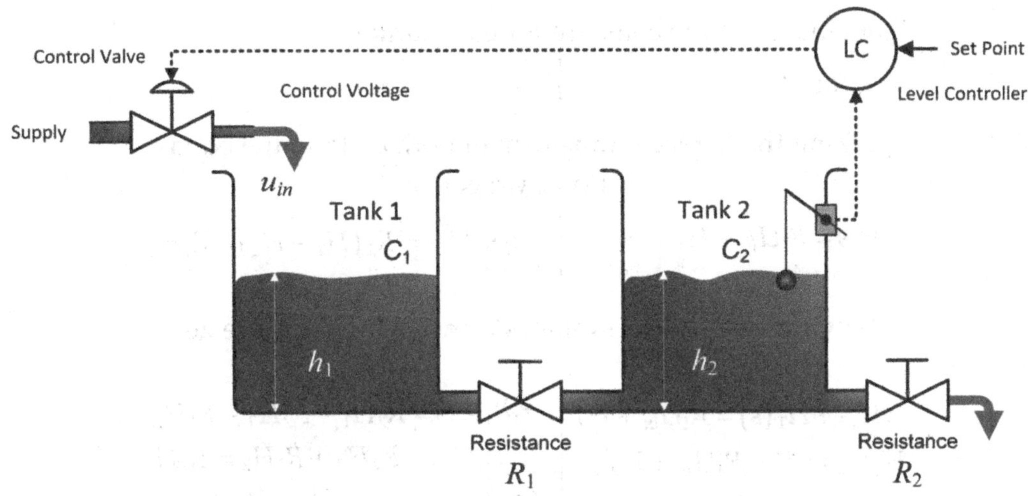

Figure 18-3. Two-tank system.

is drawn out of Tank 2 through a similar pipe. A level transmitter located in Tank 2 provides the required feedback to control the level in Tank 2. This setup allows drastic changes in Tank 1 while minimizing disturbances to Tank 2.

Consider the following variables:

$C_1 = 1$ m³/m	capacitance of Tank 1
$C_2 = 1$ m³/m	capacitance of Tank 2
$h_1(t)$ = height of liquid in Tank 1	in meters (m)
$h_2(t)$ = height of liquid in Tank 2	in meters (m)
$u_{in}(t)$ = input flow rate	
$R_1 = 1/2$ m/(m³/min)	resistance out of Tank 1
$R_2 = 1/3$ m/(m³/min)	resistance out of Tank 2
h_1 = level of Tank 1	in meters (m)
h_2 = level of Tank 2	in meters (m)

Tank 1	Tank 2
$C_1 \dfrac{dh_1}{dt} = u_{in} - q_1$	$C_2 \dfrac{dh_2}{dt} = q_1 - q_2$
$q_1(t) = \dfrac{h_1 - h_2}{R_1}$	$q_2 = \dfrac{h_2}{R_2}$
$C_1 \dfrac{dh_1}{dt} = u_{in} - \dfrac{h_1 - h_2}{R_1}$	$C_2 \dfrac{dh_2}{dt} = \dfrac{h_1 - h_2}{R_1} - \dfrac{h_2}{R_2}$
$R_1 C_1 \dfrac{dh_1}{dt} = R_1 u_{in} - h_1 + h_2$	$R_1 R_2 C_2 \dfrac{dh_2}{dt} = R_2 (h_1 - h_2) - R_1 h_2$

Note that the time constant for each tank is:

$$\tau_1 = R_1C_1 \qquad \tau_2 = R_2C_2$$

Taking the Laplace transform of both differential equations, we get:

$$\tau_1 H_1 s = R_1 U_{in} - H_1 + H_2 \qquad \tau_2 R_1 H_2 s = R_2(H_1 - H_2) - R_1 H_2$$

Solving for $\frac{H_2}{U_{in}}$, the system's transfer function, we get:

$$\tau_1 H_1 s + H_1(s) = R_1 U_{in} + H_2$$
$$H_1(\tau_1 s + 1) = R_1 U_{in} + H_2$$
$$H_1 = \frac{R_1 U_{in} + H_2}{(\tau_1 s + 1)}$$

$$\tau_2 R_1 H_2 s = R_2 H_1 - R_2 H_2 - R_1 H_2$$
$$\tau_2 R_1 H_2 s + R_2 H_2 + R_1 H_2 = R_2 H_1$$
$$(\tau_2 R_1 s + R_2 + R_1) H_2 = R_2 H_1$$
$$H_2 = \frac{R_2 H_1}{(\tau_2 R_1 s + R_2 + R_1)}$$

Two-Tank Transfer Function Derivation

Combining the equations of Tank 1 and Tank 2, we get:

$$H_2 = \frac{R_2\left(\frac{R_1 U_{in} + H_2}{(\tau_1 s + 1)}\right)}{(\tau_2 R_1 s + R_2 + R_1)}$$
$$(\tau_2 R_1 s + R_2 + R_1) H_2 = R_2 \frac{R_1 U_{in} + H_2}{(\tau_1 s + 1)}$$
$$[(\tau_2 R_1 s + R_2 + R_1) H_2](\tau_1 s + 1) = R_2(R_1 U_{in} + H_2)$$
$$\{(\tau_1 s + 1)\tau_2 R_1 s + (\tau_1 s + 1) R_2 + (\tau_1 s + 1) R_1\} H_2 = R_2 R_1 U_{in} + R_2 H_2$$
$$\{(\tau_1 s + 1)\tau_2 R_1 s + (\tau_1 s + 1) R_2 + (\tau_1 s + 1) R_1\} H_2 - R_2 H_2 = R_2 R_1 U_{in}$$
$$\tau_1 \tau_2 R_1 H_2 s^2 + \tau_2 R_1 H_2 s + \tau_1 R_2 H_2 s + R_2 H_2 + \tau_1 R_1 H_2 s + R_1 H_2 - R_2 H_2 = R_2 R_1 U_{in}$$
$$H_2\left(\tau_1 \tau_2 R_1 s^2 + \tau_2 R_1 s + \tau_1 R_2 s + R_2 + \tau_1 R_1 s + R_1 - R_2\right) = R_2 R_1 U_{in}$$
$$\frac{H_2}{U_{in}} = \frac{R_2 R_1}{\tau_1 \tau_2 R_1 s^2 + \tau_2 R_1 s + \tau_1 R_2 s + R_2 + \tau_1 R_1 s + R_1 - R_2}$$
$$\frac{H_2}{U_{in}} = \frac{R_2 R_1}{\tau_1 \tau_2 R_1 s^2 + (\tau_2 R_1 + \tau_1 R_2 + \tau_1 R_1) s + R_1}$$
$$\frac{H_2}{U_{in}} = \frac{R_2}{\tau_1 \tau_2 s^2 + (\tau_2 + \tau_1 \frac{R_2}{R_1} + \tau_1) s + 1}$$

Recall that $\tau_1 = R_1C_1$; we get:

$$\frac{H_2}{U_{in}} = \frac{R_2}{\tau_1\tau_2\, s^2 + (\tau_2 + R_1C_1\frac{R_2}{R_1} + \tau_1)s + 1}$$

$$\frac{H_2}{U_{in}} = \frac{R_2}{\tau_1\tau_2\, s^2 + (\tau_2 + C_1R_2 + \tau_1)s + 1}$$

$$\frac{H_2}{U_{in}} = \frac{R_2}{\tau_1\tau_2 s^2 + (C_1R_2 + \tau_1 + \tau_2)s + 1}$$

Substituting values:

$$\tau_1 = R_1C_1 = \frac{1}{2}\times 1 = 0.5$$

and

$$\tau_2 = R_2C_2 = \frac{1}{3}\times 1 = 033$$

and

$$\frac{H_2}{U_{in}} = \frac{0.33}{0.5\times 0.33 s^2 + (1\times 0.33 + 0.5 + 0.33)s + 1}$$

$$\frac{H_2}{U_{in}} = \frac{0.33}{0.165 s^2 + (1.16)s + 1}$$

$$\frac{H_2}{U_{in}} = \frac{2}{s^2 + 7s + 6}$$

Two-Tank Simulation

Simulating the control strategy using software such as Simulink is as simple as drawing the block diagram. The software will also auto-tune the system and provide appropriate values for the PID controller. In addition, you can use the software to fine-tune the system by modifying the PID parameters while observing the system's response. A block diagram of the system and its response is shown in Figure 18-4.

Note that the feedback element, as well as the control valve, have been omitted for simplicity. However, the example demonstrates the application of models in simulating control systems. Both will alter the behavior of the control systems, depending on their respective models.

Another advantage of simulation is its ability to analyze the signal outputs of each block. The output of the PID controller is shown in Figure 18-4. The graph for the control action shows that the valve immediately opens fully and in 2 s dips to its more or less steady value. Using simulation, many what-if scenarios can be studied, compared, and verified.

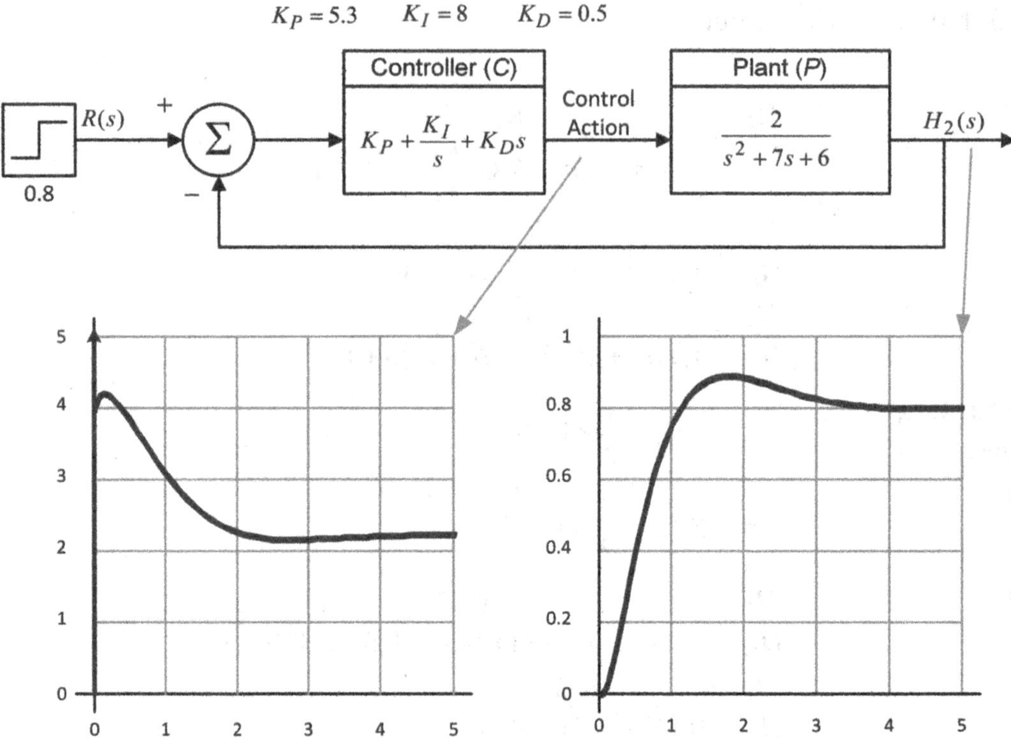

Figure 18-4. Block diagram and simulation of two-tank control system.

18.7 Summary

We have focused primarily on the development of transfer functions based on ordinary differential equations that model different aspects of a control system. The models can be used to simulate the control system, providing a better understanding of system behavior.

Process models can also be used to create virtual sensors that measure one set of variables and deduce the values of another set mathematically. Possibly the most advantageous application for process models in the chemical and petrochemical industries is constraint management. Constraints are the material and practical limitations that determine the maximum/minimum allowable values for the process variables, such as the capacity of a tank, the temperature limit of a reactor, or a pipe's maximum flow rate.

19

Model-Based Control

Model-based control is based on the concept that if an accurate model of the process can be obtained, a better controller for the process can be designed. This type of control strategy is increasingly used in industry. When a process model is available, processes with long dead time and/or interacting variables can be controlled more effectively.

Model-based control can be a complex topic, especially when applied to multiple loops in a process plant. Keeping to this book's philosophy, we will only outline some of the more fundamental concepts.

Learning Objectives

- Understand the concept of model-based control
- Understand the nature of process dead time
- Review control system design to eliminate unnecessary or unfortunate dead times
- Appreciate the use of predictor algorithms and techniques
- Recognize the benefits of Lambda tuning

19.1 Introduction

In applying feedback control, as discussed previously, no knowledge of the process is required, except for knowing whether the process requires direct or reverse action of the controller. For example, when a proportional-integral-derivative (PID) controller receives negative feedback, it can bring the process to the desired value when subjected to set point and demand changes or disturbances.

With detailed knowledge of the process in the form of a mathematical model, a prediction of process behavior to set point and demand changes or disturbances can be computed, and additional controller action can be applied to the process. This provides better control for difficult applications, such as processes with a long dead time, interacting control loops, and control loops with physical or operational constraints.

In previous chapters, we discussed control strategies based on some type of model-based control, such as in feedforward (Chapter 12) and multivariable (Chapter 13) control.

19.2 Dead Time

Dead time is one of the most difficult dynamic characteristics to control. Effectively, we are attempting to control something that has not yet been measured. Not only is it difficult, but the control of dead time also requires complex mathematics to be performed by the controller compared to a standard PID controller.

Dead time was first introduced in Chapter 5. Before discussing the control of dead time, it is helpful to get a sense of its physical outcome. Refer to Figure 19-1, which illustrates a hot-water supply system used to supply water for your morning shower. The arrangement consists of a temperature control system, therefore the installation of a temperature sensor to measure the controlled variable becomes important. The question is where should this sensor be located? It is tempting to install the sensor at the point where the water will be used: at the shower. However, this location will add

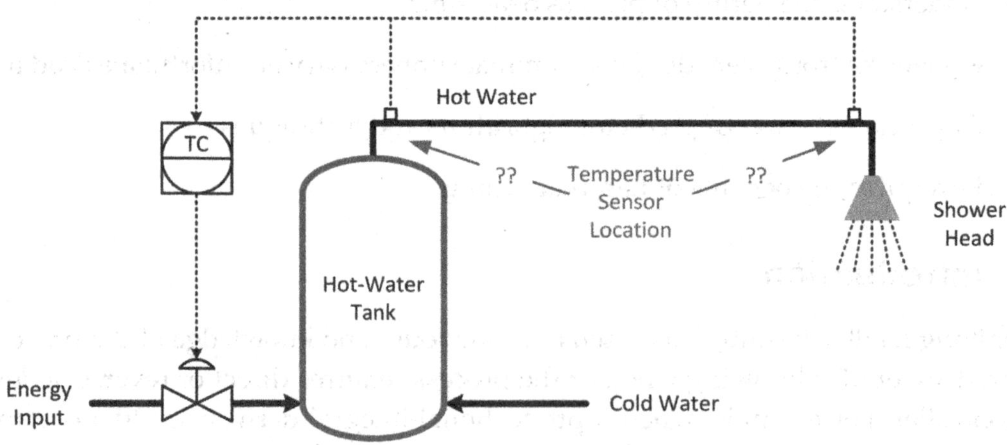

Figure 19-1. Hot-water supply system.

a significant amount of *dead time* to the control loop. The sensor should be installed where the water leaves the hot-water tank. This provides better control of temperature. If we were to install the sensor at the shower, then water would have to flow the distance from the tank to the shower, and the elapsed time would depend on the velocity of the water flowing through the pipe. The time required (dead time) would be equal to the distance divided by velocity. This type of dead time is usually called *transportation lag*. When the sensor is located at the shower, this dead time is within the control loop. With the sensor at the tank exit, the dead time is outside of the control loop, and thus the control loop will function far more quickly and effectively.

Although dead time often can be reduced by proper consideration of the sensor location, it is not always avoidable. There are many cases where a dead time is inherent in the nature of the process itself. Typical simplified examples are shown in Figure 19-2.

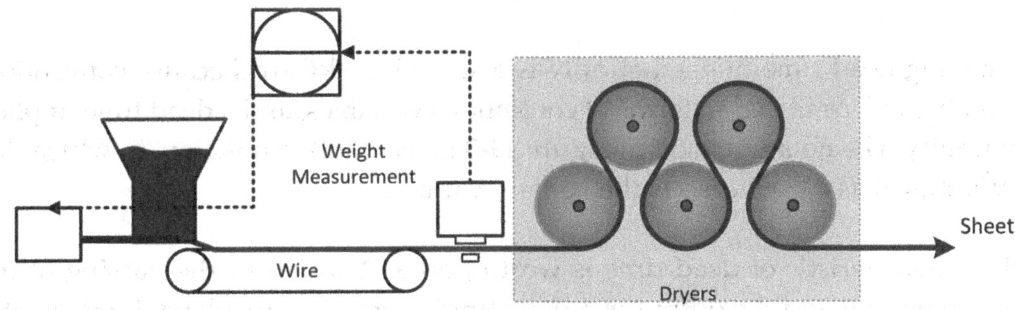

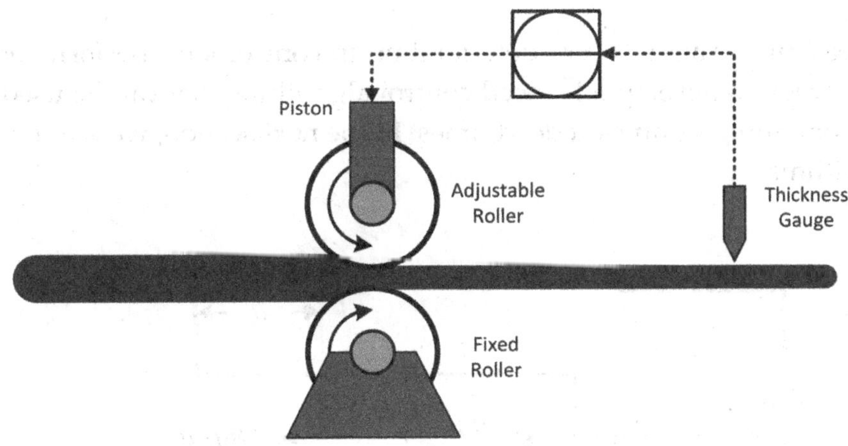

Figure 19-2. Typical dead time examples in process units.

The actual papermaking and rolling mill processes and controls are much more complex, but Figure 19-2 illustrates the dead time concept. In each case, there is a condition in which the material within the process must flow through equipment, and there is an inherent distance-velocity lag or dead time within the process equipment. Another common type of dead time is encountered in analytical equipment such as chromatographs or many, but not all, viscometers. A dead time is created from the time a controller output (CO) change takes place until the measure process value begins to respond to that change. These sample times are, of course, dead times, but from a control loop point of view, they are located in the feedback part of the loop.

There are also other situations that are either dead times in a physical sense or that create the equivalent of a dead time as far as a control loop is concerned. The mathematical form of a dead time is shown in Figure 19-3, which illustrated the graphical picture of dead time θ. You can clearly see that during the dead time (θ) there is no output (in response to the input), and thus no output information is available to initiate control or corrective action. This illustrates why dead time is so difficult to control.

Representing dead time mathematically is also a bit awkward because conventional mathematics is formulated in terms of continuous variables, and a dead time implies a discontinuity. The notation used in Figure 19-3 is the Laplace notation, in which dead time θ is shown as $e^{-\theta s}$, where s is the Laplace variable.

Another characteristic of dead time is worth noting; because of the passing of time between the input and the output of a dead time, memory is involved. Implementing memory using analog hardware is difficult. Digital hardware is better suited to implement the control calculations required.

Because dead time causes severe deterioration in control loop performance, there is a real motivation to develop advanced control algorithms that can be used to control processes containing significant dead times. In the next section, we will review one of these algorithms.

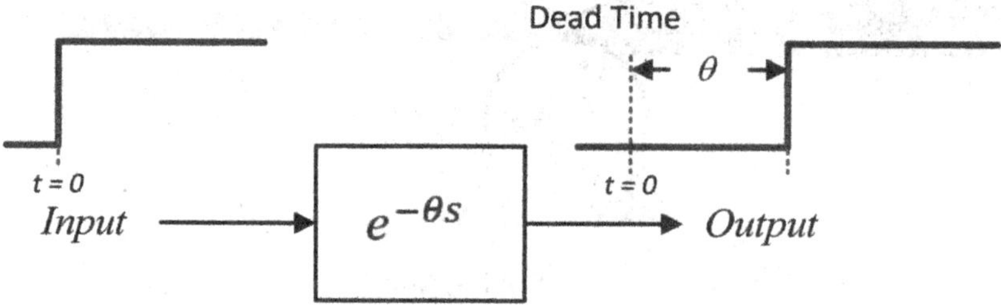

Figure 19-3. Mathematical form of dead time.

19.3 The Smith Predictor Algorithm

The Smith predictor control algorithm was developed specifically for processes with long dead times. It is also the best known of all the dead time compensation algorithms, and it is based on a mathematical model of the process to be controlled. To demonstrate its function, we will assume a process that can be represented as a first-order lag plus a dead time, as shown in Figure 19-4.

We learned previously (in Chapter 5) that a first-order lag system with dead time can be represented mathematically as follows:

$$t_d = \frac{G_P}{1+\tau s} \times e^{-\theta s} \tag{19-1}$$

where

t_d = dead time transfer function
G_P = process gain

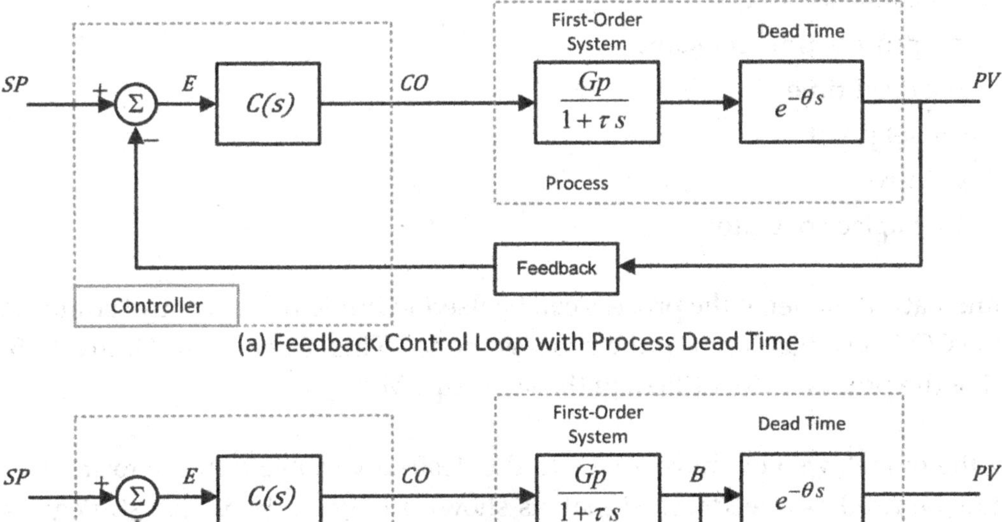

Figure 19-4. Feedback control loop with process dead time.

τ = process time constant

θ = dead time

e = error

s = Laplace operator

Figure 19-4(a) applies this to a control loop. It is convenient to split the representation of the process into two parts, a first-order lag block and a dead time block. When we move the dead time out of the loop, it defines a fictitious variable B as shown in Figure 19-4(b), which can then be fed back to the controller, also shown. The dead time is now outside the loop, and the process variable will do whatever B does but is delayed by θ. This representation cannot be implemented because B is internal to the process and cannot be measured. We can identify the following in Figures 19-4(a) and (b):

where

t_d = dead time transfer function

G_P = process gain

CO = controller output

τ = process time constant

θ = dead time

SP = set point

e = error

s = Laplace operator

A mathematical model of the process can be used to implement a control strategy such that the CO is used as input to the process model. This is shown in Figure 19-5. The model of the process is identified by the subscript M.

Using the model, we now have access to the desired variable B in the form of B_M. B_M can now be used as a feedback signal, as shown in Figure 19-5. This strategy works well, especially for set point changes, as long as our model is perfect and no disturbances enter the loop. Referring to Figure 19-5, mathematically:

if $$PV = PV_M$$

then $$B_M + (PV - PV_M) = B_M \tag{19-2}$$

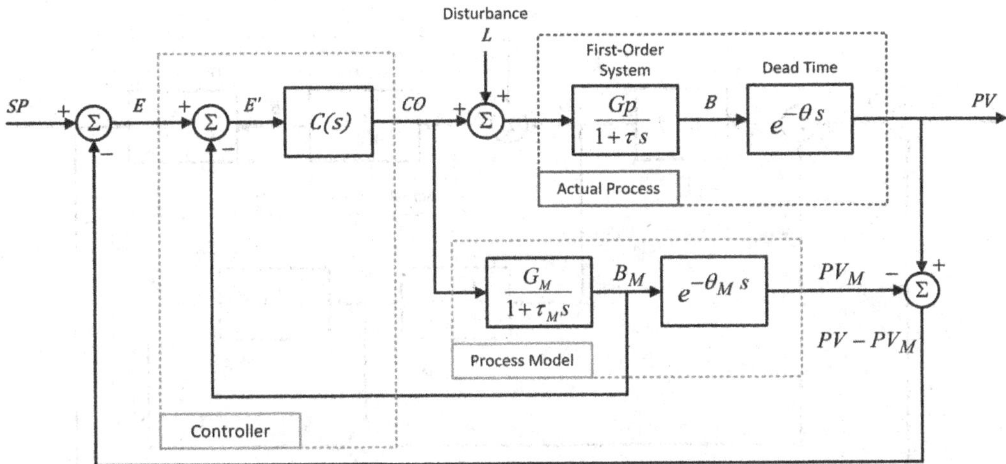

Figure 19-5. Smith predictor control algorithm using B_M for feedback.

A second feedback loop can be added, however, as shown, and this is the Smith predictor control strategy.

The model of the process without the time delay calculates the predicted effect of the control actions on the non-delayed output B_m. The controller then uses the predicted response to calculate the CO signal. The predicted process output is also delayed by the amount of the time θ as indicated by the term $e^{-\theta_M s}$. This delayed mode output is denoted by PV_M, which is then compared with the actual non-delayed output *PV*. This calculation corrects for modeling errors and for any disturbance (*L*) that entered the process. This delayed mode output is denoted by PV_M.

As shown in the block diagram of Figure 19-5, an additional internal error signal (E') is defined as:

$$E' = SP - (PV - PV_M) - B_M \tag{19-3}$$

If the process model is accurate and the disturbance *L* is zero:

Then $$PV = PV_M$$

or $$PV - PV_M = 0$$

and $$E' = SP - B_M$$

This shows that the controller responds to a change in the set point (*SP*) error signal as though no time delay were present.

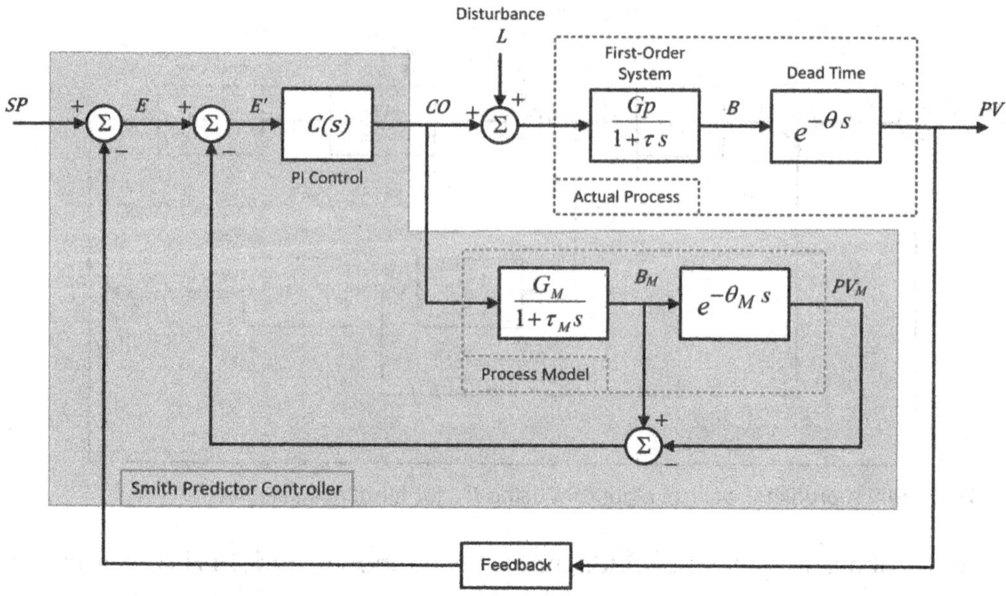

Figure 19-6. Computer control implementation of Smith predictor.

This control strategy is usually implemented in an equivalent form, shown in Figure 19-6. From Figure 19-6:

$$\begin{aligned} E' &= SP - PV - (B_M - PV_M) \\ &= SP - PV - B_M + PV_M \\ &= SP - (PV - PV_M) - B_M \end{aligned}$$

This gives Figure 19-6 the same result as Figure 19-5.

The control action C(s) in the loop can be a conventional proportional-integral (PI) controller.

A distinctive feature of the application of the Smith predictor is that results are radically affected by modeling errors. If the mathematical description of the process is inaccurate or changes during operation, it may be difficult or impractical to use the Smith predictor control strategy.

Also note that in using the Smith predictor control strategy, although it requires a model of the process, it does not use that model to tune the PI feedback controller.

There are several modifications as well as other dead time compensation strategies available, but these are beyond the scope of this book. One example of using the Smith predictor in tuning dead time follows.

Tuning Dead Time

For dead time compensators, such as the Smith predictor, tune the PID controller with the compensator turned off. Then enter the correctly identified time delay, time constant, and process gain into the compensator. If necessary, compute and update the time delay continuously. The controller gain setting identified with the compensator off is used, but a much faster integral time should be possible if the identified settings are accurate. In fact, for dead time–dominant processes, the minimum integral time approaches one-quarter of the error in the identified dead time used in the compensator.

In practice, the integral time is set to be about 20% of the sum of the identified dead time and time constant.

19.4 Lambda Tuning

Lambda tuning is a form of model-based control that provides a PI controller with the ability to generate smooth, non-oscillatory control when responding to changes in the SP. Its name is derived from the Greek letter lambda (λ), which designates a user-specified performance parameter that is based on a model of the system, and dictates how long the controller is allowed to spend on the task of bringing the PV from point A to point B.

Similar to the Ziegler-Nichols tuning method (Chapter 10), lambda tuning involves a set of tuning rules that dictate the values of the controller's PI parameters that are required to achieve the desired controller performance. The first step in applying lambda tuning is to obtain an accurate first-order model of the system (see Chapter 5, "Dynamics of Systems"). Using the model, we can determine the process gain (G_p) and the process time constant (τ), and how the process responds to a change in the set points during a bump test (see Figure 19-7).

Process gain was defined as $\frac{\textit{change in output}}{\textit{change in input}}$, and the time constant τ as the time required for the process to reach 63.2% of its final value, not including dead time (see Figure 19-7).

Using this model—either derived from measurements or determined mathematically—lambda tuning is defined by the following rules:

$$K_C = \frac{\tau}{G_P(\lambda + t_d)} \tag{19-4}$$

and

$$T_I = \tau$$

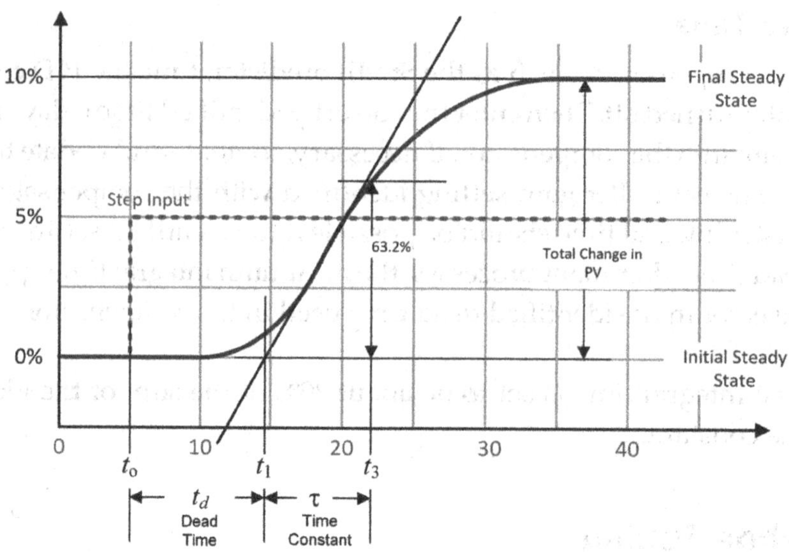

Figure 19-7. Process bump test.

where

K_C = controller gain

G_P = process gain

τ = process time constant

t_d = process dead time

T_I = controller integral time

λ = chosen response time to fit the control strategy

Note: Lambda tuning does not use derivative control action, only PI control.

The variable lambda (λ) should be chosen such that a closed-loop system with a non-oscillatory SP response will settle out in approximately 4 x λs.

Note that these tuning rules require the user to specify only one performance parameter. This not only simplifies the calculation of K_C and T_I, but it also enables the user to select the controller's desired performance in terms of the time required to complete an SP change—a physically meaningful quantity as opposed to the less intuitive concepts of proportional band and reset time.

Closed-Loop Performance

A PI controller tuned using the lambda tuning rule will theoretically complete an SP change in about 4 s and will drive the PV toward the SP gradually to guarantee that the error will continue to diminish steadily with no overshoot.

This overdamping characteristic is particularly useful in applications where the PV must be maintained near some limiting value that it must not cross. The controller will never violate such a constraint because it will never drive the PV past the SP. Nor will a lambda-tuned controller ever cause unstable oscillations in the PV because it will never need to reverse course after an SP change.

Lambda tuning guarantees an overdamping response, and stability, which is why it has become particularly popular. The absence of any overshoot also prevents interacting loops from causing the actuators to oscillate. Individual actuators (especially valves) will be subject to less wear and tear because they will never be required to reverse course unless the SP change requires it.

19.5 Summary

In this chapter we gained a better understanding of the nature of process dead time, and we applied model-based control to provide better dead time control using the Smith predictor algorithm. We also applied lambda tuning to model-based control.

Model parameters must be known within about 10% to 20% of actual values.

Exercise

19.1 Consider the following system, in which diesel oil is injected into fuel oil to control product viscosity:

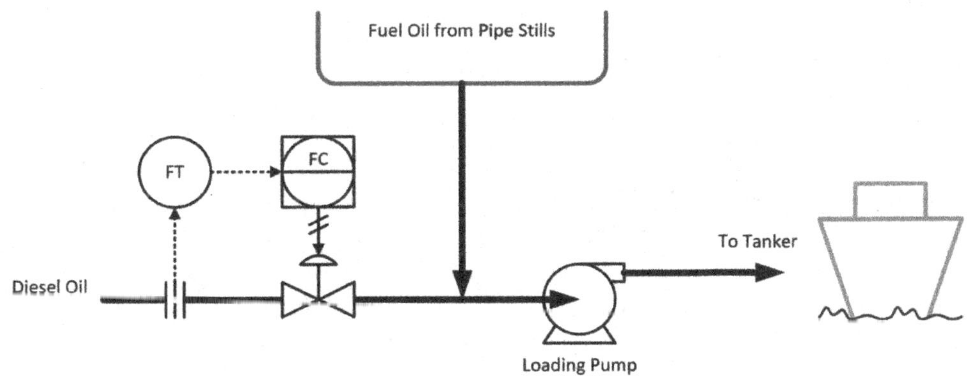

Determine a suitable location for a viscosity meter to be located to measure viscosity and then, through cascade control, to be used to provide a set point to the flow controller (FC) that controls diesel oil injection.

19.2 Discuss the dead time inherent in a steel rolling mill or in a paper machine.

19.3 Compare dead time compensators as presented in this section with conventional feedback control.

19.4 In all the algorithms presented in this section, the process model was assumed to be a first-order lag plus dead time. Can or should higher order systems be considered when developing dead time compensation algorithms?

19.5 Assume your digital control hardware vendor gives you a software program that implements an advanced dead time compensator, but you must supply the gain, the time constant, and the dead time for the process. How could you measure these experimentally?

20 Nonlinear and Adaptive Control

Practically all process control theory is taught in terms of linear systems, whereas all real applications, to some extent, have nonlinear characteristics. There are two basic approaches for dealing with this contradiction

1. either compensate for the nonlinearities, or
2. adapt to them.

We will briefly discuss both in this chapter.

Self-tuning (also called *auto-tuning* or *adaptive tuning*) is a technology that has greatly advanced in recent years. However, the fundamental concepts underlying adaptive control have been around for some time. Practically all proportional-integral-derivative (PID) controllers and control systems today include some type of self-tuning.

Equal-Percentage

- Understand the character and implications of nonlinearities
- Appreciate the use and usefulness of compensation for nonlinearities
- Understand adaptive tuning
- Appreciate general adaptive control concepts

20.1 Nonlinearity

We saw previously that most, if not all, systems can be described by differential equations (Chapters 6 and 18).

Linear systems are better understood and easier to model and predict. Nonlinear systems are much more difficult to model and predict. When something is said to be linear, we tend to think of a straight line. In other words, solutions to systems that are linear can be extrapolated. Linear differential equations have general solution techniques. But nonlinear differential equations only have specific solution techniques, for specific equations under specific conditions.

A control system is considered linear when it can be described by a linear differential equation. This implies that its performance is predictable. If we change an input by increasing or decreasing it by Δx, allow the system to reach its new steady state, and then again change the input by an additional Δx, the new output will be exactly the same as, and in addition to, the output to the first Δx input. If we reduce the input by Δx, the output will again be the mirror image of the output caused by an increase of Δx in the input. The output is a predictable linear extrapolation of earlier outputs for different amounts of Δx both in extent and shape.

When a control system is nonlinear, its response to a positive change of Δx is different from the mirror image of a response due to a negative change of Δx. The response of the system changes in time and as the load on the system changes. It must be retuned for each change in set point and for different load conditions. However, a linear system may be tuned once and stay tuned for different set points and load conditions. Linear systems behave in a predictable manner.

In general, linear systems have the following characteristics:

- All dependent variables are continuous; that is, they give a continuous response over the range of values exhibited by the system.
- All coefficients of the differential equations that describe the system are constants or functions of time only.
- When several inputs are applied at the same or different locations at the same or a different time, the total response is the sum of the individual responses (superposition).
- The magnitude of the output is directly proportional to the magnitude of the input.

Nonlinear systems lack one or more of these characteristics. Much can be done when designing control systems to cause them to be linear. When it is inevitable that nonlinearities are present—either because of inherent characteristics or because of changes in parameters in time or under varying loads—we can sometimes install compensation so that the overall result is linear.

In a control sense, it is also possible to design a system that adapts itself to varying parameters. *Adaptive control* is somewhat of an inaccurate term, but it is most often used to deal with nonlinearities that cause process conditions and parameters to change. The control strategy or controller characteristics are adapted or changed during actual operations to maintain acceptable good control of these changing conditions. Most adaptive control systems have some method of process identification, whose results are then used to adjust one or more parameters in the control algorithm. As an example, process gain is measured in real time and controller gain is adjusted so that the product of the two is always constant.

It is usually better to avoid or compensate for nonlinearities. If that cannot be done, adaptive control should then be considered.

20.2 Valve Characteristics

Control valves are a good place to begin a consideration of nonlinearities. The flow characteristics of control valves generally fall into three broad categories, as was shown in Figure 8-4 in Chapter 8. Figure 20-1 is a repeat of that figure.

These characteristic curves are somewhat misleading. Superficially, it appears that the Quick-Opening (1) and Equal-Percentage (3) curves are nonlinear, while the Linear (2) curve is linear. However, the application of a valve characteristic is more complex. To better understand control valve behavior, it is important to consider its specific dynamic characteristics when installed as a part of a total piping and control loop arrangement. The concepts involved significantly affect the valves' operating characteristics and the total performance of the feedback control system.

At this point, you might want to review Section 5-10, "Actuator Dynamics," in Chapter 5. The result of that discussion is that a linear control valve may not behave as a linear device, whereas an equal-percentage valve may operate over a range in which its behavior is effectively linear. By considering the family of valve curves and analysis of system pressure losses, one can obtain a better understanding of a specific valve's behavior. Only then can the control system be intelligently designed.

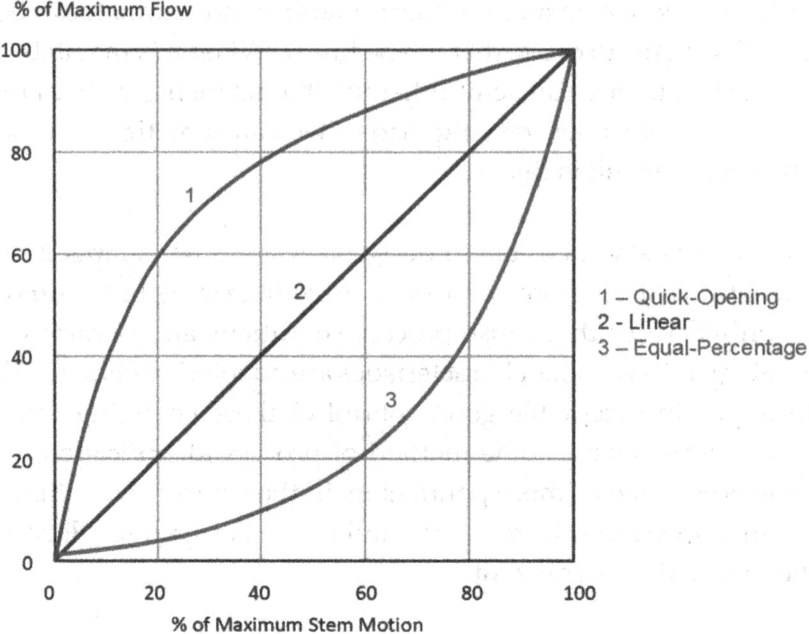

Figure 20-1. Valve flow characteristics.

20.3 Process Characteristics

Consider the heat exchanger shown in Figure 20-2. Is the process itself linear or nonlinear?

At steady state, the energy balance is as follows:

$$W C_p(T_o - T_i) = F \Delta H \tag{20-1}$$

where

W = heated liquid flow rate
C_p = heat capacity of liquid
T_o = outlet temperature
T_i = inlet temperature
F = steam flow rate
ΔH = steam enthalpy change

Solving for F gives:

$$F = \frac{W C_p(T_o - T_i)}{\Delta H} \tag{20-2}$$

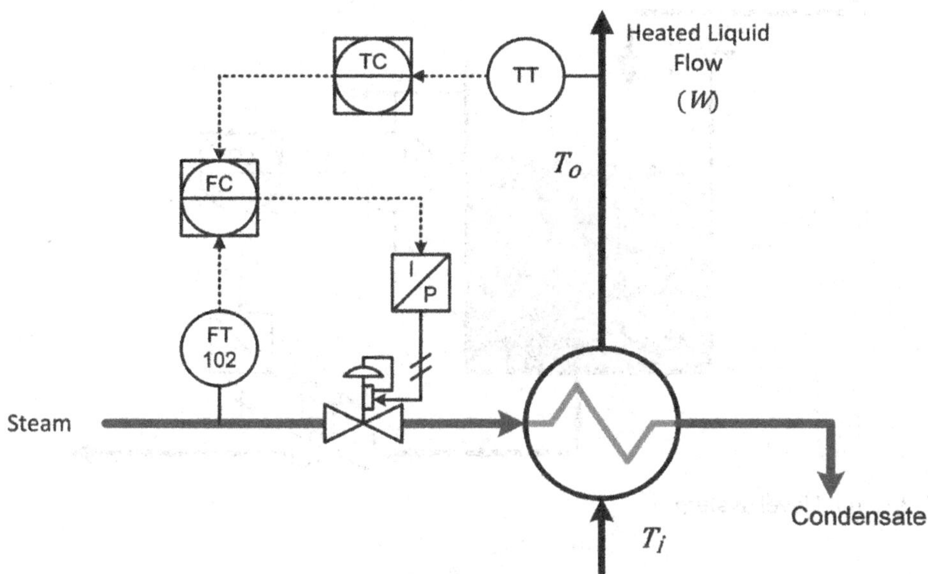

Figure 20-2. Heat exchanger system.

Equation 20-2 is nonlinear if we assume that W varies and because we know that T_o varies. But if we have a steady state involving one value of W, consider what happens if we double W. Using these two situations as example base cases, for each one, consider what happens when T_o is below set point by 5°F. The same error exists, and the same signal goes to the control valve. But in the case where W is doubled, we need twice the increase in steam. We really need an equal-percentage increase in steam, which means that we really need true equal-percentage trim and behavior from the control valve. When this happens, the total combination of the nonlinear exchanger and the nonlinear equal-percentage valve becomes linear.

Now consider the liquid level control system in Figure 20-3.

This process (tank system) is linear. When the level is lowered by 1 in, for example, we get the same signal to the valve no matter where the set point is. That is good because we need the same change in outlet flow because the tank sides are straight. The system is linear, and for the combined system to be linear, we need linear trim and behavior in the control valve.

The process itself may have nonlinear characteristics that are inherent and cannot be designed around. Consider a pH process, for example. A titration curve for a strong acid and base is shown in Figure 20-4. Even where the curve appears to be a straight vertical line (between 2 and 12 pH), a blowup of this curve in that region yields additional S-shaped regions (nonlinearities). These major nonlinearities make a pH process difficult to control, making pH control a subject of its own.

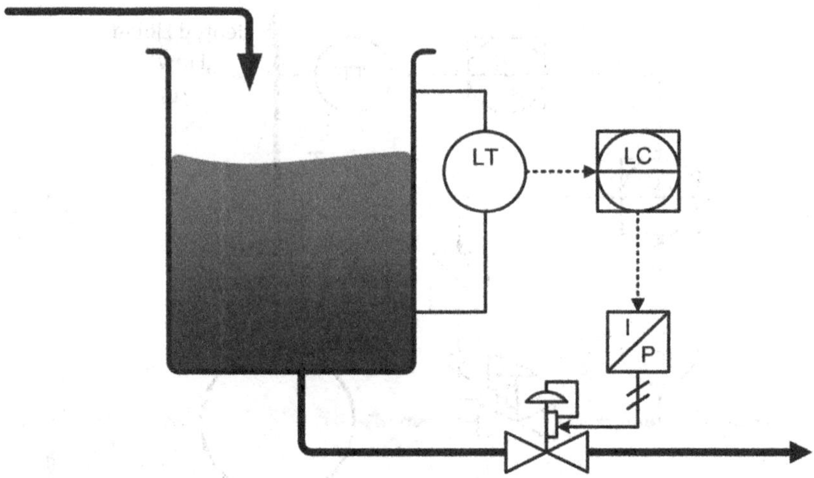

Figure 20-3. Liquid level system.

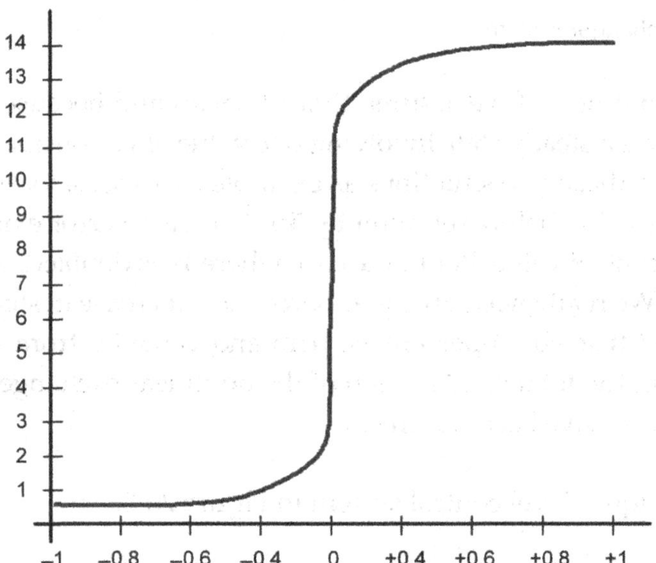

Figure 20-4. Titration curve.

In Chapters 5 and 19, we reviewed a very special type of nonlinearity, dead time, and learned that it created the need for special control considerations. Transmitters and sensors themselves can also create special nonlinear problems. For example, differential-pressure flow sensors transmit a signal that is proportional to flow squared, and in some cases, it is desirable to use a square root extractor in connection with such transmitters. Many transmitters (e.g., chromatographs) operate in definite cycles, and their output is not continuous but instead consists of a series of discrete data points.

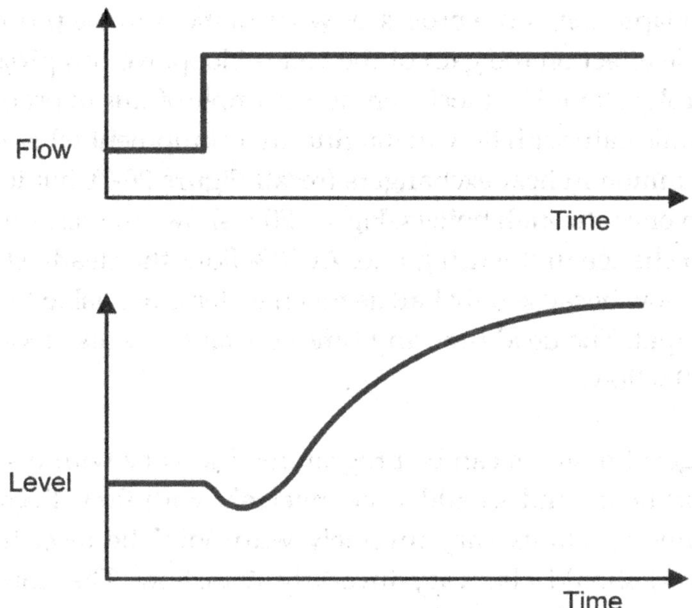

Figure 20-5. Inverse response of boiler drum level.

Some processes even exhibit a nonlinearity of an "inverse" nature, such as shown in Figure 20-5. This is the response of a boiler drum to an increase in feedwater flow. What happens is that cooler water flows into the drum, some steam bubbles are condensed, the apparent level falls, and then the level rises as more new water enters and new steam bubbles are formed.

There are many more examples. Nonlinearities can sometimes be minimized by good analytical design, by wise equipment selection and specification, or by the use of special compensators. However, sometimes they are unavoidable and must be dealt with using special control techniques.

20.4 Adaptive Control

Suppose that we compensate for nonlinearities by introducing suitable nonlinear functions at the controller input or output, for example, or by selecting a particular valve trim to offset process characteristics. Neither of these is thought of as adaptation because these functions (compensation) are fixed. Adaptive control is based on the development of a compensating adjustment through knowledge of a disturbing factor or on the basis of the loop response itself. A system that adapts itself on the basis of a measurement of a disturbing factor is referred to as a *programmed system*. A system that uses a measurement of its own performance is termed *self-adaptive*.

In programmed adaptation, those processes where a measurable process variable produces a predictable effect on the gain of the control loop, we can program compensation into the control system. The most common example of this in process control is the variation in dynamic gain with flow in longitudinal equipment where no back-mixing is evident. It is common in heat exchangers (recall Figure 20-2), but it is a particularly severe problem in once-through boilers. Figure 20-6 shows the response of steam temperature to a step change in the firing rate. At 50% flow, the steady-state gain is twice as high as a 100% flow because only half as much water is available to absorb the same increase in heat input. The dead time and time constants are also twice as great when the system is at 50% flow.

The controller algorithm gains can be programmed to vary with the fraction of full-scale flow. Proportional band should vary inversely with flow. Because the process dead time and time constants vary inversely with flow, the reset time (T_I) and the derivative time (T_D) should also vary inversely with flow. The flow-adapted three-mode algorithm becomes the following:

$$CO = \frac{100w}{PB}\left(e + \frac{w}{T_I}\int e\,dt + \frac{T_D}{w} \times \frac{de}{dt}\right) \tag{20-3}$$

where

CO = controller output
PB = proportional band
T_I = integral time
T_D = derivative time
w = flow rate
e = error signal

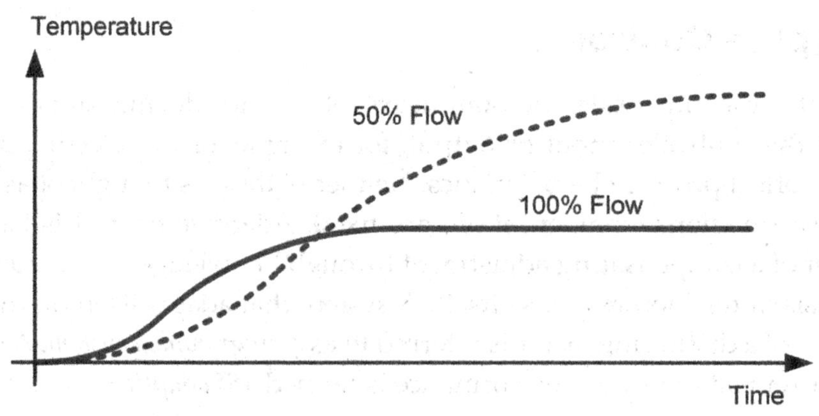

Figure 20-6. Boiler response to firing rate change.

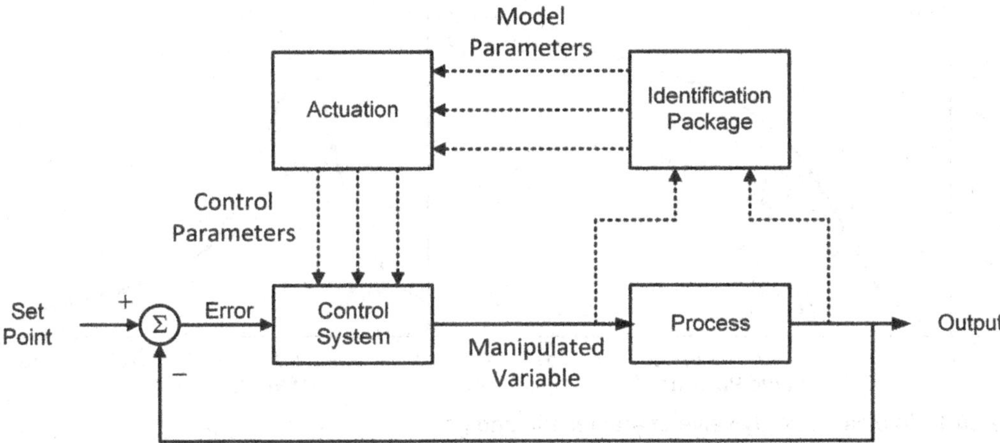

Figure 20-7. Self-adaptive system.

The principles involved in such programmed adaptive control can be extended to other process situations as well.

On the other hand, self-adaptive systems must sense variations in plant behavior. This is referred to as *identification*, and it is usually a partial measurement because total identification is either too complex or impossible. After identification, the adaptive device or system must be able to adjust some controller parameter or function. This is referred to as *actuation* and is illustrated in the schematic diagram in Figure 20-7.

The identification, actuation, and control functions of a system are limited only by the skill and ingenuity of the system designer. They may be simple or complex, linear or nonlinear, continuous or discontinuous. In a real sense, programmed adaptive control is feedforward in nature, whereas self-adaptive control is feedback.

A final note: The definition of adaptive control we have used here is broad compared to other definitions you may encounter. Some authors restrict the use of the term *adaptive control* to systems that measure, and attempt to control, some index of performance. In such cases, systems like those described in the preceding paragraphs are called *nonlinear control* systems.

20.5 Adaptive Gain Control

Adaptive gain control is probably the most commonly employed self-adaptive control technique. For example, all jet aircraft have adaptive gain control so the controls will always "feel" the same to the pilot whether the plane is at sea level or at 35,000 ft and whether the plane is empty of cargo or full. There are many different techniques for maintaining the product of all the gains around a loop as a constant.

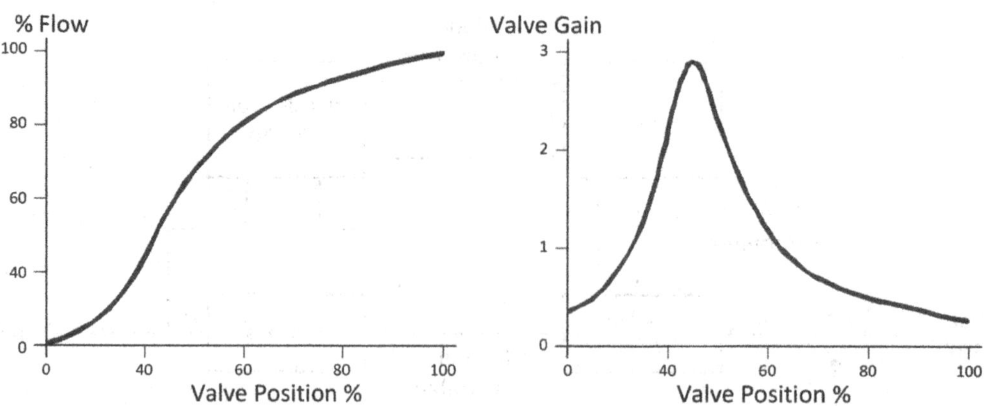

Figure 20-8. Typical butterfly valve characteristic and gain.

When it is possible to measure or infer the variable gain of a component, it becomes possible to adjust other gains to compensate for those variable gain changes. For example, control valves often have nonlinear gains, as illustrated in Figure 20-8. If such a curve is known, then the gain of the valve can be determined by measuring stem position. By neglecting valve dynamics, it is possible to change the gain in the control algorithm to keep the product of the controller gain and the valve gain constant.

Note that such compensation strategies can be computationally complex.

A prevalent technique for adaptive gain control is demonstrated in Figure 20-9. The objective is to maintain the product of the controller gain (K_C) and the process gain (G_P) at a constant value. The method is based on the following: As the loop's total gain $G_P \times K_C$ increases, the transmission of high-frequency noise around the loop will also increase. As the loop gain increases, the amplitude of low-frequency noise decreases and the amplitude of high-frequency noise increases. This in turn is reflected in the error signal. The system includes a high-pass (HP) and a low-pass (LP) filter, and they are used to change the controller gain (K_C).

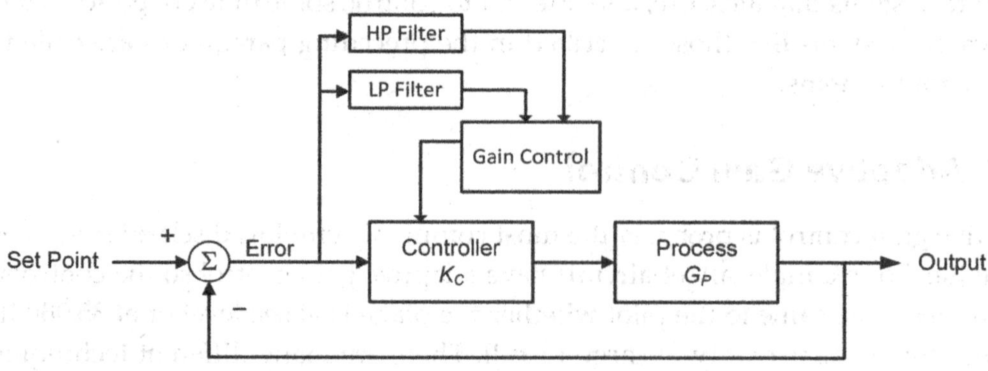

Figure 20-9. Adaptive control.

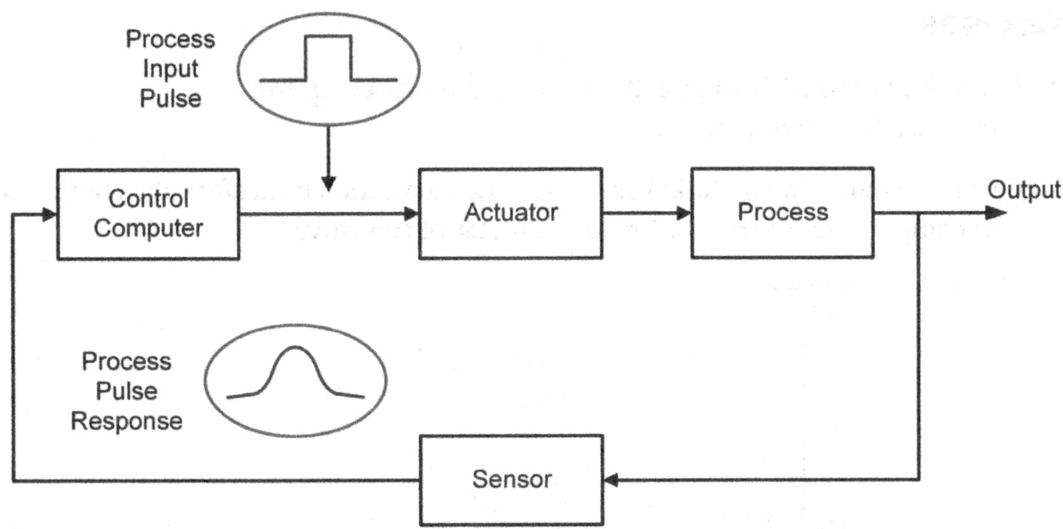

Figure 20-10. Pulse testing a process.

20.6 Three-Mode Adaptive Tuning

The concepts of adaptive tuning can be extended to adjust all three PID parameters in the controller algorithm. For example, as part of a computer control system, it is possible to have the computer periodically pulse the system, identify process characteristic coefficients, and then calculate parameters to tune the controller. This is illustrated in Figure 20-10.

If the process is characterized as a first-order lag plus dead time, the process reaction curve can be used to determine the process parameters: K, θ, and τ. These parameters can then be related to proportional gain, reset time, and derivative time by way of tuning maps.

Most control systems vendors now supply adaptive-tuning hardware and software, and their literature is a good source of current practice.

20.7 Summary

Adaptive tuning, also referred to as *self-tuning, auto-tuning*, or *adaptive tuning*, is a technology that has made rapid strides in the past few years. Although the theoretical concepts underlying it have been around for some time, it is only recently that the algorithms and the computing power required have become inexpensive enough for calculation-intensive self-tuning techniques to be economically viable in commercial products. Today, practically all control systems on the market include some type of self-tuning.

Exercises

20.1 Given Equation 20-1, what is the steady-state process gain? How does this vary with flow?

20.2 Given the following installed characteristic curve for a butterfly valve, how could you compensate for the nonlinear character of the valve?

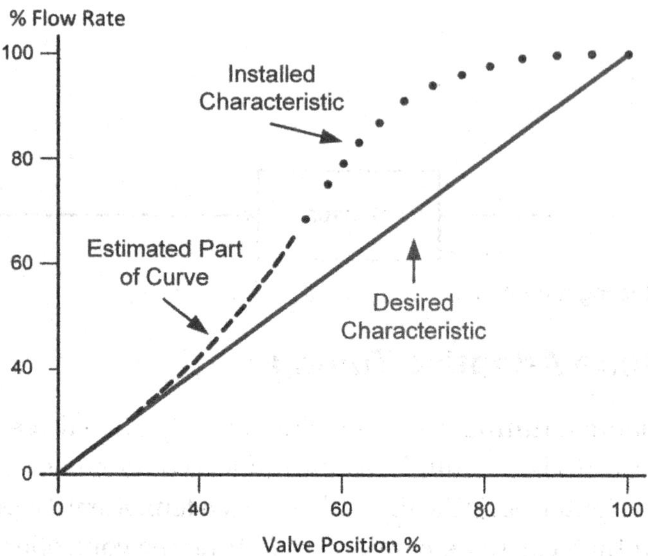

21 Control System Architecture

This chapter focuses on how individual control loops are assembled into larger control systems. The structure of these larger systems is also illustrated, and some insight into their advantages and disadvantages is provided.

Process control and automated systems have evolved from a simple to an incredibly sophisticated level. To understand the architecture and functional layout of today's process control and automated systems, it is useful to know how they have evolved over time. A brief review of the history will demonstrate how decisions made in the design of early process control systems have impacted today's systems.

Learning Objectives

- Identify the basic hardware components of control loops
- Identify the methods and structure that interconnect components into a control system
- Describe a direct digital control (DDC) system
- Describe a supervisory control and data acquisition (SCADA) system
- Describe a distributed control system (DCS)

21.1 Basic Components

Most control systems consist of five basic components:

1. Sensors and transmission components
2. Discrete devices
3. Final control elements
4. Controllers
5. Display/recording components

We will discuss each of these briefly in this section. At a minimum, a basic control loop consists of items 1, 3, and 4.

Sensors and Transmission Components

Temperature, flow, pressure, and level are the four major variables measured in process control systems. However, many other variables, such as pH, speed, position, humidity, composition, and so on, may also be measured. The sensors that make these measurements may be simple or very complex. They are connected from their field location to a display, record, and/or control location that may be nearby or remote. The transmitted signal may take many different forms, but 4 to 20 mA and digital transmission over wires, optics, Wi-Fi, or radio are the most established standards. Most variables have their own unique twisted-pair wire connection, but signals may also be multiplexed and/or transmitted via a fieldbus structure.

Many sensors are intrinsically digital, such as on-off devices. The trend for nondigital sensors is toward providing digital output capabilities. Also, many sensors are now being made "smart" with an internal microprocessor, providing a facility for remote or self-calibration and data logging. Digital sensor output can be converted back to analog for transmission on a 4 to 20 mA wire, but there is a loss of accuracy and repeatability.

Discrete Devices

Discrete devices, such as indicators and switches, are a part of practically all control installations. Indicators provide process status and alarm conditions. Switches are used to start up and shut down equipment and also to initiate proper sequencing of batch processes.

Final Control Elements

Control valves are the most prevalent final control element in most process control loops. A valve usually has a pneumatic actuator and an optional positioner to adjust the valve stem. Signals from the controller to the final control element are usually 4 to 20 mA, and these are converted to an air pressure in the range of 3 to 15 psig by a current-to-pressure (I/P) converter to operate the actuator. Most valve positioners are equipped with an I/P converter. Some valve positioners accept digital signals directly, and the trend is moving increasingly toward this direction. Other types of final control elements include variable-speed drives for pumps and motors and on-off devices. Appropriate signal conversions are needed to convert the signal being transmitted from the controller to the form needed by the specific actuator.

Controllers

Analog and pneumatic controllers were the standard control devices for many years. Many of these controllers had digital interfaces that enabled them to be switched from manual to automatic. In practically all cases, however, the tuning of the controller was done manually.

Most controllers today are microprocessor-based and are designed to handle a single loop, but some can handle multi-loop arrangements. All controllers provide standard proportional-integral-derivative (PID)-type control, but more advanced algorithms are also available. Many controllers include provision for automatic tuning and signal scaling. Interfaces to other control system equipment are typically via digital interfaces, usually operating in conjunction with a fieldbus.

Display/Recording Components

Older and mainly analog control systems were oriented toward using large panels to display and record data. Typically, these displays were located in a separate control room that also contained termination racks for most field wiring and perhaps even a separate computer console for control-related purposes. Often, the data display and recording equipment were shared among multiple loops.

Most modern control systems are highly digitally oriented and use flat panel display units to display information. The human-machine interface (HMI) graphics that are used to facilitate this display can be very detailed. Operators input data or commands to the system through a keyboard, mouse, light pen, or touch panel. Multiple video display units are often used. These digitally oriented operator stations can be centrally located or distributed throughout the plant. In some systems, parameters can be viewed through portable tables and/or smartphones.

21.2 System Components

Now that we have described some of the basic components that make up a control loop, we will look at the system components. The four basic system components are:

1. Computer systems and storage
2. Software
3. Field network
4. Local area networks (LANs)

Computers and Storage

When computers started to be used in process control, typically there was one central machine that did all computations and data storage. Today, wherever computational power is required, a computer of some sort is installed. Small, special-purpose computers are connected with larger machines over LANs. Programmable logic controllers (PLCs), first used in discrete manufacturing in the automobile industry in 1970, have increased in use in process control automation. They have replaced hardwired electromechanical relays and provide greater flexibility without the need for rewiring control relays, Most modern PLCs also have PID capabilities.

In addition to the normal storage needed by individual computers, there is also a need for bulk storage. This bulk storage is necessary for the following tasks:

- To archive plant operating records
- To store plant design data and flow sheets
- To maintain operating instructions, procedures, and recipes
- For operator training facilities and procedures
- For troubleshooting and/or design studies

The need for storage always grows, and the cost drifts lower as new technologies become available. Today there is a trend toward using the cloud for bulk storage.

Software

System operating software is usually vendor supplied. With it, the user can easily implement a control strategy, but the software itself cannot be (and should not be) modified by the user. The vendor typically supplies a programming function that allows the user to customize control applications. In addition, special software for statistical process control, artificial intelligence, and optimization is now available, as is database software. As of this writing, most software runs under the Windows operating system.

Networks

Early field networks were based on a twisted-pair copper wires, or on the same wires used for 4 to 20 mA signals. Individual wire pairs were often used for each variable, and power to operate the field device was often provided through this same pair. This structure results in the use of literally thousands of pairs of twisted wires, which can cost more than the control equipment itself. The cost is driven even higher by the need for explosion-proof or intrinsically safe installations and/or the need to prevent ground loops.

Because of the high costs of such field wiring, the increasing load placed on transmission systems, and the need to transmit digital signals directly, a standard digital communication system was developed.

Modbus, originally designed in the mid-1970s by Modicon, is considered to be the first industrial communication protocol using a bus structure to interconnect multiple devices. It was a way to link an intelligent device with PLCs using a simple client-server protocol. The idea of a bus structure and a client-server protocol enabled multiple devices to be connected using the same interconnecting wires (see Figure 21-1).

Fieldbus is the term used today to describe the digital replacement for 4 to 20 mA as the method for communicating between the elements of a distributed process control system. Fieldbus must provide for two-way communication that allows full access to intelligence located in field devices.

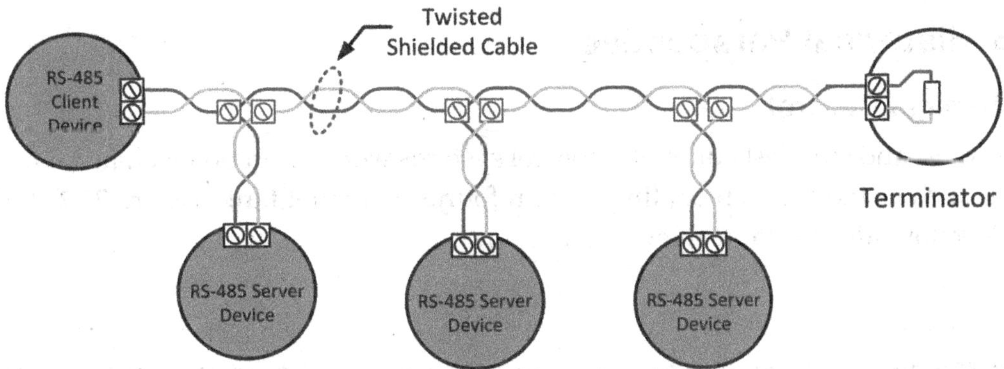

Figure 21-1. Simple bus structure.

The characteristic requirements of a digital fieldbus are:

- It must be secure, and this is usually accomplished through redundancy.
- It must have the capacity to recover from signal loss typically due to electrical noise found in manufacturing environments.
- It must be fast, with response times from microseconds to milliseconds, and it must be capable of powering field devices over long distances, although power can also be provided separately.
- It must be able to provide electrical isolation and intrinsic safety if needed.
- Finally, there must be protocol support for the elements of a real-time operating system to synchronize the time-critical functions in many devices that are only connected by a communications network.

Several proprietary fieldbuses have been on the market for some time, but there was an obvious need for standards. The International Society of Automation (ISA) started an effort in the late 1980s to develop such standards, and they are now a reality. For example, see ANSI/ISA-50.02-1992, *Fieldbus Standard for Use in Industrial Control Systems*.[1]

LAN

Multi-vendor equipment always exists in large computer control systems. This equipment is usually interconnected through a LAN. Communicating between these various vendors' systems can present problems. The solution is for communication to be based on a common protocol. Several vendor proprietary protocols exist, but today no single standard appears likely to emerge. With different LANs using different protocols, interconnecting devices such as repeaters and layered gateways are used to interconnect LANs.

21.3 Historical Perspective

Mechanical Control

The earliest and simplest automatic control systems were based on mechanical mechanisms, such as the mechanical link system found in a toilet tank. Figure 21-2 shows a simple automatic mechanical level controller.

---

1. ISA-50.02-1992, *Fieldbus Standard for Use in Industrial Control Systems – Part 2: Physical Layer Specification and Service Definition*. Withdrawn 2008. Replaced by IEC 61158.

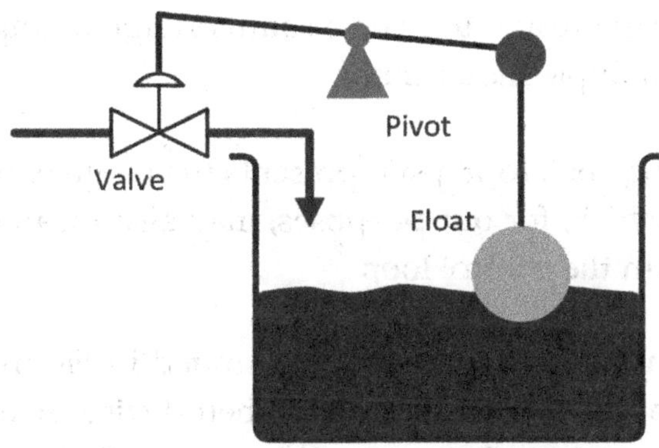

How it works

As the level in the tank increases, the float rises, causing the pivot arm to rise on the float side and lower on the valve side. The lowering of the pivot arm on the valve side causes the valve to close an amount proportional to the rise of the liquid in the tank. The closing of the valve reduces flow into the tank, which reduces the rate of rise of the liquid level. If the level goes down, the reverse takes place.

Figure 21-2. Simple mechanical control system.

This mechanism includes a float that sits in the tank in which the liquid level is being controlled. The float is attached to a mechanical arm attached to a pivot arm, which is in turn connected to a fixed pivot and then connected to a valve mechanism. This simple mechanical device does a fairly effective job of controlling the level in a tank and requires no power source. The primary disadvantage of such a control system is that it is only practical in local environments. It cannot be managed remotely by an operator.

Pneumatic Control

After basic mechanical control systems came the development of pneumatic control systems (Figure 21-3 shows a typical configuration), the first powered systems. They

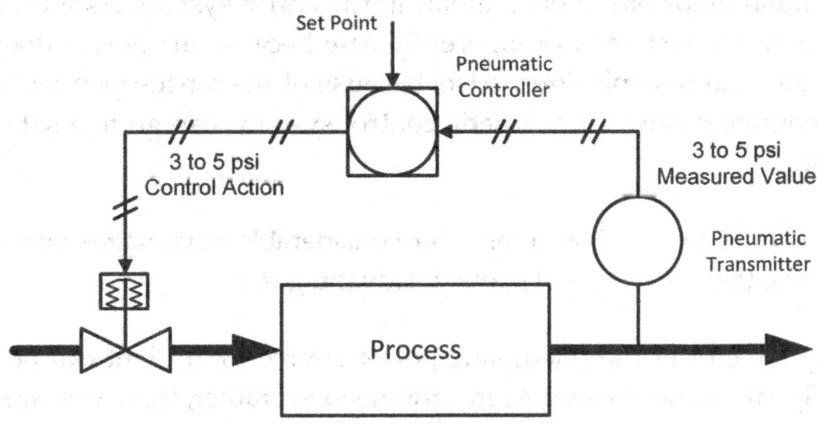

Figure 21-3. Pneumatic control.

were powered by air pressure and utilized a common signal range of 3 to 15 psi, which was agreed on across multiple industries.

This means that the range of 3 to 15 psi represents the range of both the measurement device and the valve, which, for our purposes, may also mean other devices used to adjust the process within the control loop.

Using a standard signal range like the one illustrated in Figure 21-3 enabled instrument, controller, and valve suppliers to include their devices in a pneumatic loop with equipment from other suppliers. In a sense, this was the first standard "fieldbus." The origin of PID control can also be traced to pneumatic devices.

A 3 psi signal to a flow controller means that 0% of flow is going through the measurement device. A 9 psi signal means that 50% of maximum flow is going through, and a 15 psi signal means that 100% of the possible flow is going through the device. A 3 psi signal from a controller to a valve means to close the valve completely, a 9 psi signal means to open it to 50%, and a 15 psi signal means to open it all the way. Setting the low range of the pneumatic signal at 3 psi rather than 0 psi enabled operators to distinguish between a low measurement condition and a system power failure.

Pneumatic control systems offered important advantages over the mechanical systems they were designed to replace. One of the greatest advantages was that the pneumatic signals could be transmitted over a distance; the controllers could be located away from the measurement instrument and valves, enabling centralized control panels where operators could supervise multiple control loops from a single control panel. Operators could clearly see how one loop affected another loop in the system, thereby providing better overall control by enabling them to monitor and adjust individual loops based on evaluating the entire system. As with mechanical systems, pneumatic systems are inherently safe because air power does not cause sparks that can lead to explosions. Also, because of the air-to-open and air-to-close design of pneumatic valves, pneumatic control systems can go to a safe state when power is lost.

Although pneumatic control systems offer considerable advantages over mechanical control systems, they also present some disadvantages.

For example, the cost of the pneumatic power source for a plant can be quite high. Pneumatic signals do not propagate instantaneously; rather, there is a time delay that is proportional to the length of the transmission line, and any delay in a control loop introduces difficulties in controlling the process.

Analog Electronic Control

To overcome some of the shortcomings of pneumatic control systems, the industry developed analog electronic control systems. These systems—like pneumatic systems—are analog but are based on an electrical power source rather than a pneumatic one. The accepted analog range for these control systems is 4 to 20 mA. Current is used instead of voltage for signal stability. The 4 to 20 mA signal range is an industry standard for electronic analog systems, as is the 3 to 15 psi signal range for pneumatic systems. Both standards are still used in industry today.

Analog electronic devices can perform the same basic functions as the pneumatic devices. This made it easy for an automation professional who had experience with pneumatic control systems to move to electronic control systems. Electrical signals can travel over much longer distances with effectively little to no propagation delays, making electronic analog control systems much more distributable than pneumatic systems. With these analog systems, a process operator could manage a large number of control loops from a single central control room.

On the disadvantage side, electronic analog control systems introduced electricity into areas of process plants in which a spark might cause an explosion or fire. Devices called *electrical barriers* were developed and often had to be installed into the circuit to reduce the probability of a spark in areas of the plant where that would be a safety risk. Also, because most valves were still pneumatic, plants that installed electronic analog systems typically had to support both pneumatic and electrical power sources in the plant, which resulted in additional installation and maintenance. Finally, although pneumatic systems did not offer any advantage in this area, the computational capability of both pneumatic and electronic analog systems was very limited.

As computers became more available, digital computation became viable in manufacturing operations. Unlike analog systems, digital systems use electrical circuits to simulate the digits of a number. Instead of a signal representing a range, the signal represents the value of a digit within a larger number. This resulted in much more flexible calculating engines than the analog systems could provide.

Digital (Set Point) Control

Computers were still quite expensive and tended not to be as reliable as was required in manufacturing. The first digital computer systems on the manufacturing floor only ensured that set points of analog controllers were optimally set and were able to monitor the process value. The resulting hybrid digital/analog systems were called *set point control* systems (also referred to as *supervisory control* systems or *supervisory* systems); see Figure 21-4.

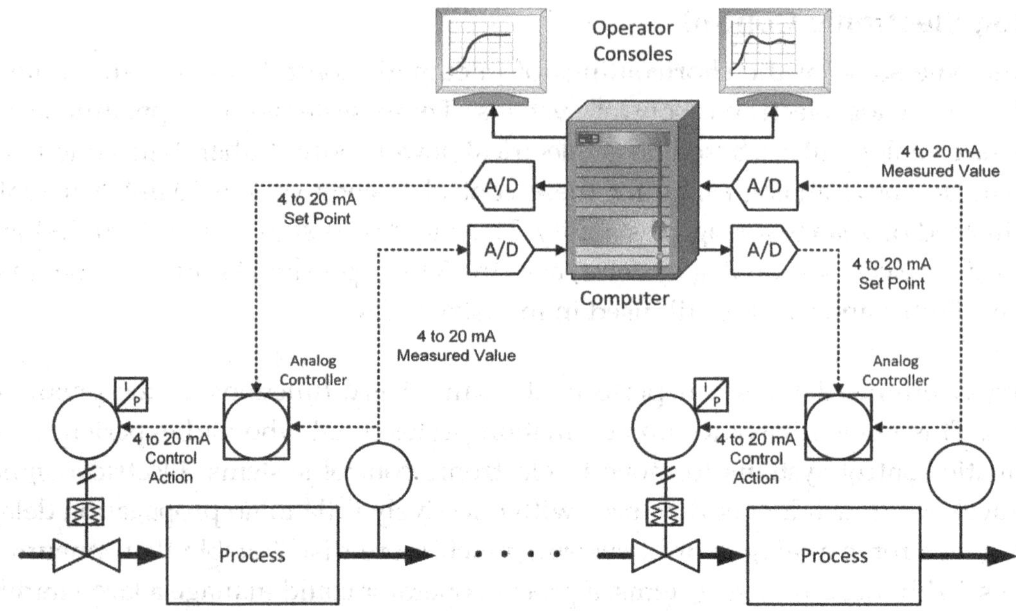

Figure 21-4. Digital (set point) control.

The computer offered the computational capability automation professionals had been looking for, and the analog controllers provided the actual feedback control mechanism. An interface between the computer and the set point input of the controllers was developed, and software in the computer was programmed to facilitate the calculation and downloading of the set points. Ensuring optimal set points enabled more effective and efficient plant operations.

The analog interfaces transferred measurement signals into the computers and output signals from the computers to the valves. The signals from the measurement devices were also analog signals that had to be converted to digital signals for use by the computer. Suppliers developed analog-to-digital (A/D) converter devices to provide this function. On the output side, because valves were also designed for analog signals, digital-to-analog (D/A) converters were developed and implemented.

These hybrid systems offered a level of distributed control performed by analog controllers, which meant the control would not fail if the computer went down. This was important because the early digital computers were not very reliable and failed frequently. Also, with computer displays and computer-based, set-point determination, operators could manage a much larger section of the plant than was previously possible.

On the downside, SPC systems were expensive because they required a digital computer and an analog control system, both of which were quite expensive by today's standards. The computer's computational capability was limited to the functionality above the control level and did not penetrate the control layer. Also, despite the computational capability of these systems, expertise and software for optimizations were in their infancy and little true set point optimization was actually implemented.

Direct Digital Control Systems

With the disadvantages of SPC systems and the development of digital PID control algorithms, many manufacturers incorporated the PID computations directly into the computer. The computer then could perform both PID control and advanced control set-point optimization functions. These systems were initially referred to as *direct digital control* (DDC) systems. DDC is illustrated in Figure 21-5.

With DDC, system software in the computer replaced the control functionality previously provided by the analog controllers. DDC systems also offered advanced calculation capability at the control and set point optimization levels, allowing for creative control approaches.

The largest downside of DDC systems was the huge consequence of a system failure. Because digital computer technology was expensive during this phase of process control system evolution, the tendency was to put as much control as possible into a single computer, despite the risk. Some manufacturers tried a backup analog control system, but this solution was also very expensive. As the cost of digital computer technology continued to decline, and reliability increased in proportion, the DDC approach became the norm for the design of control systems.

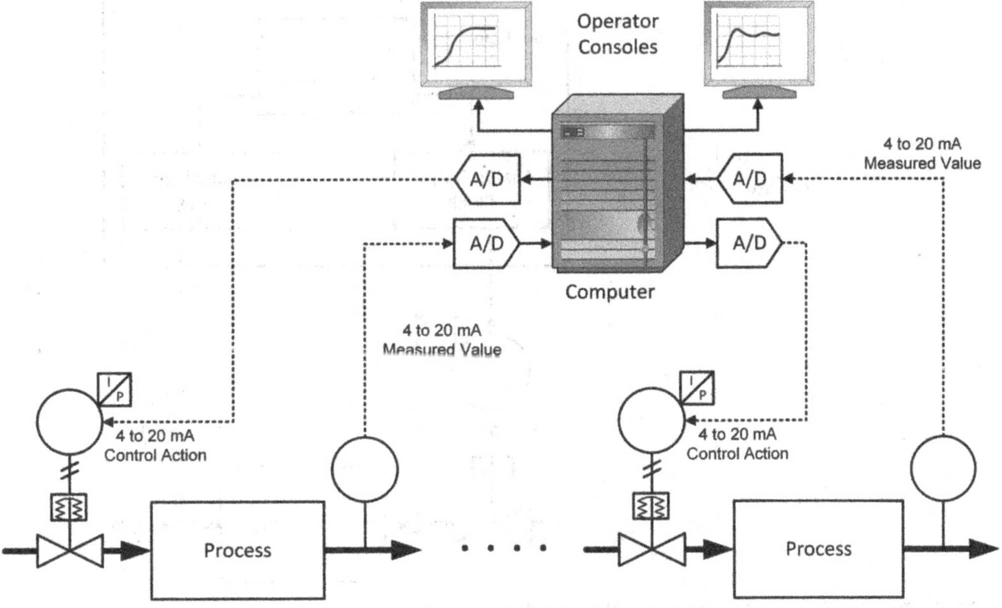

Figure 21-5. Direct digital control.

One of the major challenges in moving from pneumatic and electronic analog systems to digital computers for process control was preserving the high-value knowledge of control automation professionals while undergoing this transformation. Control automation professionals were well trained in the development and implementation of control strategies with analog control systems. Analog control systems were designed and developed through the selection of hardware components built to perform specific functions within a control strategy. For example, to develop a simple feedback control loop, an automation professional might select an electronic analog instrument, a proportional and integral analog controller, an electronic analog-to-pneumatic signal converter, and a pneumatic valve. More complicated control strategies were developed by selecting the necessary additional hardware components and connecting them in the appropriate order. Control automation professionals developed significant expertise over years of working with analog systems, based on this hardware component approach.

With digital computers, the control functionality is done using software rather than the hardware components of analog systems. As digital computer technology became more prevalent, most automation professionals were not familiar with computer programming languages. To deal with this issue, automation software developers devised a block programming approach that emulated the hardware components of controllers.

The block programs were developed for each of the major functional entities typically found in an analog control system. See Figure 21-6. Block programs enabled

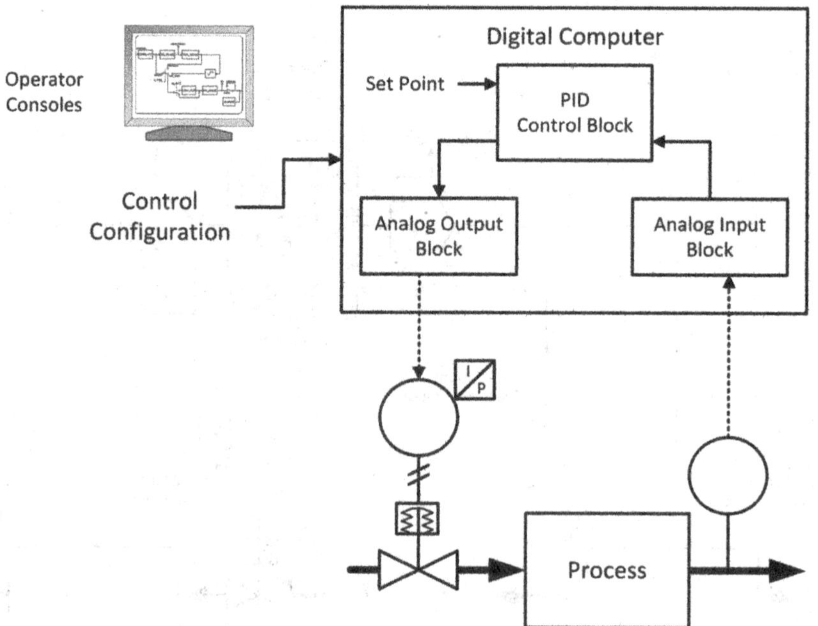

Figure 21-6. Process control software—the block concept.

automation professionals to construct control schemes using the knowledge they had developed with analog systems. They could sit at a computer terminal and work with configuration software that allowed them to build PID blocks, analog input blocks, analog output blocks, and a variety of other blocks, and connect them through software connections.

21.4 Distributed Control Systems

The introduction of digital PID controllers, PLCs, and fieldbus networks, and the operators' acceptance of using the process control block to configure computer programs, led to the evolution of a new class of systems: distributed control systems (DCSs).

DCS Architecture

The DCS architecture offers the best features of both local control and centralized control. In a DCS, PID controllers are provided locally to individual control loops, additional PLCs provide discrete control groups of equipment, and all devices are networked to one or more operator stations in a central location through digital communication links.

Honeywell introduced the first DCS, the TDC2000, in 1975, setting a new direction for control system design that would take industrial automation through the latter decades of the twentieth century. Figure 21-7 shows a typical DCS configuration.

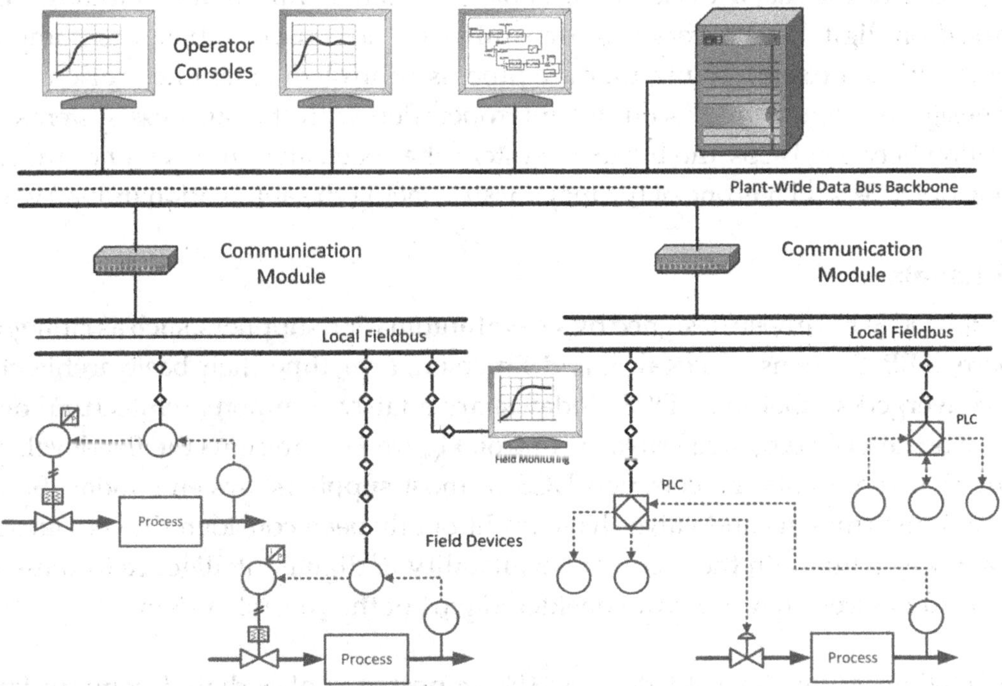

Figure 21-7. Typical DCS.

The DCS offered several advantages over the previous digital control systems. First, DCSs are much more distributable than previous digital control systems were. They offer two forms of distribution: geographic and functional.

In geographic distribution, the computer modules of the system are networked together and distributed to different locations in the plant. In functional distribution, the computer modules are designed to perform specific functions within the overall system. For example, one computer module of a DCS might be dedicated to the function of process control while another might be dedicated to the function of providing an operator interface. The combination of geographic and functional distribution provided by DCSs resulted in discrete control using PLCs and smart transmitters to monitor functional parameters. PLCs also gained PID functionality. The DCS forms an extremely flexible system design that can meet a wide variety of requirements across industries.

Most DCSs also allow much more centralized supervision over much larger sections of a process plant, often resulting in a single control room for very large plants. By distributing the functionality across multiple computer modules, the consequence of a failure in any one module is much more contained. In addition, DCS suppliers designed redundancy and fault tolerance into critical modules in the DCSs so the failure of a component would not result in any loss of functionality.

Perhaps one of the major disadvantages of DCS architecture is that, although DCSs are based on digital computers that are similar to those used in business computing systems within the same plants, they are process-control and plant-floor oriented, and their design has tended to discourage interoperation with the business systems. This separation between DCSs and business systems has been an issue ever since the introduction of DCSs, and it is one of the major issues facing the automation industry today.

DCS Levels

Although DCSs have been designed by several automation suppliers, such as Honeywell, Foxboro, ABB, Siemens, Yokogawa, and Emerson, over time their basic architectures have converged so that most DCSs today share a fairly common architectural design that can be described by five levels of control. Figure 21-8 presents the five-level architecture that represents the common DCS in most suppliers' systems today. Starting at Level 1, instruments and valves have traditionally been considered to be outside of the DCS scope, but with the increased availability of digital intelligence in these (and other) field devices, they are now considered part of the control system.

With intelligence in the field devices, the importance of a digital communication infrastructure to connect them to the DCS has become apparent. The resulting digital

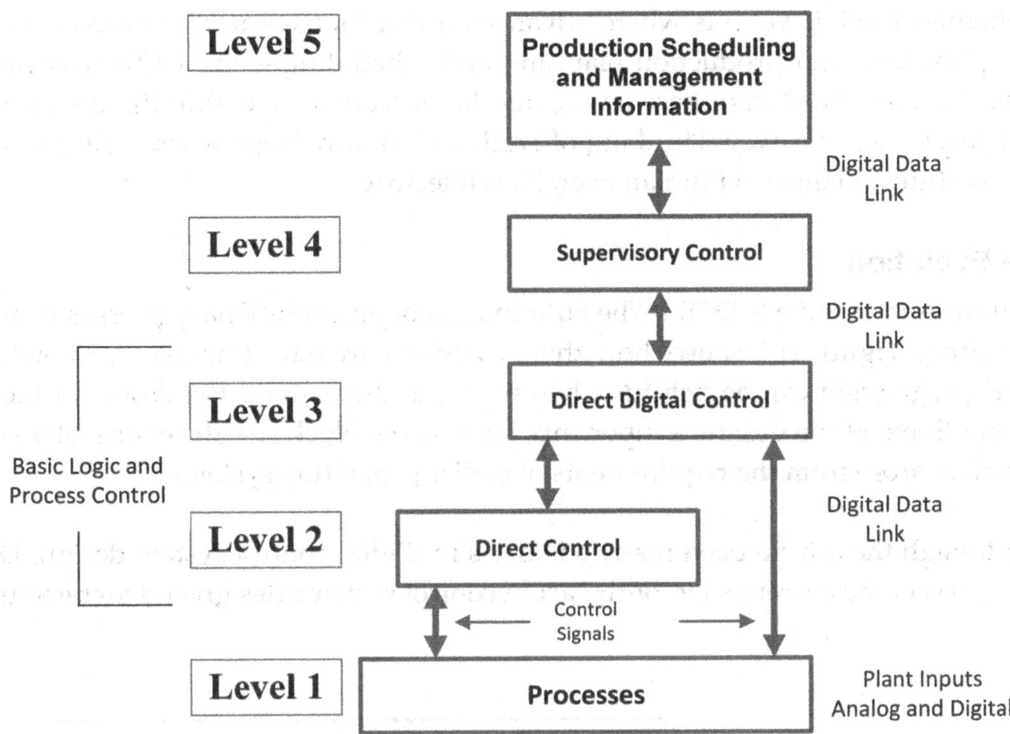

Figure 21-8. Common DCS architecture.

networks at this level have been classified as fieldbuses. Several proprietary vendor-developed fieldbuses were originally introduced to the marketplace, but they only allowed the field devices to interoperate with the systems if they were provided by the same vendor. An industry standard fieldbus was developed by the FOUNDATION Fieldbus, a digital fieldbus standardization organization, in an attempt to enable intelligent field devices from any vendor to work with DCSs from any vendor. The development of a standard digital fieldbus required considerably more effort than that required for the analog fieldbuses because the addition of intelligence to field devices meant that there was much more information of interest to the DCS.

Level 2 in the architecture is the input/output level that provides the interface to the intelligent and non-intelligent field devices. Even though there are advantages to having intelligence in the field devices, millions of devices installed in process plants prior to the availability of intelligent field devices remain—and will remain for years to come. Intelligent and non-intelligent devices must be brought into the DCS, and the input/output level addresses this.

The next level is control Level 3 where basic process and logic control occur. Above the control level is the supervisory control level, Level 4, at which the operators interact with the system through computers to supervise the operation of the plant.

The highest level, Level 5 is where advanced applications—such as process historians, optimizers, and production planning and scheduling—run. DCSs may not be physically partitioned into these levels, but the functionality within these systems is often partitioned this way. Thinking of DCSs in this way helps when trying to align required functionalities within an overall architecture.

DCS Evolution

The transition to today's DCS is the culmination of an evolutionary process over the last century. Figure 21-9 shows how this evolution may have transpired. Knowledge of this progression can be helpful when trying to understand the design of today's systems. Some of the design components, such as the block structure for system configuration, arose from the requirements of earlier generation systems.

Even though there have been many advances in digital control system design, DCSs still represent the current state-of-the-art in control systems design and architecture.

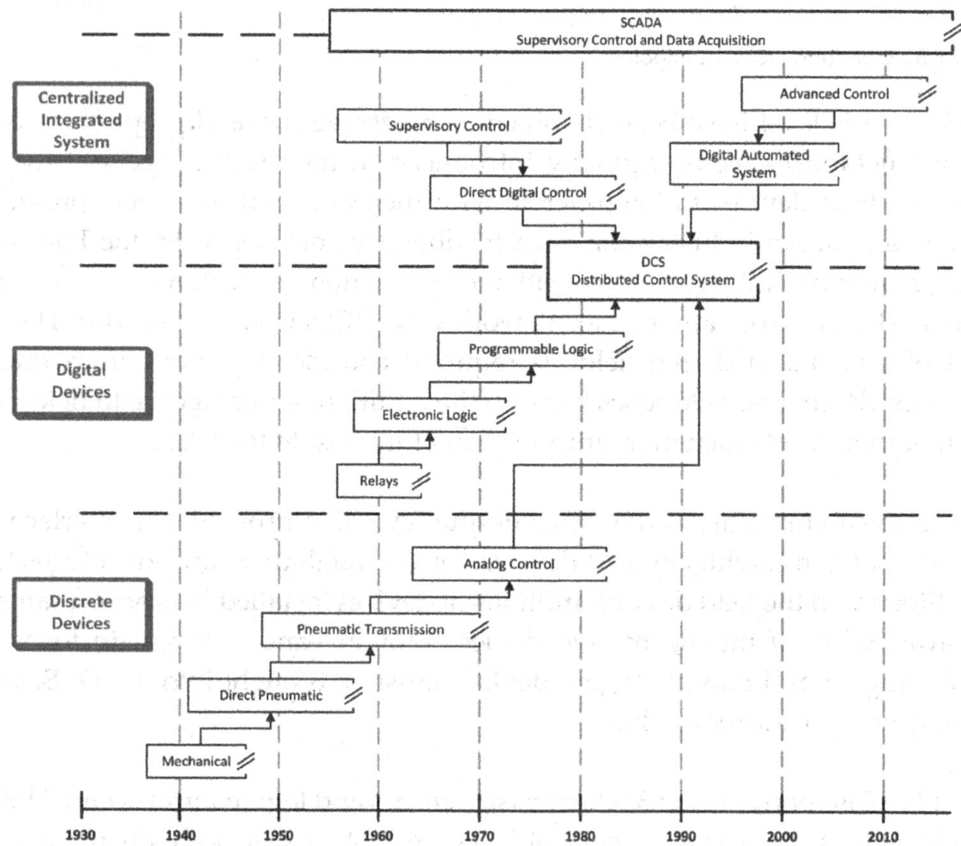

Figure 21-9. Control system evolution.

DCS Function

Control strategies of particular tasks within a manufacturing plant are usually done locally. In a DCS, each station is controlled independently but at the same time interfaced or connected to a central control. Within a DCS, we speak of primary and secondary control.

Primary control refers to the control being done locally at the point of instrumentation or actuation. Today this is usually done with a PID controller, PLC, or smart sensor.

Secondary control is what happens in a control room and depends on inputs from the various primary systems. Typical secondary control commands are:

- Change in set point (PID)
- Change in quantity (PLC counters or timers)
- Change in batch composition (multiple set points and/or PLC counters)
- Change in tolerances or quality

Basic DCS functions can be summarized as:

- Networking multiple controls to work together under the supervision of a single control
- Communication between control room and control functions
- Controlling loops (output of one controlling function may become the input or set point of a second control loop)
- PLC control (multiple units)
- Robotic control with PLC-controlled functions
- Special control functions (interfacing to standard computers for report generation, Excel, Word, etc.)
- Alarm and action (local, remote)
- Data logging and statistical process control

21.5 Supervisory Control and Data Acquisition

When you observe the conventional process industry, you see that most plants operate with large numbers of feedback controllers, and occasionally you encounter a particular loop with a specific bit of feedforward control or other sophisticated or advanced control installed. But even with such an extensive control system in place, closer inspection reveals that the supervisory function of the plant usually is only modestly automated, if at all. As an example, the individual plant operators or plant supervisory personnel determine all the set points for the feedback and feedforward controllers. They do this based on past experience and elementary calculations. Experience has shown that often operating personnel undertake this responsibility in a crude, conservative, and slow fashion. The result is that most plants are not operating at their optimal level.

The basic objective of a process operation is to produce a financial return (profit) on investment. This economic return depends on several factors, and the operating strategy is not always clearly perceived or broadly understood by all the decision makers involved. A plant is a complex, highly interactive entity, and often the optimum operating strategy can be determined only through sophisticated analysis and calculations. Clearly, the digital computer should be able to help perform such calculations in a real-time, optimal way. This concept is illustrated in Figure 21-10.

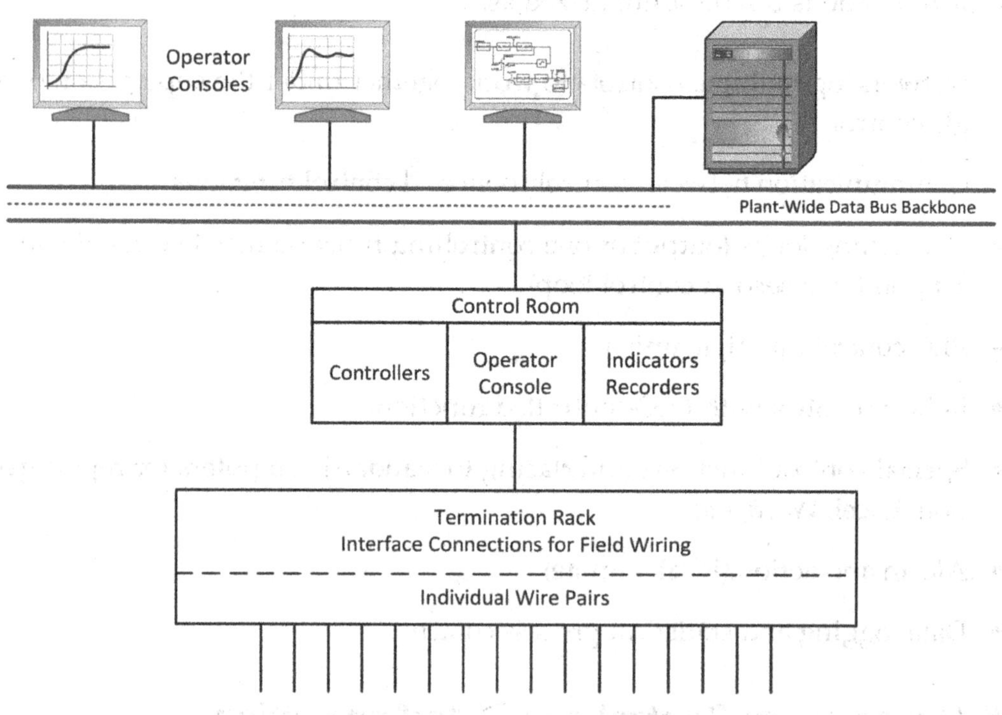

Figure 21-10. Supervisory control and data acquisition (SCADA) system.

The computer can measure significant amounts of information about actual operating conditions in the plant. Additional information may be provided to the computer, including the following:

- The cost of raw materials and utilities
- The value of products
- The composition of various raw materials, products, and intermediate streams
- The current values of variables within the process
- Constraints on the operation of the process
- Specifications on products

A model of the process can be used to relate these various factors, including providing a complete picture of plant economics. This model can then be optimized to determine the best operating strategy for the unit. The result is a specific indication of the optimum or most desirable values for the set points of all the individual controllers within the plant. These set points may be provided simply as information to the operator; that is, the supervisory control system may be open-loop. In other cases, the set points for the controllers may be set directly via a hardwired system. This is referred to as *closed-loop supervisory control.* When statistical process control is integrated into the SCADA system, the set points may be determined using statistical process control methods.

The installation of supervisory control systems has proved to be an attractive concept in process control. It illustrates the fact that the supervisory control function was not automated before digital computers entered the scene, and it often was being done less efficiently than is possible with computers. The economic justification for supervisory control systems is based on the prospect that the control system will produce improvements in process operations. There are, therefore, several situations in which supervisory control systems are prime candidates for installation. These include:

- Plants with large throughput
- Very complex plants
- Plants subject to frequent disturbances

The major difficulty in using and installing supervisory control systems is that mathematical models of plants are seldom available. Significant engineering and technical investigations are required before the system may be optimally used to improve plant operations. The incentive to do so is clearly present, because of the significant benefits that can be had.

21.6 Sequential or Batch Control

Note that up to this point, we have concentrated totally on continuous control and have largely ignored sequential batch control because sequential control is discrete and event based.

Sequential batch control is important, however, and there are several ANSI/ISA standards on the subject, for example, ANSI/ISA-88.00.01-2010, *Batch Control – Part 1: Models and Terminology*. The standards focus on protocols, however, and are procedure oriented. If readers are interested in this kind of control, they should refer to these standards for detailed treatment. It is also appropriate to note that PLCs are mostly used to implement sequential control systems. However, their hardware and software capabilities, as well as those of DCSs, are converging, so they are increasingly competitive with one another. This is especially true for larger systems.

21.7 Summary

Today's automated process control systems have their roots in the late 1960s and early 1970s as computers and microcomputer-based controls became more widespread. This then led to the conception and development of DCSs and SCADA being employed to provide central optimization of a manufacturing plant.

22

Digital Systems

Control methods have evolved from mechanical, pneumatic, and electromechanical systems to electronic and digital control systems. Electronic controls have traditionally been implemented with analog components such as resistors, capacitors, and operational amplifiers. However, with the availability of microprocessors, control systems are now being implemented mostly in digital form. The use of microprocessors in digital control systems has created new opportunities in implementing control strategies, but with them come a need for a new body of knowledge that utilizes these processing capabilities.

This new knowledge is commonly referred to as *digital signal processing* (DSP).

Most, if not all, proportional-integral-derivative (PID) controllers and process models used to control or simulate processes in industry today are microprocessor-based. This chapter will examine how computer computations, models, and simulations are implemented in a typical controller and/or computer.

Understanding the fundamental principles and computational aspect required by the control algorithm will help in the selection and use of the equipment.

Learning Objectives

- Understand the fundamental principles of DSP
- Understand the computational aspect of a process controller
- Convert differential equations to difference equations

22.1 Hardware Elements

With the performance and reliability inherent to microprocessors and microcontrollers, digital controllers now predominate. In a digital control system, the control processing and calculations are implemented with a microprocessor and/or microcontroller. However, because the actuator commands from the microcontroller are digital, they may have to be converted into analog signals using a digital-to-analog (D/A) converter and low-pass filter. Similarly, the measurements from the sensor are usually analog and will have to be converted into a digital signal using an analog-to-digital (A/D) converter, as well as an anti-aliasing filter and a sample-and-hold (S/H) circuit. The anti-aliasing filter is a low-pass filter that prevents high-frequency signals from entering the A/D converter. The Nyquist[1] criteria state that the sampling frequency of a signal should be greater than twice the maximum frequency of interest. The S/H circuit ensured that the sampled signal does not change during conversion by the A/D converter.

Note that increasingly, measuring instruments and transmitters, as well as actuators, are also microprocessor-based, and many of these devices are capable of providing and receiving signals in digital form.

Figure 22-1 shows a block diagram of a typical digital control system. The D/A converters convert the digital signal output of the microcomputer, $u(n)$, into an analog signal, $u(t)$, and the A/D converter converts the analog signal $u(t)$ into a digital signal, $u(n)$.

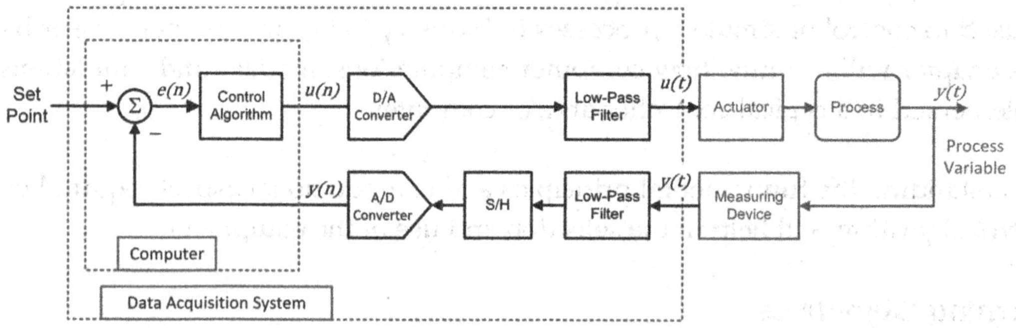

Figure 22-1. Digital control system.

1. Named after Harry Nyquist. Originally from Sweden, Nyquist was an American physicist and electrical and communications engineer who made fundamental theoretical and practical contributions to telecommunications.

Although digital controllers and digital control offer many advantages, they do have a negative side. Digital controllers sample the signal at discrete-time intervals. This limits the bandwidth that can be handled by the controller. The sampling and processing of the signal takes a finite amount of time, adding to dead time and phase delay in the system. In addition, the resolution of the signal is limited by the resolution or word length of the A/D and D/A converters and the processor used.

Digital controllers use discrete steps to represent a signal, which is limited to the word length of the processor. Coefficients or gain constants also must be represented in the limited word length. This discretization or loss of resolution is referred to as *quantization error.* In addition, results of mathematical operations must fit in a limited word length and may lose part of the result because of this limitation. This is referred to as *truncation error.* Both of these errors can cause oscillations and can lead to instability in the process.

Another problem that occurs in digital controllers is overflow of registers. Successive mathematical operations, such as integration, can cause registers to overflow. Registers in most processors wrap around, causing the result of a calculation to go from most positive to least negative, in turn causing the output to reverse directions. Most of these problems occur because their architectures are not designed for signal processing.

Most of these problems occur in processors that have 8/16-bit arithmetic logic unit (ALU)s and registers. This 8/16-bit architecture limits the accuracy of intermediate and final results and generates truncation and/or quantization errors. Lack of proper scaling to maintain the required significant bits can cause additional quantization and/or truncation errors.

Most of these problems can be eliminated with the use of proper DSP techniques and floating-point arithmetic.

In digital signal processing, most algorithms, including control algorithms, can be represented by a difference equation. A difference equation is a formula for computing an output sample at time n based on the present input samples and past output samples. In general, the equation to be evaluated multiple times is of the form:

$$a = a \times r + b \tag{22-1}$$

This is not a standard algebraic equation but rather an assignment. The equal sign in computer language is an assignment, meaning the new value of a is assigned the previous value of a times a constant r plus the value of b.

22.2 What Is Digital Signal Processing?

Digital signal processing (DSP) takes real-world signals such as sound, video, temperature, pressure, and position that have been digitized into a sequence of numbers representing the signal, and then mathematically manipulate the signal to produce the desired output.

The best way to explain DSP is by an example. Consider temperature readings taken at 1 min intervals (Figure 22-2).

From inspection of the data, it is clear that sample 3 and sample 13 are anomalies, noise, or simply incorrect readings. They could also be correct readings that did not conform to the general trend, shown as the solid line connecting the readings.

DSP is concerned with mathematically analyzing (processing) the data (signal). This processing can be identifying the trend or the average over a specific time frame, filtering the signal, removing the anomalies (as shown), or calculating a desired control action based on the signal.

A commonly used signal processing function to remove high-frequency noise is to calculate what is known as a *running average*. This function is similar to a low-pass filter. Running averages are calculated by adding samples (data points) and then dividing the total by the total number of samples used. A running average continuously changes as more samples are collected, and the calculation is repeated for each new sample. For example, a 3 point average can be represented mathematically as:

$$3\ point\ avg\ (3) = \frac{1}{3}[x(3) + x(2) + x(1)] \quad \text{For sample 3}$$

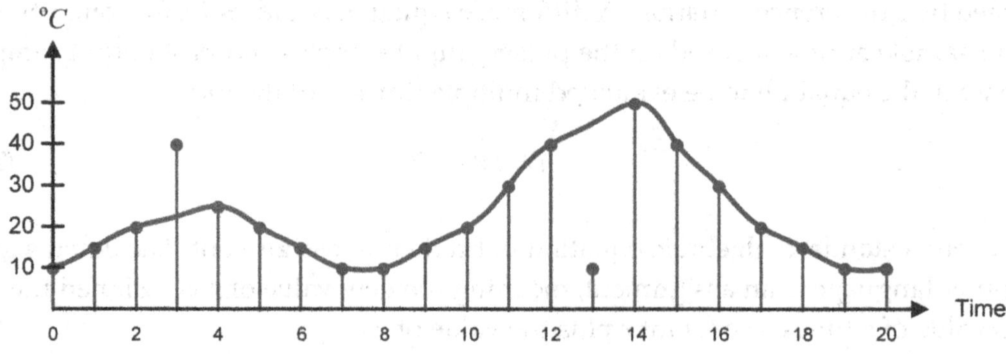

Figure 22-2. Temperature samples.

$$3\ point\ avg\ (4) = \frac{1}{3}[x(4) + x(3) + x(2)] \quad \text{For sample 4}$$

and

$$3\ point\ avg\ (5) = \frac{1}{3}[x(5) + x(4) + x(3)] \quad \text{For sample 5}$$

For the first and second sample, we must assume that the preceding values are the same as the first sample, that is:

$$3\ point\ avg\ (1) = \frac{1}{3}[x(1) + x(1) + x(1)] \quad \text{For sample 1}$$

and

$$3\ point\ avg\ (2) = \frac{1}{3}[x(2) + x(1) + x(1)] \quad \text{For sample 2}$$

In general, we can write the *n-point* average at point *i* using the following equation:

$$y(i) = \frac{1}{n}[x(i) + x(i-1) + x(i-2).\ .\ .\ .\ \ x(n-i-1] \tag{22-2}$$

Evaluating this equation for each consecutive sample can be computationally intensive, but it can be simplified, as will be shown next. Note that for each new calculated average value in a 3-point average, at least two points are always the same and will give the same 2-point average.

Consider next the *n-point* average prior to the ith point:

$$y(i-1) = \frac{1}{n}[x(i-1) + x(i-2) + x(i-3).\ .\ .\ .\ \ x(n-i] \tag{22-3}$$

Subtracting Equation 22-2 from 22-3 gives:

$$y(i) - y(i-1) = \left\{\frac{1}{n}[x(i) + x(i-1) + \cdots\ \ x(n-i-1]\right\} - \left\{\frac{1}{n}[x(i-1) + x(i-2) + \cdots\ \ x(n-1]\right\}$$

Simplifying gives:

$$y(i) - y(i-1) = \left\{\frac{1}{n}[x(i)]\right\} - \left\{\frac{1}{n}[x(n-1)]\right\}$$

or
$$y(i) = \frac{1}{n}x(i) - \frac{1}{n}x(n-1) + y(i-1) \tag{22-4}$$

where

n = number of points to average
i = i^{th} sample in the sequence
$y(i)$ = present output value
$y(i - 1)$ = previous output value
$x(i)$ = present input value
$x(n - 1)$ = previous input value

Equation 22-4 is simpler to evaluate than our original equation and is referred to as a *difference equation.*

Note how this equation conforms to the description of Equation 22-1:

$$a = a \times r + b$$

A difference equation shows the relationship between consecutive values of a sequence of values and the differences among them. Most often they are rearranged as a recursive equation that computes the output from the input signal and past outputs.

We will see that all DSP algorithms can be expressed as difference equations.

22.3 Digital Signals

Discrete-Time Signals

In this section, we will describe the fundamentals of discrete-time signals. We will start by defining the concepts and establishing the terminology. Consider the analog system example shown in Figure 22-3.

The system is a physical entity that serves a specific purpose. It might be an electrical circuit that consists of resistors, capacitors, inductors, transistors, integrated circuits, and the like, or a control system. The input to the system is the time variation of some physical quantity, such as temperature, and is usually represented by an electrical voltage or a current. In system theory, any of these input signals can be modeled with a mathematical function of time $x(t)$. The system's task is then to modify the input signal

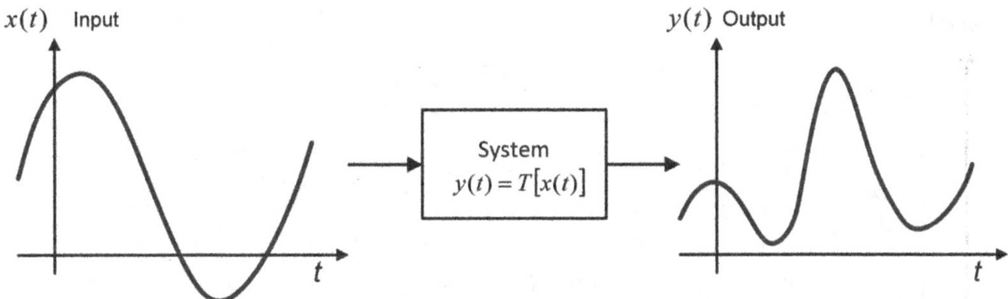

Figure 22-3. Continuous-time system working on continuous-time signals.

into an output signal, which may or may not be of the same type as the input signal. As an example, the system might produce a voltage signal at its output in response to temperature variations at its input. Another system might take the displacement of a lever as input and produce a voltage signal in response. The output signal of the system can also be modeled with a mathematical function $y(t)$.

In contrast to this analog system, now consider the discrete-time system in Figure 22-4.

In this case, the system is a digital computer that processes numbers. The input signal that is applied to the computer is an *indexed sequence* of numbers in a form suitable for use by the computer and is represented by the mathematical series $x(n)$. It is possible to draw an analogy from a spreadsheet program and to think of the discrete-time input signal as a set of numbers stored in one column of the spreadsheet. Given the index of a particular cell within that column, the value of the element stored in that cell can be obtained. For example, $x(85)$ refers to the value contained in cell 85 or, in signal-processing terms, the 85th sample of the sequence $x(n)$. Note that the term *sample* is related to the act of sampling a continuous-time signal to obtain a discrete sequence of numbers. In digital signal processing, it is customary to represent a discrete-time signal graphically as a set of vertical lines placed on a horizontal axis holding the values of the index, n (see Figure 22-5).

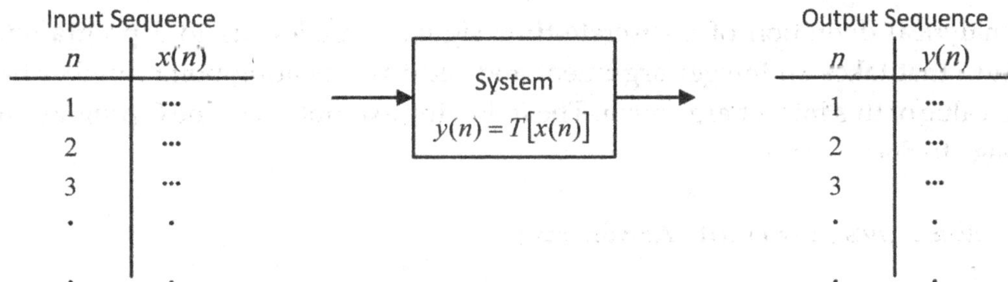

Figure 22-4. Discrete-time system working on discrete-time signals.

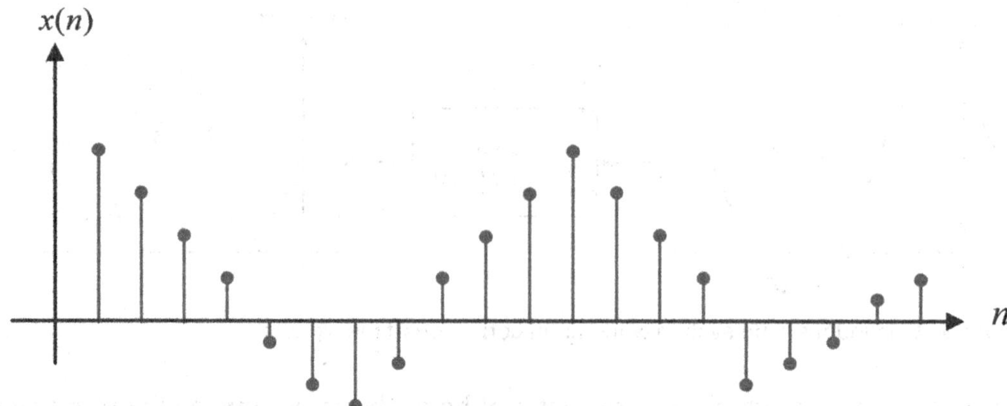

Figure 22-5. Graphical representation of a discrete-time signal.

The height of each vertical line represents the magnitude or value of the corresponding sample of the sequence. Alternative forms of descriptions are also utilized. The signal can be described by means of a mathematical formula. For example:

$$x(n) = 0.5n + sin(\pi n)$$

For any specified integer value of *n*, the corresponding amplitude can be calculated from the formula. Another method of describing a discrete-time signal is to tabulate the values of its samples. In this format, amplitudes of signal samples are listed between braces and separated from each other by commas. The location of the sample $x(0)$ is indicated by an up arrow (↑) on the line below the list of samples; that is:

$$x(n) = \{\ldots,\ x(-2),\ \ x(-1),\ \ \underset{\uparrow}{x(0)},\ \ x(1),\ \ x(2),\ \ x(3),\ \ \ldots\}$$

An ellipsis (. . .) just after the left brace or just before the right brace indicates the existence of more samples in the respective direction. The absence of an ellipsis on either side indicates that there are no more significant samples in that direction or, equivalently, all samples not shown are zero valued.

The analytical definition of a discrete-time signal is analogous to a programming structure that takes an integer argument and returns a floating-point number based on the value of this integer argument. The following example uses the C programming language to demonstrate.

```
float MySignal(int Argument)
{
    if (Argument == -1)
```

```
            return 1.7;
        else if (Argument == 0)
            return -0.6;
        else if (Argument == 1)
            return 2.3;
        else if (Argument == 2)
            return 5.6;
        else
            return 0.0;
}
```

The function **MySignal()** represents an infinite-length sequence that has only three nonzero samples. In tabular form,

```
MySignal(n) = { 1.7, -0.6, 2.3, 5.6}
                      ↑
```

Note the arrow ↑ under the value –0.6. This is the value of **MySignal** when $n = 0$.

Although the function can be called with any integer value for the argument, it will return a value of zero for all except four specific argument values. In the case of complex-valued discrete-time signals, complex values can be returned either by using the readily available complex data type of the language (e.g., FORTRAN) or by defining a complex data structure (e.g., with C, C++, C#, Pascal, and some compiled languages).

In the continuous-time system example in Figure 22-3, the actual system would be a physical entity that might consist of electrical, mechanical, electromechanical, or hydraulic components. On the other hand, our mathematical model of the system is an implementation of a signal processing circuit (usually represented with a block diagram) that transforms the function $x(t)$ into the function $y(t)$. In the discrete-time system example in Figure 22-4, the system is a digital computer, and our model is a mathematical formula or computer algorithm that transforms the sequence $x(n)$ into the sequence $y(n)$. After appropriate mathematical models are found for the system and its input and output signals, the analysis becomes independent of the physical nature of the problem. In other words, the mathematical functions $x(n)$ and $y(n)$ and the functional relationship between them form the basis for analysis in systems theory. The fact that $x(t)$ might represent a voltage or that the system might be an electric motor has little bearing on the theoretical analysis from this point on. The same was true when we were using the Laplace transform. This does not mean, however, that practical considerations regarding the physical nature of the problem are unimportant.

In real-world problems, physical constraints of the problem at hand and deviations of physical phenomena from idealized mathematical models must be taken into consideration in parallel to theoretical analysis. Systems theory simply provides a common framework for the analysis of vastly different physical systems.

One detail that must be elaborated on is the distinction between a discrete-time signal and a digital signal. The time variable is discrete in both signals; that is, both are defined as functions of an integer index. A discrete-time signal consists of samples that are continuous in amplitude. The amplitude of a sample can take any value. On the other hand, samples of digital signals are quantized and can only take amplitude values from a predefined set, and the set depends on the A/D converter used. We will work with discrete-time signals, rather than digital signals, because on paper, any amplitude can be represented with infinite precision.

When we represent signals in a computer, however, they become discrete signals. Computers represent numbers with fixed word lengths and therefore have finite precision.

Elementary Discrete Signals

Many basic discrete-time signals appear often in digital signal-processing literature and are typically used as building blocks in describing other, more complicated signals.

The discrete *unit-impulse* sequence, also referred to as a *unit-sample* or *Dirac delta* sequence in some textbooks, is defined as:

$$\begin{aligned} \delta(n) &= 1, \ \text{if } n = 0 \\ &= 0, \ \text{otherwise} \end{aligned} \tag{22-5}$$

and is shown in Figure 22-6.

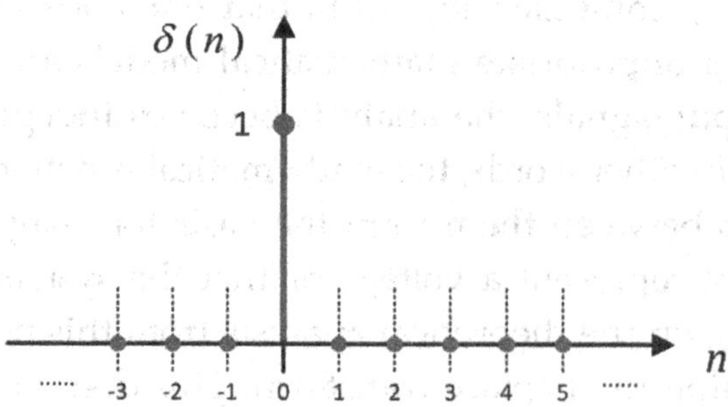

Figure 22-6. Discrete unit-impulse sequence.

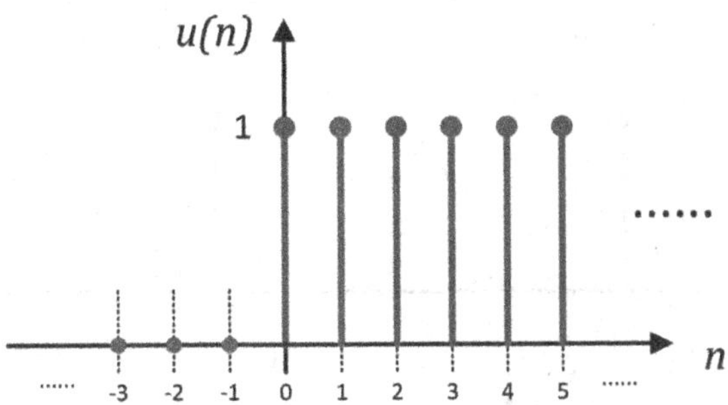

Figure 22-7. Unit-step sequence.

It is important to realize that $\delta(n)$ does not consist of just one sample; rather, it is an infinite-length sequence in which all samples except one happen to be equal to zero in amplitude.

The *unit-step* sequence is defined as:

$$u(n) = 1,\ if\ n \geq 0 \\ = 0,\ otherwise \tag{22-6}$$

and is shown in Figure 22-7.

It is not difficult to see the similarity between these signals and their analog equivalent, namely the unit impulse, and the step function.

Deterministic and Random Signals

Discrete-time signals can also be characterized as either deterministic or random. A signal is deterministic if it can be expressed in an analytical form, which means as a mathematical expression or equation.

Random signals, on the other hand, do not have analytical descriptions. Instead, they are described through their statistical (probabilistic) properties. Figure 22-8 shows an example of a random sequence.

22.4 Mathematical Operations on Discrete Signals

In this section, we will define several arithmetic operations for manipulating discrete-time sequences. These operations enable sequences to be analytically expressed in terms of simpler sequences.

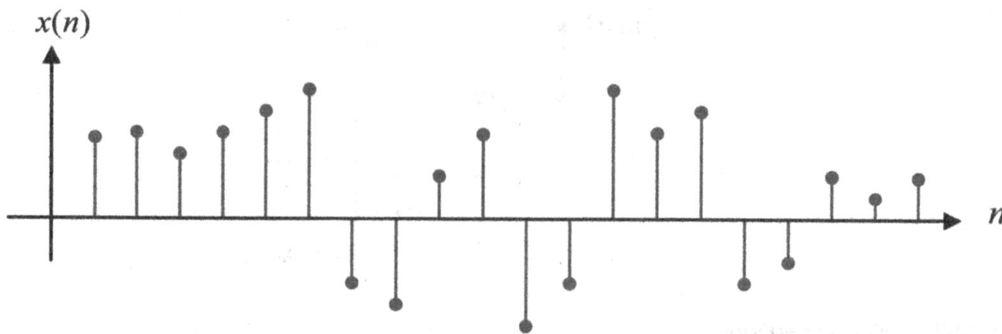

Figure 22-8. Random sequence.

Addition

Two sequences are added by adding the values of their corresponding samples for each value of the index n. As a result, the sum of two sequences is a new sequence where each sample is equal to the sum of the corresponding samples in the two sequences being added; that is:

$$y(n) = x_1(n) + x_2(n) \quad \text{For all values of } n.$$

Multiplication

The product of two sequences is a new sequence, each sample of which is equal to the product of the corresponding samples in the two sequences being multiplied; that is:

$$y(n) = x_1(n) \times x_2(n) \quad \text{For all values of } n.$$

Scalar Addition

Adding a constant scalar offset to a sequence means adding the scalar to each sample of the sequence; that is:

$$y(n) = x_1(n) + a \quad \text{For all values of } n.$$

Multiplication by a Constant

Multiplying a sequence by a constant scalar means multiplying each sample of the sequence by the constant; that is:

$$y(n) = a \times x(n) \quad \text{For all values of } n.$$

Time Shifting

Replacing integer argument n of a sequence with $n-k$ causes the sequence to be delayed (shifted to the right) by k samples if k is positive. If k is negative, then the shift is to the left, corresponding to an advance.

To understand the mechanism that causes the shift, consider the programming example given previously. Recall that the function *MySignal(n)* returns the values 0.7, –0.6, 5.6, and 2.3 when the argument n is equal to –1, 0, 1, or 2 respectively. We can define a new function *NewSignal(n)* as follows:

```
float NewSignal(int Argument)
{
  return MySignal  (Argument  -  5)
}
```

The relationship between the two signals represented by these two functions is:

$$NewSignal(n) = Mysignal(n-5)$$

When the function *NewSignal(n)* is called with the index value 6, it calls the function *MySignal(n)* with the index value 1, thus causing the value of –6 to be returned. The function *NewSignal(n)* returns 1.7, –0.6, and 2.3 when its argument is 5, 6, and 7, respectively. It is thus a five-sample delayed version of *MySignal(n)*.

Time Reversal

Replacing the integer argument n of a sequence by $-n$ causes the sequence to be reversed in time, that is, flipped around the time origin $n = 0$. It is possible to justify this by using a similar logic as in the previous time-shifting example.

Signal Representation

Up to now, we have represented discrete signals as sequences of numbers. It is this sequence that the computer stores and that the program operates on. The relation of the sequence and time is not apparent inside the computer. The operations performed on the sequence, as we will see shortly, depends on the relationship of the numbers in the sequence in relation to its time. We will next see how sequences can be represented as a function of time.

Discrete signals are generally represented in one of three ways.

Consider the following continuous signal, a ramp represented by the following equation $f(t) = 2t$. When this signal is sampled at 0.1 s intervals, we get the following sequence:

t	$f(t)$
0.0	0
0.1	0.2
0.2	0.4
0.3	0.6
0.4	0.8

This sequence can be represented as:

$$f(t) = \{0.0,\ 0.2,\ 0.4,\ 0.6,\ 0.8,\ \ldots\ldots\} \quad \text{or as}$$

$$f(nT) = 2t \quad for \quad T = 0.1 \quad \text{or as}$$

$$f_n = \{0.0,\ 0.2,\ 0.4,\ 0.6,\ 0.8,\ \ldots\ldots\}$$

In the first representation, it is assumed that the interval for each n is 0.1 s.

In the second representation, the interval is specified by the sampling interval T.

In the third representation, only the index is provided, and the interval is either assumed or not required.

Realize that inside the computer, the actual sequence:

$$\{0.0,\ 0.2,\ 0.4,\ 0.6,\ 0.8,\ \ldots\ldots\}$$

is stored in exactly the same way. The different forms are used interchangeably.

Flow Graphs

Flow graphs are typically used to represent difference equations pictorially. They are composed of the symbols shown in Figure 22-9. The symbols represent summation (addition of all the input elements to produce an output), multiplication by a constant, and one-unit, discrete-time delay.

Using these symbols only, any difference equation can be represented graphically.

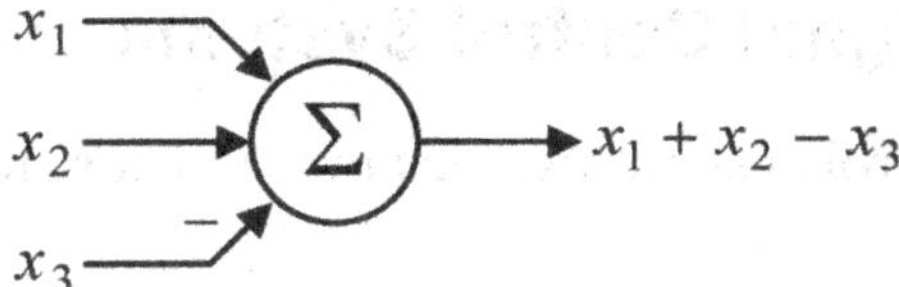

Summation

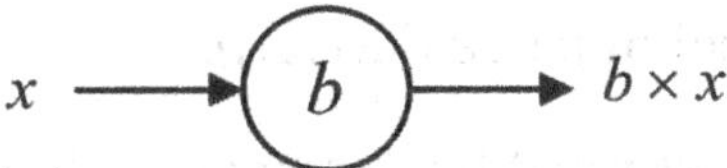

Multiplication

$x(n)$ → z^{-1} → $x(n-1)$

Unit Time Delay

Figure 22-9. Signal flow symbols.

As an example, we will represent the following difference equation:

$$Y(n) = 0.2X(n) + 0.8Y(n-1)$$

This equation represents a first-order system with a time constant of 0.4 s sampled at T = 0.1 s, which will produce the flow graph in Figure 22-10.

Difference Equation

As we saw in previous chapters, a differential equation describes the behavior of a continuous system. The solution to a differential equation establishes the system time or frequency response to a given input. To model a discrete system, we identify the corresponding difference equation. The solution of the difference equation then establishes the discrete-time response of the system.

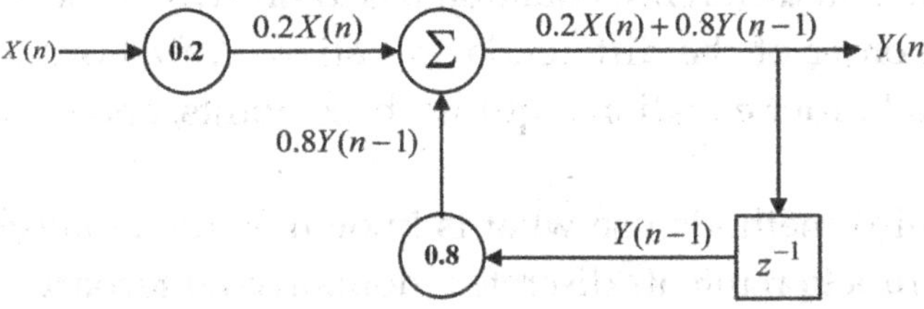

Figure 22-10. Flow graph of $Y(n) = 0.2X(n) + 0.8Y(n-1)$.

22.5 Design of Digital Control Systems

Designing a digital control system involves three major steps:

1. The process must be put into a mathematical form so its behavior can be analyzed and evaluated. This is done by defining the system using differential equations and then taking the Laplace transform to identify the system's transfer function (i.e., a model must be derived).
2. An appropriate controller must be designed so that the plant gives the desired response under the influence of the control system.
3. The system must be converted into its discrete form. This conversion can be done in two different ways.
    a. The design of the controller can be carried out entirely in analog form in the s-domain and then converted into discrete form at the final stage for implementation.
    b. The design of the controller may be carried out entirely in discrete form. In this case, the plant model must be converted into its discrete form, using the z-transform. The z-transform is considered to be the discrete (digital) version of the Laplace transform. This will be discussed in more detail later.

Discretization of Analog Systems

There are several ways to convert existing continuous or analog control systems into discrete or digital systems. However, the conversion from the time domain or s-domain to the discrete domain causes some distortion. We will investigate the three most popular methods and later discuss when to use each one.

A continuous function, $f(t)$, can be represented digitally as $F(nT)$, where n is the nth sample and T is the sampling period.

The first method is the simplest and is based on the fact that a differential equation can be approximated with a difference equation based on two or more points or samples, depending on the order of the differential equation. A first-order equation requires two points, a second-order equation requires three points, and so on.

The second and third methods use what is known as the *z-transform*, converting an s-domain transfer function into its discrete z-domain counterpart to then be converted to its required difference equation. This method is also known as *zero-order hold*.

The zero-order hold method is the most accurate method of designing digital control systems, and all the work is done in the digital domain. It is also the most difficult.

The third method is known as the *bilinear transformation*. This method transforms an analog solution into a digital approximation.

Approximating a Continuous System

The following procedure can be used to convert a continuous system expressed as a transfer function or differential equation into a digital system. The digital system will result in a difference equation, suitable to be processed by a computer.

1. Obtain the analog transfer function of the control expression (Laplace transform) and determine the required sampling time (T).
2. Convert the transfer function to a differential equation by taking the inverse Laplace transform.
3. Convert the differential equation to a difference equation using the following relations:

$$\begin{aligned} y(t) &= Y(n) \\ x(t) &= X(n) \\ dy(t) &= Y(n) - Y(n-1) \\ dt &= t(n) - t(n-1) = T \end{aligned}$$

   where

$$\begin{aligned} n &= \text{present value} \\ n-1 &= \text{the previous value} \\ T &= \text{the sampling time of the digitized system} \end{aligned}$$

4. Solve the expression for the present value of the output in terms of the present input, past inputs, and past outputs as needed.

The following example demonstrates the method.

Step 1.

The following transfer function describes a simple first-order system with a time constant of $\tau = 0.4$ s.

$$\frac{Y(s)}{X(s)} = \frac{1}{1 + \tau s}$$

Step 2.

To convert the transfer function to a difference equation, we first cross-multiply the transfer function and multiply through:

$$Y(s)\times(1+\tau s)=X(s)$$
$$Y(s)+\tau s\,Y(s)=X(s)$$

Next, we take the inverse Laplace transform, by replacing the *s* operator with its inverse Laplace transform and expressing the input and output in the time domain.

substituting
$$s\rightarrow\frac{d}{dt}$$
$$Y(s)\rightarrow y(t)$$
$$X(s)\rightarrow x(t)$$

then $Y(s)+tsY(s)=X(s)$ becomes

$$y(t)+t\frac{d}{dt}y(t)=x(t)$$

Step 3.

Digitizing the equation by making the following replacements, as previously stated, we get:

$$Y(n)+\tau\frac{Y(n)-Y(n-1)}{T}=X(n)$$

Step 4.

Solve for the present output, Y(n):

$$Y(n)+\tau\frac{Y(n)-Y(n-1)}{T}=X(n)$$
$$Y(n)+\frac{\tau}{T}Y(n)-\frac{\tau}{T}Y(n-1)=X(n)$$
$$Y(n)\left(1+\frac{\tau}{T}\right)-\frac{\tau}{T}Y(n-1)=X(n)$$
$$Y(n)\left(1+\frac{\tau}{T}\right)=X(n)+\frac{\tau}{T}Y(n-1)$$

Solving for output $Y(n)$:

$$Y(n) = \frac{X(n) + \frac{\tau}{T}Y(n-1)}{\left(1 + \frac{\tau}{T}\right)}$$

$$= \frac{X(n) + \frac{\tau}{T}Y(n-1)}{\left(\frac{T+\tau}{T}\right)}$$

$$Y(n) = \frac{T}{T+\tau}X(n) + \frac{T}{T+\tau} \times \frac{\tau}{T}Y(n-1)$$

$$Y(n) = \frac{T}{T+\tau}X(n) + \frac{\tau}{T+\tau}Y(n-1)$$

The sampling interval, as stated in the beginning, should be at least twice the frequency of interest. In the case of a first-order system, this is the time constant, τ. A good rule of thumb is to set $f_{sample} > 4\tau$ or $T \le \tau/4$. In our case, this would be $T = 0.1$ s.

Substituting: $\tau = 0.4$ and $T = 0.1$ we get our final difference equation.

$$Y(n) = \frac{0.1}{0.1+0.4}X(n) + \frac{0.4}{0.1+0.4}Y(n-1)$$

$$Y(n) = 0.2\ X(n) + 0.8\ Y(n-1)$$

Note that the two coefficients add up to exactly 1. This is not a coincidence but rather the property of a first-order system. The difference equation describes a weighted average or low-pass filter.

Figure 22-11 shows the difference equation calculated at 0.01 s intervals. The response is clearly that of a first-order system with a step input $X(n)$ of 1.

This example enables us to appreciate that a computer can approximate the solution to differential equations by using a difference equation, provided that an appropriately designed data acquisition system is available.

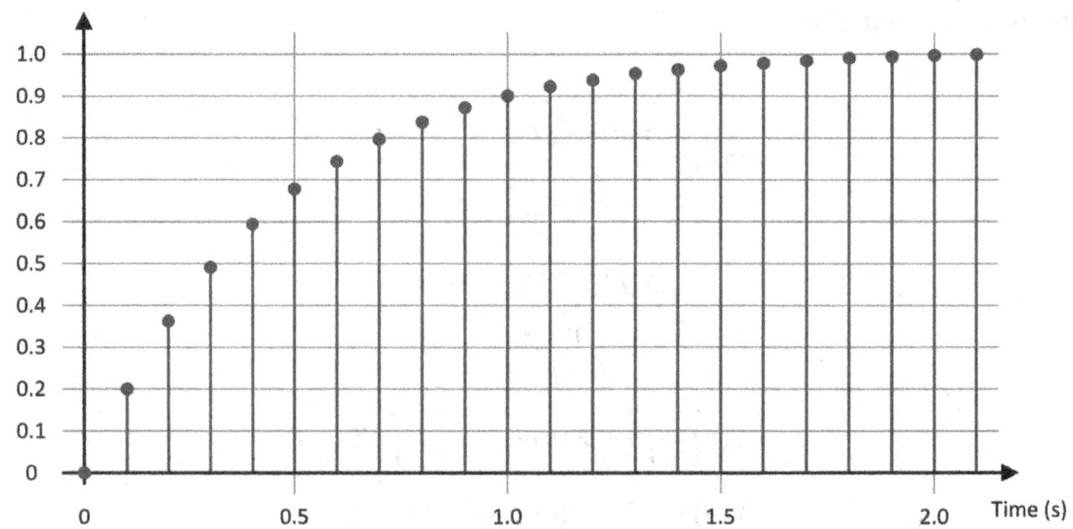

Figure 22-11. Time response of: $Y(n) = 0.2\ X(n) + 0.8\ Y(n-1)$.

22.6 Design and Implementation of a PID Controller

This section describes the design and implementation of a PID controller. PID is a commonly used technique in classical control. In designing controllers, just minimizing a term proportional to the error is often not sufficient. Including the integral of the error term reduces the steady-state error to zero because it represents the accumulated error. To further improve stability and plant dynamics, a differential of the error term is introduced. This term represents the error rate.

Two different approaches are used to convert PID into discrete form: rectangular and trapezoidal approximations. For the rectangular approximation, the design is done in the analog domain and then converted into the z-domain. For the trapezoidal approximation, the design can be done in the analog domain or directly in the z-domain.

The PID algorithm in the time domain is given by:

$$u(t) = K_P\ e(t) + K_I \int e(t)\,dt + K_D \frac{de(t)}{dt} \tag{22-7}$$

where

K_p = proportional gain
K_I = integral gain
K_D = differential gain
$e(t)$ = error signal

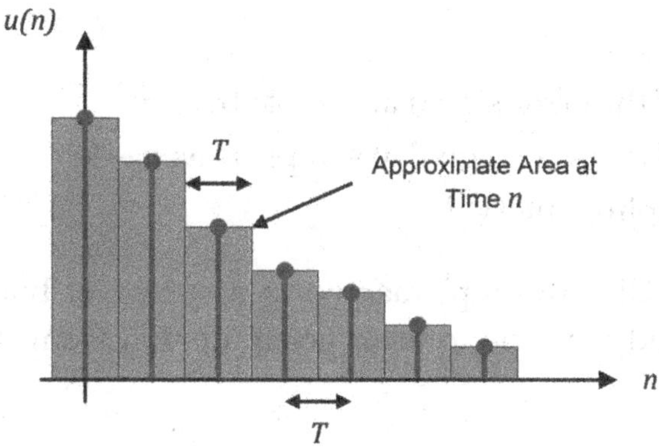

Figure 22-12. Rectangular approximation.

Rectangular Approximation (First Approximation)

To convert this equation to its discrete form, the integral term $\int e(t)\,dt$ is approximated by the summation of rectangles as shown in Figure 22-12.

Mathematically, this translates as follows:

$$\int e(t)\,dt = \sum_{n=0}^{N} e(n)\,T \qquad (22\text{-}8)$$

where

T = the sampling interval

$e(n)$ = the value of the error signal at sample time n

N = number of samples

The differential term $(de(e))/dt$ can be approximated as before by:

$$\frac{de(e)}{dt} = \frac{e(n) - e(n-1)}{T} \qquad (22\text{-}9)$$

where

$e(n)$ = value of the error signal at sample time n

$e(n-1)$ = value of the error signal at sample time $n-1$

T = the sampling interval

The PID algorithm can now be approximated in discrete form by:

$$u(n) = K_P\, e(n) + K_I\, T \sum_{n=0}^{N} e(n) + \frac{K_D}{T}[e(n) - e(n-1)] \qquad (22\text{-}10)$$

where

$e(n)$ = value of the error signal at sample time n

$e(n-1)$ = value of the error signal at sample time $n-1$

T = the sampling interval

This equation can be directly implemented in a computer but doing so is computationally intensive and is prone to error accumulation from the integral term (i.e., $\Sigma e(n)$).

To reduce this equation into a difference equation, we first calculate the output at time $n-1$.

$$u(n-1) = K_P\ e(n-1) + K_I\ Ts\sum_{n=0}^{N-1} e(n) + \frac{K_D}{Ts}[e(n-1) - e(n-2)]$$

Subtracting the $u(n-1)$ from $u(n)$ gives:

$$\begin{aligned} u(n) - u(n-1) &= K_P[e(n) - e(n-1)] + K_I\ T[e(n)] \\ &\quad + \frac{K_D}{T}[e(n) - 2e(n-1) + e(n-2)] \end{aligned}$$

Note that for the integral term, all terms except $e(n)$ remain, and that

$$[e(n) - e(n-1)] - [e(n-1) - e(n-2)] = [e(n) - 2e(n-1) + e(n-2)]$$

Multiplying through and combining similar terms gives:

$$\begin{aligned} u(n) &= u(n-1) + K_P\ e(n) - K_P\ e(n-1) + K_I\ T\ e(n) \\ &\quad + \frac{K_D}{T}e(n) - 2\frac{K_D}{T}e(n-1) + \frac{K_D}{T}e(n-2) \\ &= u(n-1) + e(n)\left[K_P + K_I\ T + \frac{K_D}{T}\right] + e(n-1)\left[-K_P - 2\frac{K_D}{T}\right] \\ &\quad + e(n-2)\left[\frac{K_D}{T}\right] \end{aligned}$$

If we let:

$$\alpha = K_P + K_I\, T + \frac{K_D}{T}$$
$$\beta = -K_P - \frac{2K_D}{T}$$
$$\gamma = \frac{K_D}{T}$$

our equation simplifies to:

$$u(n) = u(n-1) + \alpha\, e(n) + \beta\, e(n-1) + \gamma\, e(n-1) \qquad (22\text{-}11)$$

which is again a standard difference equation, and less computationally intensive.

Trapezoidal Approximation (Second Approximation)

In this approximation, the integral at time n is calculated as:

$$T \times \frac{e(n-1) + e(n)}{2}$$

The trapezoidal area (the average height of two successive points) gives a more accurate approximation of the integral term (see Figure 22-13).

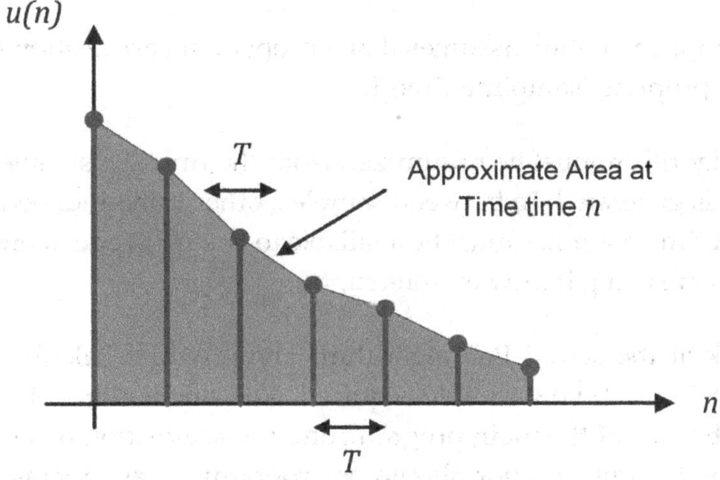

Figure 22-13. Trapezoidal approximation.

The PID algorithm can now be approximated in discrete form by:

$$u(n) = K_P\ e(n) + \frac{K_I\ T}{2} \sum_{n=0}^{N} e(n) + \frac{K_D}{T}[e(n) - e(n-1)] \quad (22\text{-}12)$$

This equation can also be reduced into a difference equation, in the same way as before, giving us the following:

$$u(n) = u(n-1) + \alpha\, e(n) + \beta\, e(n-1) + \gamma\, e(n-1) \quad (22\text{-}13)$$

where

$$\alpha = K_P + \frac{K_I\ T}{2} + \frac{K_D}{T}$$

$$\beta = -K_P + \frac{K_I\ T}{2} + \frac{K_D}{T}$$

$$\gamma = \frac{K_D}{T}$$

Note the change in the α, β, and γ coefficients from the previous approximation.

22.7 Computer Implementation

We saw that the PID control action could be implemented using the following difference equation:

$$u(n) = u(n-1) + \alpha\, e(n) + \beta\, e(n-1) + \gamma\, e(n-1)$$

Our computer implementation assumes that a proper data acquisition system is being utilized, with appropriate sampling time *T*.

The computer algorithm must be minimized from the time the sample is taken to the time the output is generated. In between samples, other (time-independent) tasks can be performed. A time reference must be available to ensure proper sampling intervals. Usually, this is done using hardware interrupts.

First, let us look at the actual PID algorithm. The program calculates a new value of control action for each new sample acquired. The calculations of alpha, beta, and gamma should be part of the main program, and the acquisition of a new sample and the outputting of the control action should be programmed as separate functions. The program should also have an appropriate function to acquire a new sample. Figure 22-14 shows a fragment of a computer program that could implement the PID function.

```
PID()       /* C version of PID algorithm */
{
            int         sample, dummy;
            float       measured_value, cntrl_out;
            Float       setpoint;
            float       alpha, beta, gamma;
            float       cntrl_action[3];
            float       error_signal[3];

            for (sample =1; sample >=0; sample --)
            {
            /* make room for present conrol action and error */
                  cntrl_action[sample+1] = cntrl_action[sample];
                  error_signal[sample+1] = error_signal[sample];
            }
      /*    read measured value */
            measured_value = a2d();
            error_signal[0] = setpoint - measured_value;
      /* calculate new control action */
            cntrl_action[0] = cntrl_action[1]
                              + alpha * error_signal[0]
                              +  beta * error_signal[1]
                              + gamma * error_signal[2];
      /*    send new control action to output
            cntr_out = cntrl_action[0];
      /* output cntrl_action to d2a */
            dummy = d2a(cntr_out);
}
                  /* end of PID calculation
```

Figure 22-14. C language implementation of the PID algorithm.

22.8 The Z-Transform

Introduction

We saw that the equation for a first-order system and the PID algorithm are both difference equations. To analyze or design a digital system, a difference equation of the system itself is also required. We get this difference equation by using the z-transforms of the sampled signals. Just as the transfer functions in an analog time domain system are obtained from the Laplace transforms of signals, system transfer functions of sampled data are obtained using the z-transform. Then, just as for Laplace transforms, a transfer function can be obtained using the ratio of the output over the input signals of the system, both signals being z-transformed.

The transfer function must be the system description, because by the preceding definition, if you multiply it by the z-transform of any input signal, you get the z-transform of the output signal. We use capitals to represent the z-transform of signals.

Consider the following transfer function in the time domain:

$$y(t) = \frac{y(t)}{x(t)} \times x(t) = h(t) \times x(t)$$

Then there exists the following relationship in the z-domain:

$$Y(t) = \frac{Y(z)}{X(z)} \times X(z) = H(z) \times X(z)$$

In the preceding equations, $x(t)$ is the input signal, $y(t)$ is the output signal, and $H(z)$ is the digital transfer function of the system, because if you multiply it by the z-transform of the input, you get the z-transform of the output.

The inverse z-transform can then be used to return to the sampled time domain and solve for the difference equations. The systematic conversion of a difference equation to a computer program then forms the basis for computer simulation of physical systems.

Example

Consider input signal $f(nT)$:

$$\begin{aligned} f(nT) &= 2n,\ 0 \le n < 4 \\ &= 0,\ n \ge 4 \end{aligned}$$

This can be graphically illustrated as shown in Figure 22-15.

Derive the z-transform of the function.

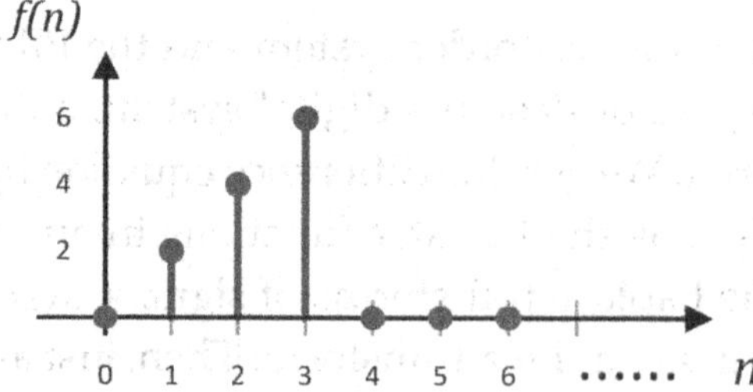

Figure 22-15. Sampled function: $F(z) = 2z^{-1} + 2z^{-2} + 2z^{-3}$.

> ***Solution***
>
> The function generates the following geometric series:
>
> $$u(n) = 0u(n) + 2u(n-1) + 4u(n-2) + 6u(n-3) + 0 + 0 + \cdots$$
>
> The z-transform of the function can then be written as:
>
> $$F(z) = 0 + 2z^{-1} + 2z^{-2} + 2z^{-3} + 0 + 0 + \cdots$$
>
> The z^{-1} notation implies a one-sample delay, z^{-2} implies a two-sample delay, and so forth.
>
> By removing all noncontributing samples of zero, it can be rewritten as:
>
> $$F(z) = 2z^{-1} + 4z^{-2} + 6z^{-3}$$

Let us next look at a somewhat more complicated example, namely the z-transform of a unit step input, *u*(*t*).

The discrete version of this function is given by:

$$u(nT) = 0, \; n < 0$$
$$u(nT) = 1, \; n \geq 0$$

The graphical implications of the sampled function are shown in Figure 22-16.

This generates the following geometric series:

$$u(n) = 1 + 1u(n-1) + 1u(n-2) + 1u(n-3) + \ldots$$

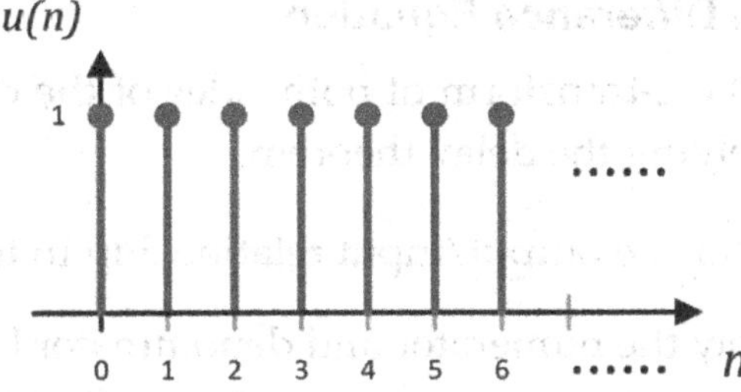

Figure 22-16. Unit-step function.

Taking the z-transform gives:

$$U(z) = 1 + 1z^{-1} + 1z^{-2} + 1z^{-3} + \cdots$$

The summation goes on forever. However, if we multiply both sides by z^{-1}, we get:

$$z^{-1}U(z) = 1z^{-1} + 1z^{-2} + 1z^{-3} + 1z^{4} \cdots$$

We next subtract the second equation from the first equation, giving us:

$$U(z) - z^{-1}U(z) = (1 + z^{-1} + z^{-2} + z^{-3} + \cdots) - (z^{-1} + z^{-2} + z^{-3} + z^{4} \cdots)$$

This gives:

$$U(z) - z^{-1}U(z) = 1$$

Solving for $U(z)$ the z-transform of $u(nT)$ gives:

$$U(z)(1 - z^{-1}) = 1$$

$$U(z) = \frac{1}{1 - z^{-1}}$$

This form of the function is referred to as the *closed form* of the geometric series.

We will not concern ourselves with the derivation of many more complex functions, leaving this to the mathematicians, but will avail ourselves of the results in our work with discrete signals.

The Inverse Z-transform

The inverse z-transform is usually obtained by using partial-fraction expansion.

Transfer Function from Difference Equation

Step 1 Take the z-transform of both sides of the difference equation by applying the delay theorem.

Step 2 Solve for the output/input relationship in terms of z.

Step 3 Multiply the numerator and denominator by the highest positive order of z to remove all terms with negative orders of z.

This is best illustrated by the following example.

$$u(k) = u(n-1) + 0.3\ e(n) - 0.4\ e(n-1)$$

Step 1
$$U(z) = z^{-1}U(z) + 0.3\ E(z) - 0.4\ z^{-1}\ E(z)$$

Step 2
$$U(z) - z^{-1}U(z) = E(z)(0.3 - 0.4\ z^{-1})$$
$$U(z)(1 - z^{-1}) = E(z)(0.3 - 0.4\ z^{-1})$$

Step 3
$$\frac{U(z)}{E(z)} = \frac{0.3 - 0.4\ z^{-1}}{1 - z^{-1}} \times \frac{z}{z}$$
$$= \frac{0.3 - 0.4}{z - 1}$$

Difference Equation from Transfer Function

Step 1 Multiply the numerator and denominator of the transfer function by the highest negative order of z to remove all terms with positive orders of z.

Step 2 Cross-multiply and take the inverse z-transform.

Step 3 Solve for the present output in terms of the past samples (output and input).

This is best illustrated by the following example.

$$\frac{U(z)}{E(z)} = \frac{z^2 + z}{z^2 - 1.2z + 0.5}$$

Step 1
$$\frac{U(z)}{E(z)} = \frac{z^2 + z}{z^2 - 1.2z + 0.5} \times \frac{z^{-2}}{z^{-2}}$$
$$= \frac{1 + z^{-1}}{1 - 1.2z^{-1} + 0.5z^{-2}}$$

Step 2
$$U(z) - 1.2\ z^{-1}U(z) + 0.5z^{-2}U(z) = E(z) + z^{-1}E(z)$$
$$u(n) - 1.2u(n-1) + 0.5u(n-2) = e(n) + e(n\quad 1)$$

Step 3
$$u(n) = 1.2u(n-1) - 0.5u(n-2) + e(n) + e(n-1)$$

Zero-Order Hold

This technique of reconstructing a digitized signal assumes that the controller is preceded by a zero-order hold (ZOH), using a sample-and-hold (S/H) circuit followed by

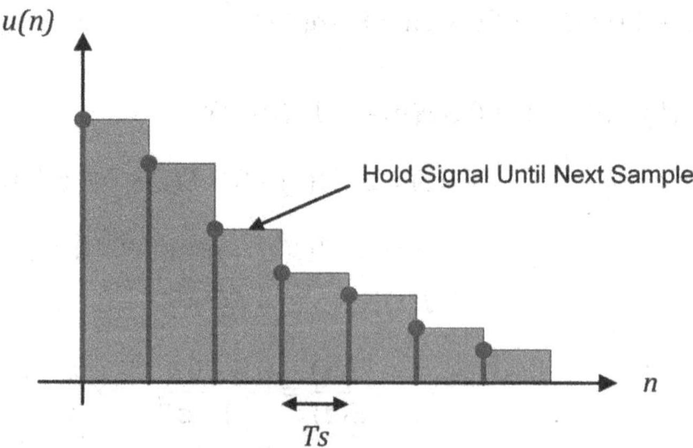

Figure 22-17. S/H circuit.

an A/D converter, so that both the input and output of the resulting system are digital. Both the ZOH and A/D converter are included in the conversion method. Figure 22-17 clearly shows the effect that the S/H circuit has on the original signal. Notice that the resulting signal is, in effect, lagging behind the original signal. It is interesting to compare this signal with that in Figure 22-12.

The conversion is given by the following equation:

$$H(z) = (1 - z^{-1}) \times Z\left[L^{-1}\left(\frac{H(s)}{s}\right)\right] \tag{22-14}$$

It is assumed that the inverse Laplace transform (L^{-1}) will be split up using partial fractions and that both the inverse Laplace transform and the *z*-transform will be looked up in tables.

The following example demonstrates this method. Consider the same system as in the previous example:

$$\frac{Y(s)}{X(s)} = \frac{1}{1 + \tau s} = H(s)$$

Using Equation 22-14, we get:

$$L^{-1}\left(\frac{H(s)}{s}\right) = L^{-1}\left(\frac{1}{s(1 + \tau s)}\right)$$

Using partial fractions to calculate the inverse Laplace transform:

$$\frac{1}{s(1+\tau s)} = \frac{1}{s} + \frac{A}{s} + \frac{B}{1+\tau s}$$
$$= \frac{A + A\tau s + Bs}{s(1+\tau s)} = \frac{A + s(A\tau + B)}{s(1+\tau s)}$$

Solving for A and B:

$$A + s(A\tau + B) = 1 + 0s$$

Giving: $A = 1$ and $A\tau + B = 0$

Then: $B = 0 - A\tau = -\tau$

Substituting back gives:

$$\frac{1}{s} + \frac{-\tau}{1+\tau s} = \frac{1}{s} - \frac{1}{\frac{1}{\tau} + s}$$

Taking the inverse Laplace transform gives:

$$1 - e^{-T/\tau}$$

Substituting this in ZOH Equation 22-14 gives:

$$H(z) = (1 - z^{-1}) \times Z\left[1 - e^{-T/\tau}\right]$$
$$= (1 - z^{-1})\left(\frac{z}{z-1} - \frac{z}{z - e^{T/\tau}}\right)$$

Substituting $\tau = 0.4$ and $T = 0.1$, we get our final difference equation:

$$H(z) = (1 - z^{-1})\left(\frac{z}{z-1} - \frac{z}{z - e^{-0.25}}\right)$$
$$= (1 - z^{-1})\left(\frac{z}{z-1} - \frac{z}{z - 0.79}\right)$$
$$= \frac{(z-1)}{(z-1)} - \frac{(z-1)}{z - 0.79}$$
$$= 1 - \frac{(z-1)}{z-0.79} = \frac{z - 0.79 - z + 1}{z - 0.79}$$
$$H(z) = \frac{0.21}{z - 0.79} = \frac{0.21z^{-1}}{1 - 0.79z^{-1}}$$

We can now identify the governing difference equation:

$$\frac{Y(z)}{X(z)} = \frac{0.21}{z-0.79} = \frac{0.21z^{-1}}{1-0.79z^{-1}}$$
$$Y(z)\left(1-0.79z^{-1}\right) = 0.21z^{-1}\,X(z)$$
$$Y(z)\left(1-0.79z^{-1}\right) = 0.21z^{-1}\,X(z)$$

Taking the inverse *z*-transform:

$$y(n) - 0.79y(n-1) = 0.21x(n-1)$$

Solving for output $y(n)$, we get:

$$y(n) = 0.21x(n-1) + 0.79y(n-1)$$

Note that this equation is similar to our original difference equation, $Y(n) = 0.2X(n) + 0.8Y(n-1)$, but the input $x(n-1)$ is one sample behind. This gives a transient response that is delayed by one sample. Depending on the system being simulated, we can correct that by using $x(n)$ in the equation:

$$y(n) = 0.21x(n) + 0.79y(n-1)$$

22.9 Bilinear Transform

This technique, also called the *Tustin* or *trapezoidal approximation*, uses the relationship:

$$s = \frac{2}{T} \times \left(\frac{z-1}{z+1}\right)$$

The term *bilinear* refers to the fact that both the numerator and the denominator are linear in z.

This method starts with a transfer function in the s-domain (Laplace), and then using the above relationship, a transfer function in the z-domain is obtained.

This method warps the frequency response at the critical frequencies of the system. To overcome this problem, the critical frequencies of the original s-domain are *pre-warped* so that the critical frequencies of the z-domain system end up where they belong. The critical frequency, ω, is pre-warped to another frequency by the following transformation:

$$\omega^* = \frac{2}{T}tan\left(\frac{\omega T}{2}\right)$$

where T is the sampling period, pre-warping only applies to second- or higher order systems. First-order systems will give the same critical frequency when warped.

For our first-order system, the critical frequency is $\tau = 0.4$ and $T = 0.1$. Using the above pre-warped transformation, we get:

$$\begin{aligned}\tau^* &= \frac{2}{0.1}tan\left(\frac{0.4\times 0.1}{2}\right)\\ &= 20tan(0.02) = 20\times 0.02\\ &= 0.4\end{aligned}$$

The following example demonstrates the bilinear transformation application.

Again, consider the following first-order system:

$$\frac{Y(s)}{X(s)} = \frac{1}{1+\tau s}$$

Substituting $\tau = 0.4$ and $T = 0.1$ and using the bilinear transformation, we get:

$$\begin{aligned}\frac{Y(s)}{X(s)} &= \frac{1}{1+0.4\times\frac{2}{0.1}\times\left(\frac{z-1}{z+1}\right)}\\ &= \frac{z+1}{z+1+0.4\times\frac{2}{0.1}\times(z-1)}\\ &= \frac{z+1}{z+1+8z-8}\\ &= \frac{z+1}{8z-6}\end{aligned}$$

Multiplying the numerator and denominator by z^{-1} gives:

$$\begin{aligned}\frac{Y(s)}{X(s)} &= \frac{z+1}{8z-6}\times\frac{z^{-1}}{z^{-1}}\\ &= \frac{1-z}{8-6z^{-1}}\end{aligned}$$

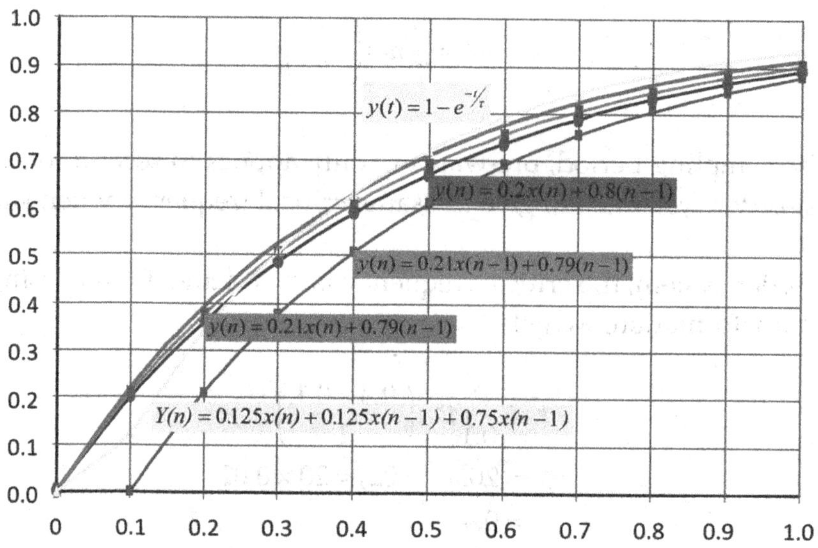

Figure 22-18. Comparison of different methods of obtaining difference equations.

Converting to difference form and solving for $y(n)$ gives:

$$\begin{aligned}
&8Y(z) - 6z^{-1}Y(z) = X(z) + z^{-1}X(z)\\
&8y(n) - 6y(n-1) = x(n) + x(n-1)\\
&8y(n) = x(n) + x(n-1) + 6y(n-1)\\
&y(n) = 0.125x(n) + 0.125x(n-1) + 0.75y(n-1)
\end{aligned}$$

Figure 22-18 plots this equation against our original difference equation and its theoretical response.

Looking at the response, we see that the start is slower than before. The reason is that the difference equation uses as its input the average; namely:

$$\frac{x(n-1) + x(n)}{2}$$

22.10 Summary

We have seen three different methods for deriving a difference equation based on an s-domain (analog) transfer function. The first method gives a relatively good approximation without any special mathematics. The second method, ZOH, and the third method, bilinear transformation, both require the use of the z-transform and result in noticeably better difference equations, although they may require some modification based on prior knowledge of the system. The ZOH is best used to provide simulation of and actual process, and the bilinear transformation is best used for generating control action (lead or lag) other than PID.s

Appendix A: System of Units

Process control involves taking measurements and understanding them. A consistent way of measuring size, mass, level, temperature, time, energy, power, speed, and other parameters requires a consistent system of units.

A consistent system of units consists of fundamental units, which represent base dimensions, and derived units. Most units exist in more than one system of units.

Standards of Measurement

In the process industry, taking accurate measurements is an essential part of process control. A review of units of measurement is in order to ensure we all speak the same language when quantifying our measurements.

Measurements are expressed in both US customary (USC) units (also referred to as *British*, *English*, or *Imperial*) and SI (also referred to as *metric*[1]) units. However, not all USC units are the same as Imperial units. In fact, there can be slight differences among USC units, Imperial units, and English units. The SI units are named after the French Système International d'Unités, or International System of Units. Most scientific work is done using the SI system of units. However, in North America and in the process industry, USC units are still in popular use.

1. There is a technical difference between the SI system and the metric system.

In the SI system, there are seven base units of measurement: length, mass, time, temperature, electric current, light, and the amount of substance. All other units of measurement are expressed in terms of these fundamental units and are known as *derived units of measurement.*

Length

Length in the SI system is expressed in meters (m).

The USC unit for length is the standard yard, which is equal to 0.914 of the standard meter.

Mass

Mass in the SI system is expressed in kilograms (kg).

The USC unit for mass is the pound. Note that in the English engineering units, the pound-mass and the force are distinct units.

The newton (N) is the SI unit of force; 1 N will give 1 kg of mass a speed of 1 meter per second (m/s). 1 N equals 0.22 lb of force.

For a more detailed explanation, see the "Some Basic Concepts" section at the end of this appendix.

Time

Time is expressed in terms of the second, in all systems of measurement.

Temperature

Temperature is expressed in terms of the kelvin (K). The thermodynamic or kelvin scale of temperature used in the SI system has its origin or zero-point at absolute zero and has a fixed point at the triple point of water, defined as 273.16 K. The Celsius (°C) scale is derived from the kelvin scale. The triple point of water is defined as 0.01°C on the Celsius scale, which is about 32.02°F on the Fahrenheit scale. This applies to all systems.

Note that the Celsius and Fahrenheit scales are denoted in degrees (°), but the kelvin scale is not.

Electric Current

Electric current is expressed in terms of the ampere (A), which is named after André-Marie Ampère, a French mathematician and physicist. This applies to all systems.

Light

Light, or luminous intensity, is expressed in terms of the candela (cd). It is defined as the luminous intensity of 1/600,000 of a square meter of a radiating cavity at the temperature of freezing platinum (2,042 K). In the US it is common to express light intensity in foot-candles. One foot-candle is equivalent to one lumen per square foot).

Summary

Quantity	Basic Unit	Symbol
Length	meter	m
Mass	gram	g
Time	second	s
Electric Current	ampere	A
Temperature	kelvin	K
Luminous Intensity	candela	cd

Most units are expressed in scientific and/or engineering notation using prefixes. All engineering prefixes are expressed as powers of 3.

Prefix	Symbol	Quantity
pico	p	10^{-12}
nano	n	10^{-9}
micro	μ	10^{-6}
milli	m	10^{-3}
centi	c	10^{-2}
deci	d	10^{-1}
hecto	h	10^{+2}
kilo	k	10^{+3}
mega	M	10^{+6}
giga	G	10^{+9}
tera	T	10^{+12}

SI and USC Units

One of the principal advantages of the SI system is its use of only one unit (standard of size) for each parameter (or the thing that is being measured). The SI units are interrelated in such a way that only five fundamental standards are required. These fundamental standards are identified with an asterisk (*) in the following tables.

Mechanics

Parameter	SI Unit	Symbol	USC Units
Time duration, delay	second*	s	second, minute, hour, day
Frequency	hertz	1/s (Hz)	cycle per second
Length, distance, displacement	meter	m	inch, foot, yard, rod, chain, furlong, mile
Velocity, speed, rate of change of position	meter per second	m/s	foot per second mile per hour, knot
Acceleration: rate of change of velocity	meter per second^2	m/s^2	foot per second squared
Mass: quantity of material	kilogram*	kg	ounce, pound, slug, ton
Force: push, pull, or weight	newton	kg m/s^2	pound-force
Impulse, force times time	newton second	kg m/s	pound-force second
Momentum: mass times velocity	kilogram meter per second	kg m/s	pound foot per second
Work, energy: force times distance	joule = 1 newton meter	kg m^2/s^2	foot pound-force
Power: rate of doing work	watt = 1 joule per second	kg m^2/s^3	foot pound-force per second horsepower
Area: size of a surface	square meter	m^2	square foot, square yard acre, square mile
Pressure: force per unit area	pascal (Pa) = newton per square meter	kg/m s^2	pound-force per square inch
Volume, capacity	cubic meter	m^3	pint, quart, gallon cubic foot, cubic yard
Density, heaviness	grams per cubic centimeter	g/cm^3	pounds per cubic foot

Heat

Parameter	SI Unit	Symbol	USC Units
Heat energy	joule calorie[2] 1 calorie = 4.186 joules	J	British thermal unit, Btu
Temperature (change)	kelvin* 1°C = 274 K	K °C	degree Fahrenheit, °F

2. Although the joule is the proper SI unit for energy, the calorie (cal) or the kilocalorie (kcal also denoted as Calorie, note the capital) are widely used by health professionals and the food industry.

Light

Parameter	SI Unit	Symbol	USC Units
Luminous intensity	candela*	cd	Same
Luminous flux	lumen, one candela produces 4π lumens	lm	Same
Illumination	lux, lumen per square meter	lx	Same
Focus	diopter, reciprocal meters	m^{-1}	Same
Astronomical distance	parsec	pc	Same

Note: Light bulbs (the incandescent type) have traditionally been rated in watts. However, newer types of light bulbs such as light-emitting diode (LED) bulbs are rated in units of absolute temperature kelvin (K).

Lord Kelvin noticed that during his research on heated blocks of carbon, the color changed progressively as the temperature increased, from a dim red to a brighter yellow and eventually a bright blue-white glow at the highest temperatures. Today, light bulb color is represented in units of kelvin rather than in watts: 2700 K indicates a warm incandescent, 3000 K is a warm white halogen, and 3500 K indicates a household fluorescent.

Electricity and Magnetism

Parameter	Metric Electrical Unit	Symbol	Notes
Electric charge	coulomb*	C	96,500 coul = 1 faraday faraday = 1 mole of electrons
Electromotive force (emg)	volt	V	joule per coulomb
Capacitance	farad	F	coulomb per volt
Electric current	ampere (coulomb per second)	A	—
Electric energy	joule (watt second)	J kWh = 3,600,000 joules	newton meter or kg m^2 s^2
Electric power	watt (joule per second)	W	—
Electrical resistance	ohm (volt per ampere)	Ω	—
Electrical conductivity	siemens (coulomb per joule)	S	ampere per volt
Electrical field strength	volts per meter	V/m	—
Electromagnetic inductance	henry (plural henrys)	H	—
Magnetic field intensity	oersted	Oe	formerly the gauss
Magnetic flux	weber	Wb	10^8 maxwells
Magnetic flux density	tesla	T	10,000 gauss
Note: There are no Imperial electrical units.			

Some Basic Concepts

Charge (Coulomb)

By definition, the electron is the smallest unit of charge.

Electrical charge is measured in coulombs and is designated as *Q*.

$$1\text{C} = 6.25 \times 10^{18} \text{ electrons}$$

Current (Amperes)

Current is the *rate* of flow of electrons, which is designated as *I* and measured in amperes (A).

$$I = \frac{Q}{t} \quad \text{or} \quad 1\text{ A} = \frac{1\text{ C}}{1\text{ s}}$$

Force (Newton)

Force is a quantity that produces motion or modifies the motion of a body. It is designated as *F* and expressed in newtons.

One newton is the force required to accelerate 1 kg of mass at a rate of 1 m/s.

$$F = m \times a \quad \text{or} \quad 1\text{ N} = 1\text{ kg} \times 1\frac{\text{m}}{\text{s}^2}$$

Pound-Force (lbf)

The unit of force in the USC gravitational system of units is the force that a mass of 1 pound exerts on whatever it is resting on on the Earth; that is, the weight of a mass of 1 pound, and it is designated as *lbf* to distinguish it from the symbol *lb* for the pound.

The weight of a mass of 1 lb varies from place to place; it weighs less at the equator than at the poles and it weighs less at high altitudes than at sea level. The pound-force, however, does not change; it is defined as the weight (a force) that a body with a mass of 1 lb would exert at a location where the acceleration due to gravity is exactly 32.1740 ft/s^2, which is the value at sea level at a latitude of 45°.

Mass versus Weight

Mass is defined as the amount of matter an object has. One of the qualities of mass is that it has inertia. Inertia is a quantitative measure of an object's resistance to change its state of motion or rest. Inertia is proportional to an object's mass. Newton's first law of motion is:

> Every object will remain at rest or in uniform motion in a straight line unless compelled to change its state by the action of an external force.

As an example of inertia, any resting object requires a certain amount of force to set it in motion. The greater the mass of the object, the more force is required to move the object. The same is true if the object is moving and needs to be stopped. The more massive the object is, the more force will be required to stop the motion of the object. Mass is a measure of how much inertia an object has.

The weight of an object on Earth depends on the force of attraction (gravity) between the object and Earth.

Work and Energy (Joules)

Work is done when a force acting on a body causes it to move a distance. The amount of work done is equal to the amount of energy that it took. In other words:

- Work is done when energy has been utilized.
- Work and energy are expressed in joules (J).

$$\text{work} = \text{force} \times \text{distance} \quad \text{or} \quad 1\,\text{J} = 1\,\text{N} \times 1\,\text{m}$$

Note: The work done is equal to the energy used, but energy can have many forms, such as kinetic (work done) or potential (no work done).

Work and energy have the same unit because the work done on an object to move it from point A to point B is the difference in the kinetic energy it has at the two points; this is referred to as the *work-energy theorem*. In simplified form, it says that to do work, you must use energy; you can never do work without using energy. Heat should be mentioned here as well. When energy is used to do work, some of that energy is converted to heat by friction, so heat also has the same unit as energy.

Work, energy, and heat are not the same thing, but they are related.

By extension, the theorem lets us define other forms of energy that can be converted to work: chemical energy, electrical energy, nuclear energy, gravitational energy, and so on.

Note that 1 J is also the amount of heat dissipated when a current of 1 A passes through a resistor of 1 Ω for 1 s.

Using USC units, work is in foot pound-force.

Voltage

Voltage or potential is the amount of energy used to move an amount of charge. It is designated as E or V and is measured in volts.

$$V = \frac{\text{energy}}{\text{charge}} \quad \text{or} \quad 1\ \text{V} = \frac{1\ \text{J}}{1\ \text{C}}$$

The symbol *E* is often used to designate electromotive force (emf).

Power

Mechanical Power

Power is the work done in a unit of time (the rate of doing work); it is designated as *P* and expressed in watts.

$$\text{power} = \frac{\text{work}}{\text{time}} \quad \text{or} \quad 1\ \text{W} = \frac{1\ \text{J}}{1\ \text{s}}$$

Using USC units, work is expressed in foot pound-force per second.

Electrical Power

We defined voltage (V) as:

$$\text{V} = \frac{\text{energy}}{\text{charge}}$$

then

$$\text{energy} = \text{voltage} \times \text{charge}$$

Because work is done by using energy,

$$\text{work} = V \times Q$$

Using the equation of power, we can define electrical power as:

$$\text{power} = \frac{\text{work}}{\text{time}} = \frac{V \times Q}{\text{time}}$$

We saw that

$$I = \frac{Q}{t}$$

Therefore

$$\text{power} = P = \frac{V \times Q}{\text{time}} = V \times \frac{Q}{t} \quad \text{(in watts)}$$
$$P = V \times I$$

The Metric System as a Superset of the SI System

The metric system is primarily based on powers of 10 relating units.

The SI system retains the concept of prefixes for handling more convenient sizes within one category of measure, but the SI system is coherent. *Coherence* means that all specially named derived units are expressed as products and quotients of the base units and no numeric factors come into play except the value 1.

For example, volume is a three-dimensional length, so it is a cube of length. The base unit of length is the meter (m), so the coherent unit of volume is SI is m^3. This unit has no special name. Next, consider the liter. The liter (L) is defined to be a cubic decimeter, which in terms of the coherent SI unit of volume is given by $\text{L} = \text{dm}^3 = (0.1\ \text{m})^3 = 0.001\ \text{m}^3$. Note that there is a numeric factor other than 1. That means the liter is not coherent with SI and is therefore not a SI unit.

Appendix B: Graphic Symbols for Process Measurement and Control

Process control systems are usually documented using the following four major types of drawings:

1. Block diagrams
2. Process flow diagrams (PFDs)
3. Piping and instrumentation diagrams (P&IDs)
4. Loop diagrams

Block diagrams are primarily used in analysis. PFDs are used preliminary in design and system conception. P&IDs, also commonly referred as *process control instrumentation diagrams* and *process/piping and instrumentation diagrams,* are an important part of understanding process control because they show schematically how all the control components and the process are interconnected. Instrumentation detail varies with the degree of design complexity. Loop diagrams are used in construction and maintenance.

The basis for the graphical presentation of process control and measurement information is ISA-5.1-2009, *Instrumentation Symbols and Identification*. The material presented in this appendix consists of excerpts from that standard and is used with permission. For full detail, however, the ISA standard should be consulted.

The standard defines how each symbol is constructed using graphical elements, alpha and numeric identification codes, abbreviations, function blocks, and connecting lines.

Figure B-1 shows a typical detailed P&ID diagram.

Deciphering the Symbols

ISA-5.1 defines four graphical elements.

1. Discrete instrument
2. Computer system and software
3. Alternative or safety instrumented system
4. Primary choice or basic process control system

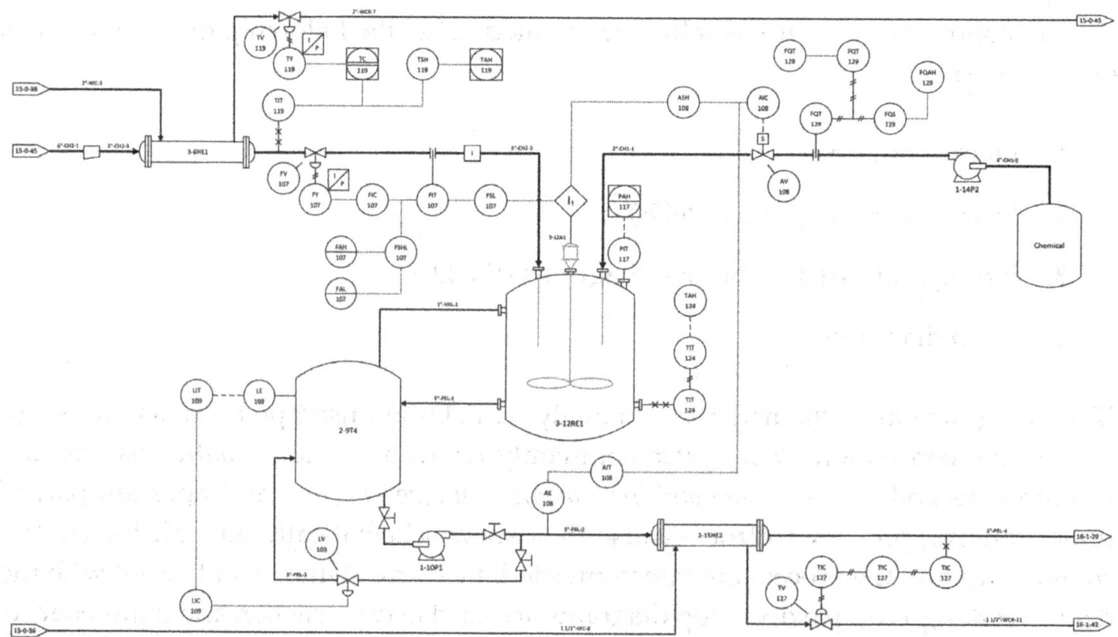

Figure B-1. Typical detailed P&ID.

These elements are grouped into five locations defined by accessibility.

1. Located in the field
2. Located in or on the front of central or main panel
3. Located in the rear of the central or main panel
4. Located in or on the front of the secondary or local panel
5. Located in the rear of the secondary or local panel or console

Discrete instruments are indicated by circular elements. Shared control/display elements are circles surrounded by a square. Computer functions are indicted by a hexagon. See Figure B-2.

Adding a single horizontal bar across any of the four graphical elements indicates the function resides in the primary location category. No line indicates the device or function is in the field. A double line indicates an auxiliary location. Devices located behind a panel board in some other inaccessible location are shown with a dashed horizontal line.

		Operator Accessible		Behind Panel	
	Field-Mounted	Control Room	Auxiliary Location		Auxiliary Location
Discrete					
Shared Display/Control					
Alternate Choice or Safety Instrumented System					
Computer Systems and Software					

Figure B-2. Instrumentation device and function symbols from ISA-5.1.

Tag Numbers and Instrument Designations

The ISA-5.1 standard contains tables and examples for creating letter abbreviations that are used for instruments on P&IDs. Most P&IDs follow this standard to some degree. Letter and number combinations appear inside each graphical element, and letter combinations are defined by the ISA-5.1 standard. Numbers are user assigned, and schemes vary with companies. Some companies use sequential numbering that ties the instrument number to the process line number, and others adopt unique and sometimes unusual numbering systems. The instrument identification is referred to as *tag numbers.*

However, the first letter always defines the measured or initiating variables, such as Analysis (A), Flow (F), Temperature (T), and so on, with succeeding letters defining readout, passive, or output functions, such as Indicator (I), Record (R), Transmit (T), and modifiers (see Table B-1 at the end of this appendix).

First letter = variable Second letter(s) = function

As an example, PV is a pressure control valve and FIT is a flow-indicating transmitter (see Table B-2 at the end of this appendix).

Figure B-3 shows a typical instrument and its designation.

Figure B-4 shows an example loop and its tag numbers. TIC 101 indicates temperature indicator control for loop number 101. All instruments (including the valve) in the same loop will have the same 101 loop number. Note that the primary control strategy is temperature, controlled by the steam flow to the heat exchanger. Therefore, the control valve is designated as TV 101, indicating it is controlling temperature, not flow.

Specific line symbols are also used to indicate connections such as instrument-to-instrument, process-to-instrument, and equipment-to-equipment. Line symbols are identified in Table B-3.

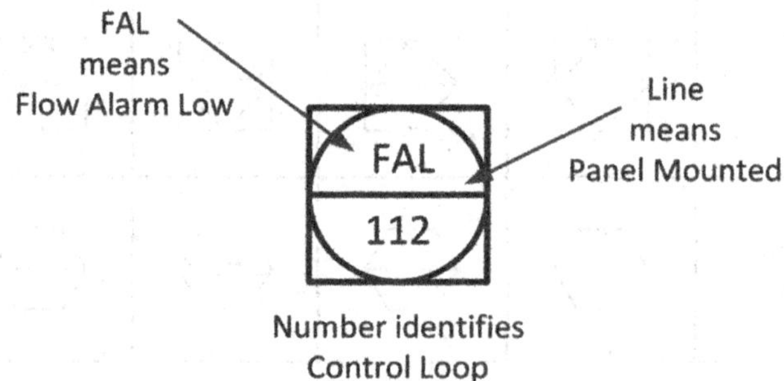

Figure B-3. Instrument designation (tag number).

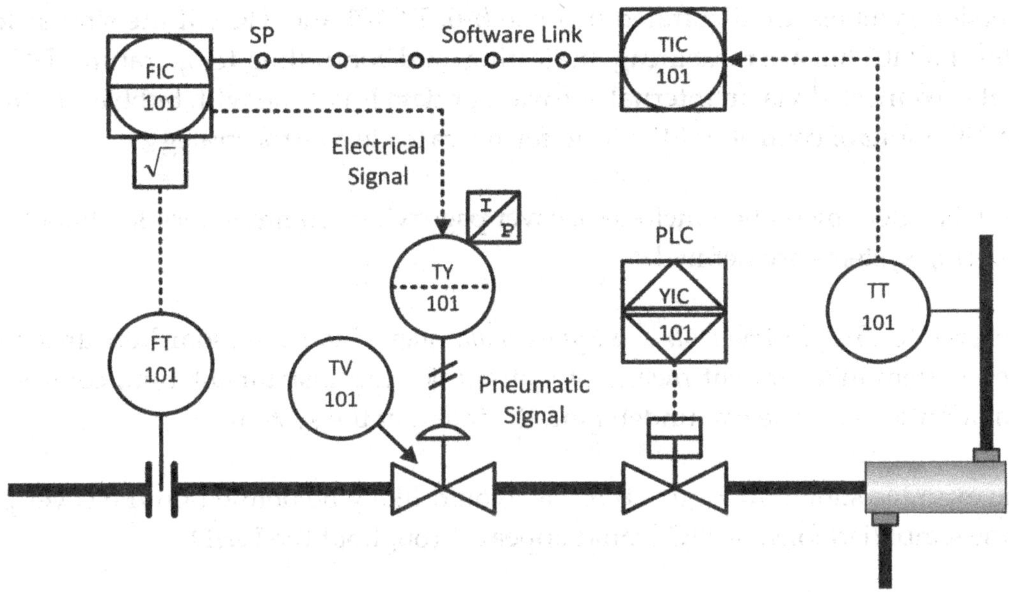

Figure B-4. All instruments in a single loop have the same loop number.

Typical valve symbols are illustrated in Table B-4.

Referring to the example P&ID in Figure B-4, FT 101 represents a field-mounted flow transmitter connected via electrical signals (dotted line) to a flow-indicating controller (FIC 101) located in a shared control/display device (solid line). A square root extraction of the input signal is applied as part of FIC 101's functionality. Signal processing symbols are summarized in Table B-5. The output of FIC 101 is an electrical signal to TY 101 which is located in an inaccessible or behind-the-panel location (dotted line). The output signal from TY 101 is a pneumatic signal (line with double forward slash marks) making TY 101 an I/P (current to pneumatic transducer) as indicated. Signal

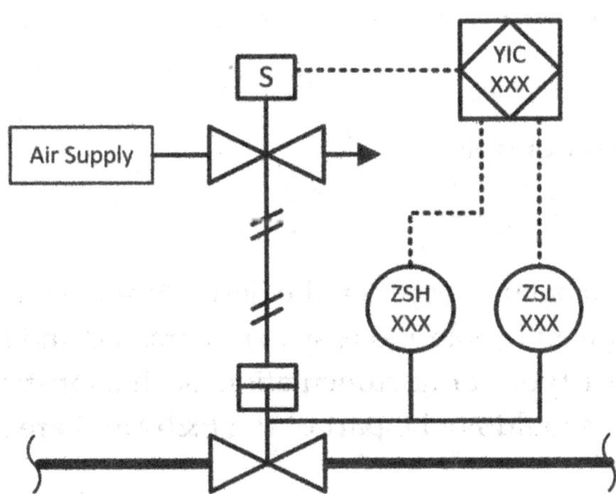

Figure B-5. Example of a typical PLC (YIC) control.

conversion symbols are illustrated in Table B-6. TT 101 and TIC 101 are similar to FT 101 and FIC 101 but are measuring, indicating, and controlling temperature. TIC 101's output is connected via an internal software or data link (line with bubbles) to the set point (SP) input of controller FIC 101 to form a cascade control strategy.

Often P&ID documentation includes a cover page where common terms, symbols, and numbering systems are defined.

Adherence to ISA's *Instrumentation Symbols and Identification* standard ensures a consistent, system-independent means of communicating instrumentation, control, and automation so everyone can understand the intent of the system.

In the example shown in Figure B-5, YIC would likely be defined on the cover page, and the simplified form of YIC would appear throughout the P&ID.

YIC indicates an on-off valve controlled by a solenoid valve and fitted with limit switches to indicate open (ZSH) and closed (ZSL) positions. All inputs and outputs are wired to a PLC that is accessible to the operator (diamond in a square with a solid horizontal line). The letter *Y* indicates an event, state, or presence. The letter *I* means the device provided indication, and the letter *C* means control takes place in this device.

Good Practice in Preparing P&ID Documents

P&ID documents are normally divided into five sections:

1. General
2. Format
3. Equipment
4. Piping
5. Instruments and Controls

General

A good P&ID provides a balance between the desire to show all relevant data with the need to make the document legible and easy to understand and read. Most details that are available from other types of documentation, such as instrument loop diagrams and vessel data sheets, should not be part of the P&ID and are not recommended for inclusion on the P&ID.

Format

The P&ID should be formatted with suitable layout and symbology. The following guidelines are general recommendations and depend on the specific P&ID.

- The top of a horizontal line and the left side of a vertical line is the top of a pipe.
- The bottom of a horizontal line and the right side of a vertical line is the bottom of a pipe.
- The drawing size is normally 17 in × 11 in (size B) or 22 in × 34 in (size C).
- Lay out each P&ID to avoid clutter and allow for future modifications, show no more than three pieces of major equipment per P&ID.
- Show the primary flow on each P&ID from left to right.
- Show flow through equipment relative to actual arrangement (e.g., cooling water supply in the bottom of the exchanger tube bundle and the cooling water return on the top of the exchanger).
- Show primary process lines heavier than secondary and utility lines. See Table B-3.
- Show off-page connector arrows for primary, secondary, and instrumentation lines entering the P&ID horizontally 0.25 in from the left inside borderline and exiting 0.25 in horizontally from the right inside.
- Utility connectors may be shown at any convenient location on the P&ID.
- Show the center point of a pressure safety element (PSE) identification bubble 0.5 in above a horizontal line or left of a vertical line and 0.5 in away from the rupture disc or equipment.
- Show all equipment, piping, and instrument symbols per the ISA standard when possible.
- Show a connection to an equipment nozzle when the connection is welded.
- Show normally closed manual valves using a darkened solid symbol.
- When darkened in valves cannot be used because of the symbol type (e.g., butterfly valve), use the abbreviation for normally closed (NC) directly below the valve in a horizontal line or to the right of the valve in a vertical line.
- Show on-off valves in the normal operating position. Do not show control valves or relief valves normally closed.

Industry Codes and Standards

Organizations

American National Standards Institute (ANSI)

American Society of Mechanical Engineers (ASME)

International Organization for Standardization (ISO)

International Society of Automation (ISA)

Standards

ANSI-ISA-50.00.01-1975 (R2002)
Compatibility of Analog Signals for Electronic Industrial Process Instruments

ANSI-ISA-51.1-1979 (R1993)
Process Instrumentation Terminology

ANSI/ISA-5.1-2009
Instrumentation Symbols and Identification

ISA-5.2-1976 (R1992)
Binary Logic Diagrams for Process Operations

ISA-5.3-1983
Graphic Symbols for Distributed Control/Shared Display Instrumentation, Logic and Computer Systems

ISA-5.4-1991
Instrument Loop Diagrams

ISA-5.5-1985
Graphic Symbols for Process Displays

ASME-B31.3-2018
Process Piping

Table B-1. Identification letters.

	Identification Letters				
	First Letter		Succeeding Letters		
	Measured or Initiating Variable	**Modifier**	**Readout or Passive Function**	**Output Function**	**Modifier**
A	Analysis		Alarm		
B	Burner, combustion		User's choice	User's choice	User's choice
C	User's choice			Control	
D	User's choice	Differential			
E	Voltage		Sensor (primary element)		
F	Flow rate	Ration (fraction)			
G	User's choice		Glass, viewing devices		
H	Hand				High
I	Current (electrical)		Indication		
J	Power	Scan			
K	Time, time schedule	Time rate of change		Control station	
L	Level		Light		Low
M	User's choice	Momentary			Middle intermediate
N	User's choice		User's choice	User's choice	User's choice
O	User's choice		Orifice, restriction		
P	Pressure, vacuum		Point (test) connection		
Q	Quantity	Integrate, totalizer			
R	Radiation		Record		
S	Speed, frequency	Safety		Switch	
T	Temperature			Transmit	
U	Multivariable		Multifunction	Multifunction	Multifunction
V	Vibration, mechanical analysis			Valve, damper, louver	
W	Weight, force		Well		
X	Unclassified	X-axis			
Y	Event, state, or presence	Y-axis		Relay, compute, convert	
Z	Position, dimension	Z-axis		Driver, actuator, final control element	

Source: ISA-5.1-2009.

Table B-2. Instrument component designations.

Flow	
FAL	Flow alarm low
FAH	Flow alarm high
FE	Flow element
FI	Flow indicator
FIC	Flow-indicating controller
FIT	Flow-indicating transmitter
FS	Flow switch
FT	Flow transmitter
FV	Flow control valve

Level	
LAL	Level alarm low
LAH	Level alarm high
LC	Level controller
LCV	Level control valve
LE	Level element
LG	Level glass
LIT	Level-indicating transmitter
LR	Level recorder
LS	Level switch
LT	Level transmitter

Pressure or Vacuum	
PAL	Pressure alarm low
PAH	Pressure alarm high
PC	Pressure controller
PCV	Pressure control valve
PDI	Pressure differential indicator
PE	Pressure element
PI	Pressure indicator or manometer
PIT	Pressure-indicating transmitter
PIV	Pressure safety relief valve
PT	Pressure transmitter

Temperature	
TAL	Temperature alarm low
TC	Temperature controller
TE	Temperature element,
TI	(thermocouple, resistance bulb,
TIC	or thermopile)
TIS	Temperature indicator
TIT	Temperature-indicating controller
TR	Temperature-indicating switch
TS	Temperature-indicating
TT	transmitter
TV	Temperature recorder
TW	Temperature switch
TV	Temperature transmitter
	Temperature control valve
	Temperature well
	Temperature interlock solenoid
	valve, relay, or converter

Miscellaneous	
BC	Burner controller
HC	Manual (hand) controller
HV	Manual (hand) control valve
II	Current indicator
ZI	Position indicator
A	Analysis
I	Current
J	Power
K	Time, program, or counting
Q	Quantity
WI	Weight indicator

Table B-3. Instrument line symbols.

Description	Line Type	Explanation
Process		A solid thick line shows the actual process, or the raw material that will produce the final product.
Capillary tubing (filled system)		A solid line with X marks indicates a tube that contains fluid, which expands and contracts when subjected to temperature changes.
Connection to the process		A thinner solid line indicates that an instrument is connected to the process or supply line.
Undefined signal		Indicates an undefined signal.
Pneumatic signal		A solid line with a double slash marking indicates a pneumatic signal (usually air). An added back slash indicates that the signal is a binary signal (on-off).
Electric signal		This signal can be current or voltage. An added back slash indicates that the signal is binary (on-off). The previous ISA standard used a solid line with a triple slash marking to indicate an electrical signal.
Hydraulic signal		A solid fine line with italic *L*s indicates a hydraulic signal (usually oil).
Wireless or guided		A sine waves indicates an electromagnetic or sonic transmitted signal. A line indicates the signal is guided.
System link		This indicates a communication link between devices of a shared display
		This indicates a communication bus connecting two or more devices
		This indicates a communication link, a fieldbus, or intelligent devices.
		This indicates a communication link between a device and a remote calibration adjustment device or system.

Table B-4. Valve symbols.

General Valve Symbols			
 General	 Rotary	 Globe	 Relief
 Butterfly	 Angle	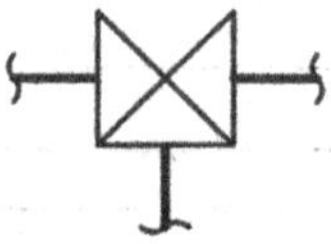 Three-Way	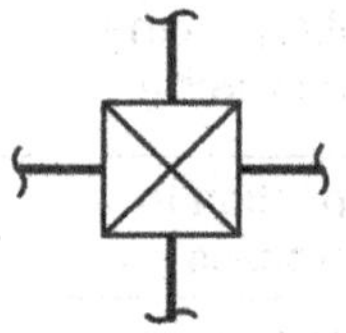 Four-Way
Alternate Symbols			
 Butterfly 	 Angle 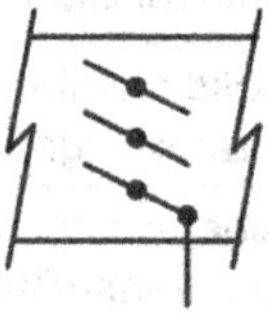	 Three-Way 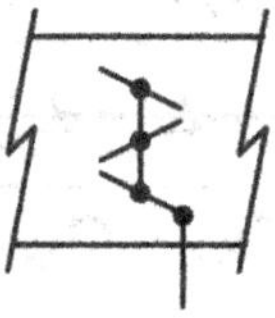	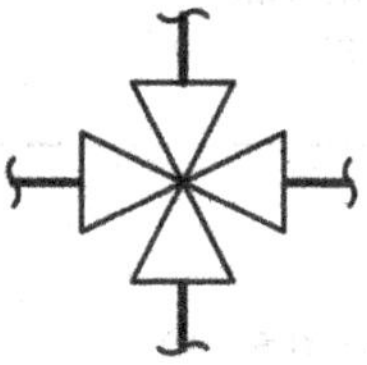 Four-Way 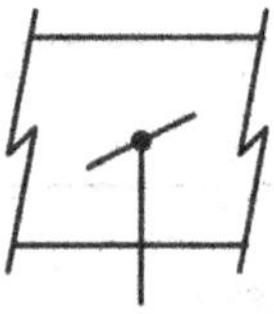
Diaphragm	Dampers or Louvers		
Control Valve Actuator Symbols			
Manual	 Diaphragm	 Pressure Balance Diaphragm	 Cylinder Piston
 Motor Operated	 Solenoid	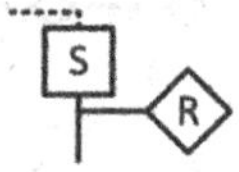 Single Solenoid with Manual Reset	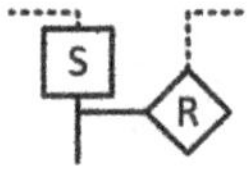 Single Solenoid with Remote Reset

Table B-5. Signal processing symbols from ISA-5.1.

Instrument Function Symbol	Description	Instrument Function Symbol	Description
Σ	Summation	>	High Select
Σ/n	Average	<	Low Select
Δ	Difference	$\not>$	High Limit
$\times$	Multiplication	$\not<$	Low Limit
$\div$	Division	$\pm$	Bias
$\sqrt{\ }$	Square Root		User Defined
X^n	Exponential	REV	Invert the Signal
$f(x)$	Unspecified Function		

Table B-6. Instrument (signal) conversion.

<table>
<tr><th>Instrument Function Symbol</th><th colspan="4">Description</th></tr>
<tr><td rowspan="6">x / y</td><td colspan="4">Conversion x/y = *Input*/*Output*</td></tr>
<tr><td>A</td><td>Analog</td><td>I</td><td>Current</td></tr>
<tr><td>B</td><td>Binary</td><td>O</td><td>Electromagnetic, sonic</td></tr>
<tr><td>D</td><td>Digital</td><td>P</td><td>Pneumatic</td></tr>
<tr><td>E</td><td>Voltage</td><td>R</td><td>Resistance (electric)</td></tr>
<tr><td>H</td><td>Hydraulic</td><td></td><td></td></tr>
<tr><td colspan="5">Example:</td></tr>
<tr><td>I / P</td><td colspan="4">Current-to-Pressure Conversion</td></tr>
<tr><td>E / I</td><td colspan="4">Voltage-to-Current Conversion</td></tr>
</table>

Appendix C: The Laplace Transform

Introduction

The Laplace transform was developed early in the nineteenth century by a French mathematician and astronomer named Pierre Simon de Laplace. It was originally strictly a mathematical exercise and was not applied to the field of electronic and control technology until the twentieth century. Chapters 6 and 18 explained that components and systems can be expressed using differential equations.

Solving differential equations can be complex and time-consuming. The Laplace transform enables us to transform these differential equations into simpler algebraic equations. Then by solving the algebraic equation, we can obtain system and control component transfer functions. With the obtained transfer functions, we can then compute the system or component's response to inputs. In other words, the Laplace transform is a mathematical model that describes systems in time and frequency response.

Frequency response, although not as intuitive as time response, should be familiar to anyone who has used the equalizer on a home stereo or computer. The equalizer display indicates the frequency distribution and magnitude of sound.

The Laplace transform takes us from a domain called the *time domain* to another domain called the *frequency domain*. In the time domain, the signal is given or determined for each instant of time. We say that the input and output signals are functions of time, and we append a (*t*) to the function name to indicate that time is the independent variable.

The time-domain equation of a component defines its output as a function of time and the input signal.

In the frequency domain, the magnitude and the phase angle of the signal are given or determined for each value of a frequency usually designated by the letter s, where s has the units of 1/s or s^{-1}. We say that the input and output signals are functions of frequency and append an (s) to the variable name to indicate that frequency is the independent variable. The frequency-domain equation of a component defines the magnitude and phase angle of the output signal as a function of both frequency and the input signal.

We use the Laplace transform to obtain the frequency-domain equation from the time-domain equation.

Lowercase letters are used to represent signals in the time domain, and uppercase letters are used for signals in the frequency domain.

Although deriving the Laplace transforms of functions gives us a better understanding of what is taking place, it is not necessary to apply the Laplace transform in automation analysis. The derivations provided at the end of this appendix are for demonstration only. The reader is by no means required to be able to demonstrate this type of mathematical competency.

The Laplace transform designated by the letter *L* is given by the following integral:

$$F(s) = L[f(t)] = \int_0^\infty f(t)e^{-st}dt \tag{C-1}$$

where

$F(s)$ = the Laplace transform of $f(t)$

t = time

s = a complex number of the form $a + j\omega$

and $j = \sqrt{-1}$

This equation represents the Laplace transform of a time function $f(t)$[1].

1. Some textbooks use the letter p instead of s. In that case, it is referred to as the *Heaviside operator.*

The time integration uniquely transforms our time function into a new function of the variable *s*. If we let:

$$s = j\omega$$

and
$$\omega = 2\pi f \tag{C-2}$$

where

ω = frequency in radians
f = frequency in hertz

then our new function is in the frequency (f) domain. The variable *s* is referred to as the *Laplace operator.* Equation C-1 represents a one-sided Laplace transform, as it assumes that all times are positive. This is a good assumption for practical systems.

The actual transformation can be quite involved mathematically, but this we leave to mathematicians. Automation professionals make ample use of tables that provide all the required transformations.

We can also make ample use of MATLAB in calculating the Laplace transform.

Figure C-1 graphically illustrates the procedure.

Properties of the Laplace Transform

This section will outline some properties (see Table C-1) of the Laplace transform.

The transfer function of a component is the Laplace transform of its unit impulse response.

Looking at entry 1 in table C-2, we note that the input to this system $X(s)$ is the unit impulse (Dirac delta function), and then $X(s) = 1$ and $H(s) = Y(s)$. This means that the impulse response of a component is its transfer function $H(s)$.

A component transfer function is obtained from the component differential equation by applying the Laplace transform and ignoring initial conditions.

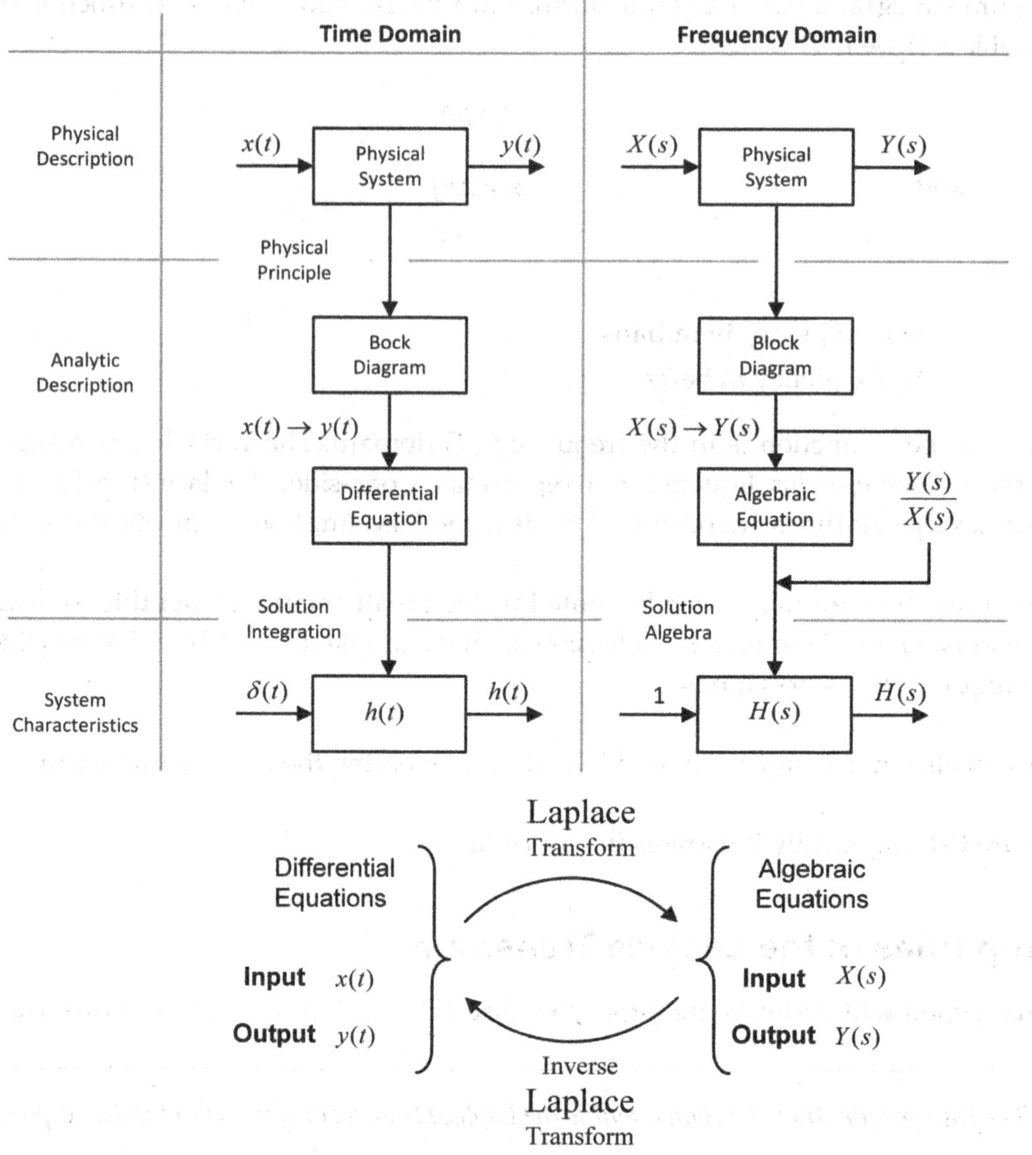

Figure C-1. Graphical illustration of the Laplace transform.

Example 1

The following differential equation defines the output *y*(*t*) with respect to the input *x*(*t*) for a specific component.

$$y(t) + 0.2\frac{d}{dt}y(t) = x(t)$$

Identify the corresponding transfer function in the frequency domain.

Solution

Applying the Laplace transforms to the equation gives:

$$Y(s)+0.2sY(s)=X(s)$$

Next, solve for the output-to-input relationship:

$$Y(s)[1+0.2s]=X(s)$$

$$\frac{Y(s)}{X(s)}=\frac{1}{1+0.2s}$$

Table C-1. Properties of the Laplace transform.

Name	Time Domain $f(t)$	Frequency Domain $F(s)$
Linearity	$K \times f(t)$	$K \times F(s)$
	$f_1(t)+f_3(t)-f_3(t)$	$F_1(s)+F_3(s)-F_3(s)$
Note	$L[af_1(t)+bf_3(t)]=aL[f_1(t)]+bL[f_3(t)]$	
Derivative	$\frac{df(t)}{dt}$	$sF(s)$
	$\frac{d^2f(t)}{dt^2}$	$s^2F(s)$
Integral	$\int f(t)dt$	$\frac{F(s)}{s}$
Final Value	$\lim_{t\to\infty} f(t)=\lim_{s\to 0} sF(s)$	
Shifting	$L[f(t-a)]=e^{-as}F(s)$ for a step function delayed (shifted) *a* unit in time	

The converse implies that a component differential equation can be generated from the component transfer function by replacing s^n with $\frac{d^n}{dt^n}$.

The time response of a component is obtained by taking the inverse Laplace transform of the component transfer function ***times*** *the required input.*

Table C-2. Selected Laplace transforms.

	Name	Graph	Time Domain $f(t)$	Frequency Domain $F(s)$
1	Dirac Delta Impulse	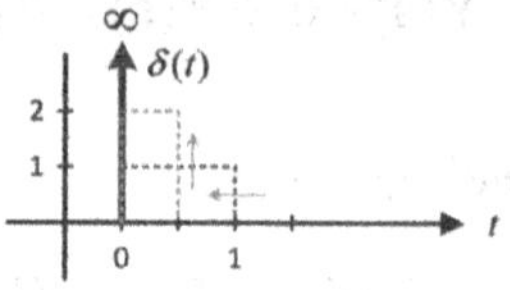 ∞; $\delta(t)$; 2; 1; 0; 1; t	$\delta(t)$	1
2	Unit Step	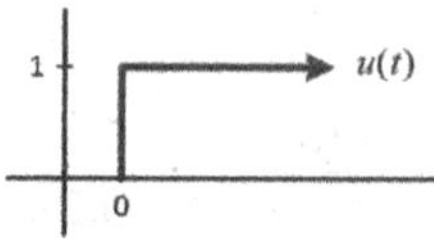 1; $u(t)$; 0; t	$u(t)$	$\frac{1}{s}$
3	Ramp of Slope K	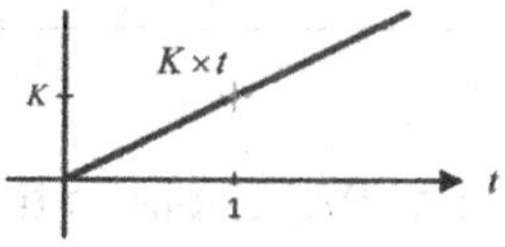 $K \times t$; K; 1; t	$K \times t$	$\frac{K}{s^2}$
4	Exponential Decay	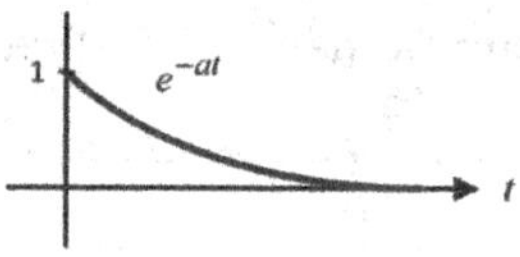 1; e^{-at}; t	e^{-at}	$\frac{1}{s+a}$
5	Exponential Growth	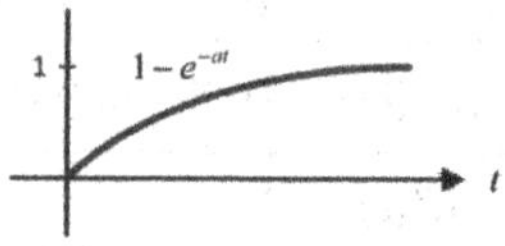 1; $1-e^{-at}$; t	$1 - e^{-at}$	$\frac{1}{s} \times \frac{1}{s+a}$
6	Sine	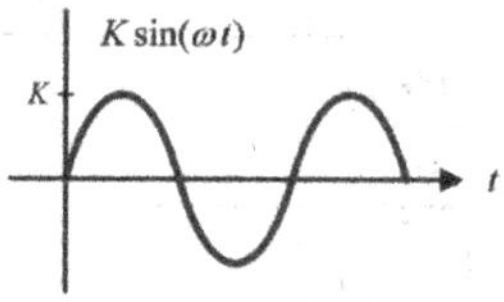 $K\sin(\omega t)$; K; t	$K \sin \omega t$	$\frac{K\omega}{s^2+\omega^2}$
7	Cosine	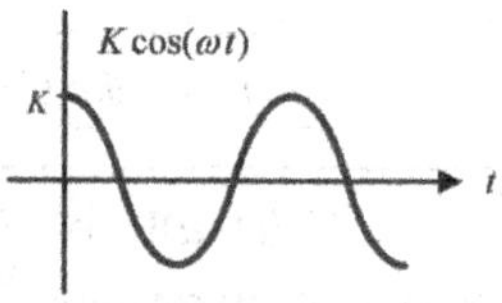 $K\cos(\omega t)$; K; t	$K \cos \omega t$	$\frac{Ks}{s^2+\omega^2}$
8	Damped Sine	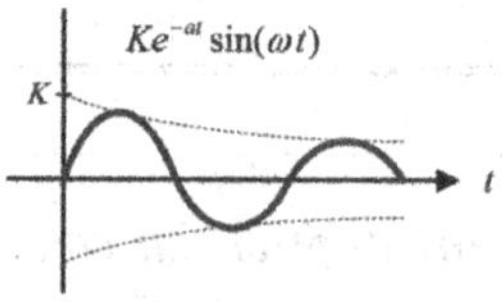 $Ke^{-at}\sin(\omega t)$; K; t	$Ke^{-at} \sin \omega t$	$\frac{K\omega}{(s+a)^2+\omega^2}$
9	Damped Cosine	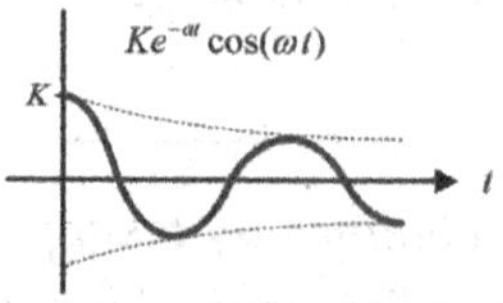 $Ke^{-at}\cos(\omega t)$; K; t	$Ke^{-at} \cos \omega t$	$\frac{K(s+a)}{(s+a)^2+\omega^2}$

Mathematically, this is stated as:

$$y(t) = L^{-1}[X(s) \times H(s)]$$

Example 2

If the input to the component in Example 1 is a step function of unit height, calculate the component time response.

Solution

The step function input in represented by $\frac{1}{s}$

Then $y(t) = L^{-1}\left[\frac{1}{s} \times \frac{1}{1+0.2s}\right]$

Applying the inverse Laplace transform, we obtain the required time response:

$$y(t) = 1 - e^{-0.2t}$$

Angular Velocity and Frequency

Linear velocity is defined as the distance traveled per unit time.

$$v = \frac{d}{t}$$

where

v = average velocity

d = distance traveled

t = time

Assume that an object is moving in a circular rather than a linear path of motion. Rather than specifying distance, it is more fitting to specify its angle of travel (θ) or rotation. We can then define *angular velocity* as:

$$\omega = \frac{\theta}{t}$$

where

ω = average angular velocity

θ = angle traveled

t = time

One revolution of angular travel equals 2π radians.

Therefore, when $\theta = 2\pi$ radians, we will have traveled one complete revolution or

$$\omega = \frac{2\pi}{t} \quad \text{and} \quad \frac{\omega}{2\pi} = \frac{1\,\text{rev}}{t}$$

The term $\frac{\text{rev.}}{t}$ or $\frac{cycles}{t}$ is defined as ***frequency*** and designated f.

Then $$f = \frac{cycles}{t} = \frac{\omega}{2\pi}$$

where

ω = average angular velocity in radians per second
f = frequency in cycles per second (also referred to as *hertz*)
t = time in seconds

$$\text{then } \omega = 2\pi f$$

Appendix D: Exercise Solutions

Chapter 2: Basic Control Concepts

2.1 Consider an electric oven in a typical modern kitchen. Identify the controlled variable (CV), the manipulated variable, and the disturbances.

Answers:

Controlled variable: oven temperature

Manipulated variable: electric current

Disturbances: ambient temperature, oven contents, endothermic and exothermic reactions leaks, door opening and closing

2.2 Consider an automatic, gas-fired, home hot-water tank. Identify the controlled variable, the manipulated variable, and the disturbances.

Answers:

Controlled variable: water temperature

Manipulated variable: gas flow

Disturbances: water usage, Btu content of gas, ambient temperature, inlet water temperature

2.3 Imagine you own a backyard swimming pool. Describe a manual control system to measure the pH of the pool water and to add an acidic solution to adjust the pH. Define the controlled variable, the manipulated variable, and the disturbances.

Answers:

Manual system:

1. Take a sample of the pool water.
2. Use litmus paper or an equivalent to measure the pH.
3. Add acidic solution as necessary.

For this manual system:

Controlled variable: pool water pH

Manipulated variable: acidic solution added

Disturbances: variation in strength of acidic solution, evaporation rate, pH of any makeup water added to pool

2.4 Describe how to automate the control of your swimming pool. Assume you have a tank of acid solution to pump into your pool to control pH; use feedback control.

Answers:

A feedback automatic control system might appear as follows:

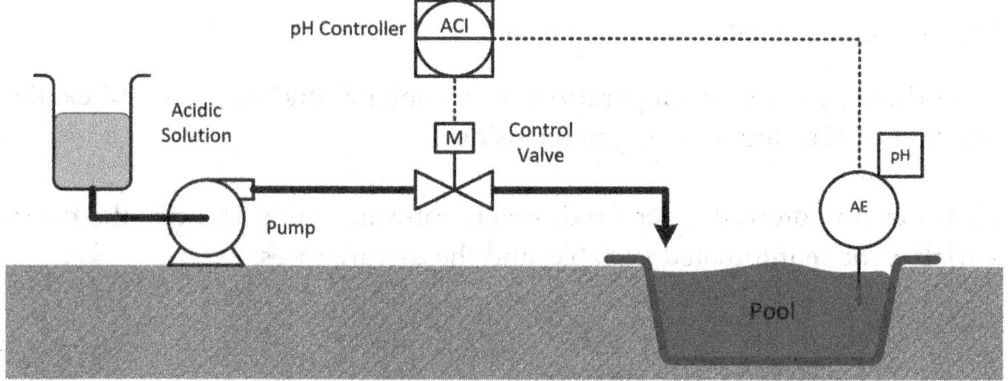

2.5 The cruise control feature used to control speed in an automobile is a good example of feedback control. Outline its operation in terms of feedback control.

Answers:

Set point: desired speed

Controlled variable: actual speed of vehicle

Manipulated variable: accelerator

Disturbances: grade of road, weight, engine performance

Note: For the technical purist, the intake manifold pressure is measured and controlled.

Chapter 3: The Control Loop

3.1 Develop a functional layout of the basic feedback control loop for the simple level control system shown.

Answers:

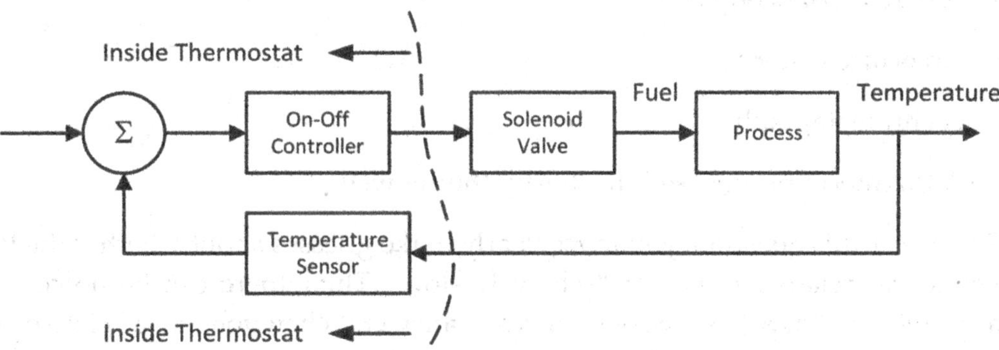

Chapter 4: Functional Structure of Feedback Control

4.1 A temperature sensor has a range of 0°F to 200°F and transmits a 3to 15 psig air signal based on the measurement. What is its sensitivity?

Answers:

$$\frac{(15-3)\text{ psi}}{(200-0)\text{°F}} = \frac{12\text{ psi}}{200\text{ °F}} = 0.06\text{ psi/°F}$$

4.2 The input elements of an electronic controller accept a set point of 0 to 60 gpm and produce a 4 to 20 mA signal to the comparator. What is the sensitivity of these input elements?

Answers:

$$\frac{(16-0)\text{ mA}}{(60-0)\text{ gpm}} = \frac{16\text{ mA}}{60\text{ gpm}} = 0.267\text{ mA/gpm}$$

Chapter 5: Dynamics of Systems

5.1 Estimate whether the dynamic characteristics of the following items of hardware would be significant in the total dynamics of a feedback control loop:

- The input elements in a controller
- An electronic transmission system
- A large, pneumatically operated control valve
- A 2000 ft pneumatic transmission system
- A pneumatic controller
- An electronic controller
- A bare thermocouple
- An orifice meter
- A chromatograph
- A thermocouple encased in a heavy thermowell

To answer this question, you must clearly make guesses about whether the listed items are "relatively fast" or "relatively slow." Thus, there can be no certainty as to what is "right" or "wrong" in your answer. When you ask a relative question, then you must ask, "Compared to what?" In process control, the answer is "Compared to the other elements around the loop." This leads to an appreciation of relative speed, and that is the important point at this stage of your education.

Answers:

Input elements: no

Electronic transmission: no

Large pneumatic valve: yes, probably

2000-ft pneumatic transmission system: yes

Pneumatic controller: no

Electronic controller: no

Bare thermocouple: no

Orifice meter: no

Chromatograph: yes, probably

Thermocouple in thermowell: yes, probably

5.2 In general terms as well as for typical industrial applications, rank the following processes as (a) responding very rapidly to input changes, (b) responding at a moderate rate, or (c) responding very slowly:

- Liquid flow in a line
- Gaseous flow in a line
- Liquid level in a small tank
- Liquid pressure in an enclosed tank
- Gaseous pressure in a large tank
- Composition in a large distillation column
- Temperature in a liquid-filled tank
- Your coworkers at 5:00 p.m.

The comment at the end of Exercise 5.1 about making guesses applies here as well.

Answers:

Liquid flow: very rapid

Gaseous flow: less rapid than liquid flow

Liquid level in a small tank: moderately rapid

Liquid pressure in an enclosed tank: rapid

Gaseous pressure in a large tank: slow

Composition in a large distillation column: slow

Temperature in a liquid-filled tank: slow

Your coworkers: you be the judge!

5.3 Refer to the immersion of the bare bulb in the bath in Figure 5-15. How would the process response curve be different if the thermometer bulb were not bare but instead encased in a protective thermowell?

Answers:

The thermowell causes more lag and slows down the temperature response. The curve in Figure 5-15 might be more S-shaped if a thermowell were used:

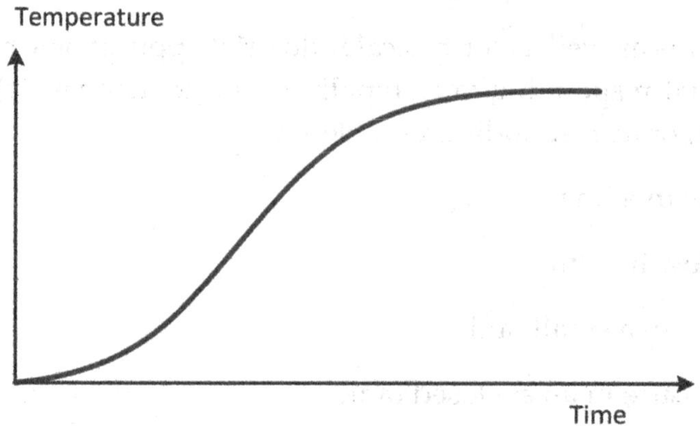

5.4 A process chromatograph is used as a sensor to measure the composition of a liquid stream. It operates on a discrete basis, that is, it takes a sample and analyzes it and establishes a signal for feedback purposes. It then takes a new sample and starts its cycle over again. What effect does this mode of operation have on loop operation as compared to a continuous sensor, such as a thermocouple or an orifice meter?

Answers:

The use of the sampling will introduce a dead time that is roughly equivalent to one-half of sampling interval *T*.

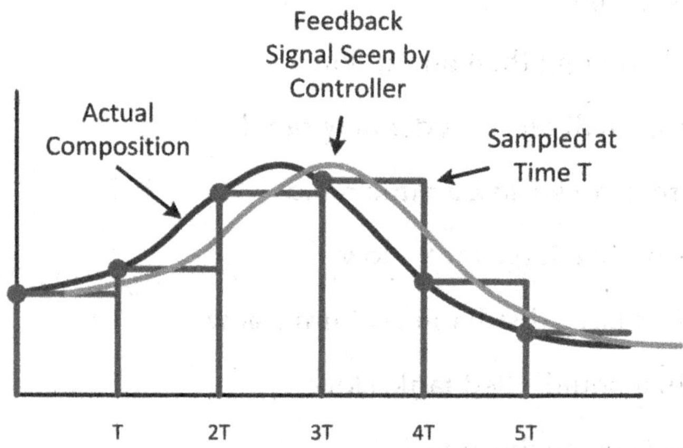

Chapter 7: Sensors, Transducers, and Transmitters

7.1 Our entire presentation of sensors and transmission systems in this unit was structured around the idea of feeding a signal back to the comparator. What would be involved (how would the hardware be different) if the process variable being measured was not a controlled quantity but was instead simply being sent to a central control room for indication and recording?

Answers:

The distinctions between transmission to a board-mounted (central control room) controller and to a board indicator and/or recorder are minor. The transmission systems are virtually identical. The only difference is in the usage of the signal once it gets to the control room.

7.2 A process chromatograph is used as a sensor to measure the composition of a liquid stream. It operates on a discrete basis, that is, it takes a sample and analyzes it and establishes a signal for feedback purposes. It then takes a new sample and starts its cycle over again. What effect does this mode of operation have on loop operation as compared to a continuous sensor such as a thermocouple or an orifice meter?

Answers:

The use of the sampling will introduce a dead time that is roughly equivalent to one-half of sampling interval *T*.

7.3 It was stated that sensor repeatability is more important than accuracy. How can this possibly be true?

Answers:

If there are errors in accuracy, they will not introduce any instability or disturbance into the process, but they may cause the operator to readjust the set point. This is a minor irritant but not nearly as serious as instabilities caused by lack of precision or reproducibility.

7.4 How far can you transmit a pneumatic signal if you can tolerate a time constant lag of 3 s and a dead time of less than 1 s?

Answers:

Refer to Figure 7-6. A time constant of 3 s dictates a maximum pneumatic transmission distance of 700 ft.

Chapter 8: Actuators and Final Control Elements

8.1 An equal-percentage control valve has 10 gpm flowing through it. When the air-to-valve signal increases by 0.5 psi, flow through the valve increases by 1.5 gpm. Much later, the base flow through the valve is operating at 20 gpm when the air-to-valve signal increases by 0.5 psi. How much will flow increase through the valve?

Answers:

A 0.5 psig air-to-valve signal increase produces a flow increase of 15%. Therefore, at a base flow of 20 gpm, the increase is 3 gpm.

8.2 Repeat Exercise 8.1 but assume the valve has a linear trim.

Answers:

For a linear valve, the gain of the valve does not change. Therefore, at 20 gpm, a 0.5 psi increase produces a change of 1.5 gpm.

8.3 A control valve controlling the flow of water has a C_v of 20 and an assigned pressure drop of 10 psi. How many gpm will flow through the valve?

Answers:

$$M_L = C_v\sqrt{\frac{\Delta P}{G}} = 20\sqrt{\frac{10}{1}} = 20 \times 4.16 = 63.3 \text{ gpm}$$

8.4 A control valve controlling the flow of air has a C_v of 32, an assigned pressure drop of 10 psi, an expansion factor of 0.8, and a specific weight of 3.103 lb/ft_3. Calculate the flow through the valve.

Answers:

$$M_g = 63.3\ C_v\ Y\sqrt{\Delta P\ \gamma_1} = 63.3 \times 32 \times 0.8\sqrt{10 \times 3.103} = 9.027 \text{ lb/h}$$

8.5 Given an equal-percentage valve installed with 10% of the dynamic pressure drop across the valve, assume the valve operates about 50% to 70% open under normal conditions. Does the valve act more like an equal-percentage valve or more like a linear valve?

Answers:

Refer to Figure 7-8. With 10% of the dynamic drop across the value and the valve operating in the range of 50% to 70% open, the valve characteristic curve is effectively linear.

8.6 Consider several actual feedback loops, such as a home heating system, an oven, and the air compressor controller at a service station. For each loop, decide how you would like the final element to "fail" in the event of power loss.

Answers:

For most situations, we prefer the valve to fail closed if it is adding energy to the system (e.g., a steam valve) but to fail open if it is removing energy from the system (e.g., a cooling water valve).

Chapter 9: Controllers

9.1 For a proportional controller, answer the following questions:

A. What gain corresponds to a proportional band of 200%?

B. What proportional band corresponds to a gain of 0.2?

Answers:

A. 0.5

B. 500%

9.2 Proportional-only controllers exhibit offset at steady state. How is this possible? (Offset implies an error; therefore, why doesn't the controller output eliminate the error?)

Answers:

At steady state, there is an unchanging error. This error produces a controller output that produces an unchanging amount of manipulated variable. All the manipulated variable so produced is used to maintain the current state, and there is no "surplus" or incremental amount of the manipulated variable available to produce a change from the current steady state. In effect, in maintaining the status quo, the manipulated variable output is consumed.

9.3 In this chapter, the error is shown as the input to the controller. Does it make any difference if the error is caused by a change in the SP in one case or by a disturbance in another?

Answers:

It makes no difference if the error is caused by a change in disturbance. The dynamics of the response will be similar either way. Of course, the transient response of the process will lead to a new final value if there has been a change in the set point.

9.4 Proportional control is often described by the equation $m = K_C \times e + M_r$, where M_r is an adjustable "manual reset" that can be used to bias the output so as to eliminate offset at a specific operating point. Will this eliminate offset at other operating points? Explain your answer.

Answers:

If manual reset is used to bias the controller output to eliminate offset at a specific operating point, there will be no offset in evidence until there is either a change to a new set point or there is a change in some of the disturbances at work on the system. Because disturbances and set points change frequently, eliminating offset by manual reset is not a long-term operating solution.

9.5 It has been proposed that a controller might be designed to check the magnitude of the error. If the error is large, the controller would operate as a proportional-only controller; if the error is small, it would operate as a reset-only controller. What would be the advantage(s) of this type of controller?

Answers:

The proposed controller would have good dynamic response for large errors, and when the error is small, the reset-only controller could be used to trim out offset. The significant advantage is that the dynamic lag penalties inherent with the reset would not be present except when that mode is in active use.

9.6 In practice, all controllers include some filtering of the error signal before or within the rate mode. Why is this necessary?

Answers:

Filtering is necessary on the rate mode because otherwise very small and rapid changes in the error signal might produce very large outputs from the rate mode. This would be deleterious to loop dynamics.

Chapter 10: Tuning Control Loops

10.1 How is it physically possible in two-mode and three-mode controllers (such as proportional-integral or proportional-integral-derivative) for there to be more than one set of controller settings that give a 1/4 decay ratio?

Answers:

For two-mode and three-mode controllers, there can be many response curves with a decay ratio of 1/4 but there will be oscillations at different frequencies.

10.2 Why is it that all liquid-level control loops do not have the same tuning parameters, all flow control loops do not have the same tuning parameters, and so on?

Answers:

Tuning parameters are not a function of loop type, such as a liquid-level loop or a flow loop. Rather, tuning constants depend on the individual dynamic parameters and the individual gains (or sensitivities) of the various components that make up the loop. Because these vary from one loop to the next, the desirable tuning parameters also vary from loop to loop.

10.3 For a level control loop whose ultimate sensitivity is 0.3 psi/ft and whose ultimate period is 3 min, calculate the settings for the various types of controllers using the ultimate method. (Sensitivity is a gain with dimensions or engineering units; gain is dimensionless.)

Answers:

Proportional only:

$$S_C = 0.5S_u = 0.5 \times 0.3\frac{psi}{ft} = 0.15psi/ft$$

Proportional-plus-reset:

$$S_C = 0.45\,S_u = 0.45 \times 0.3\frac{psi}{ft} = 0.135\,psi/ft$$

$$T_I = \frac{P_u}{1.2} = \frac{3}{1.2} = 2.5\,min$$

Proportional-plus-derivative:

$$S_C = 0.6\,S_u = 0.6 \times 0.3 = 0.18\,psi/ft$$

$$T_D = \frac{P_u}{8} = 0.375\,min$$

Proportional-plus-reset-plus-derivative:

$$S_C = 0.6\,S_u = 0.6 \times 0.3 = 0.18\,psi/ft$$

$$T_I = 0.5\,P_u = 0.5 \times 3 = 1.5\text{ min}$$

$$T_D = \frac{P_u}{8} = \frac{3}{8} = 0.375\text{ min}$$

Chapter 11: Cascade Control

11.1 Given the following heat exchanger, sketch a cascade arrangement for TC-FC (temperature control-flow control) in which a secondary flow loop has been provided for the steam.

Answers:

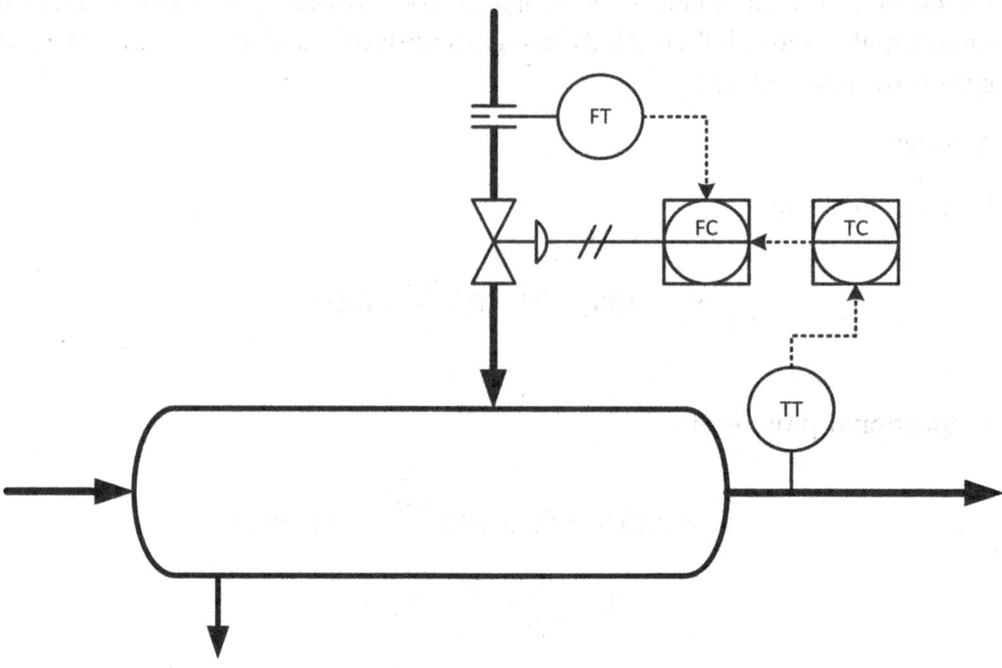

11.2 Redo Exercise 11-1 but provide TC-PC (temperature control-pressure control) with a secondary pressure control loop on the exchanger shell.

Answers:

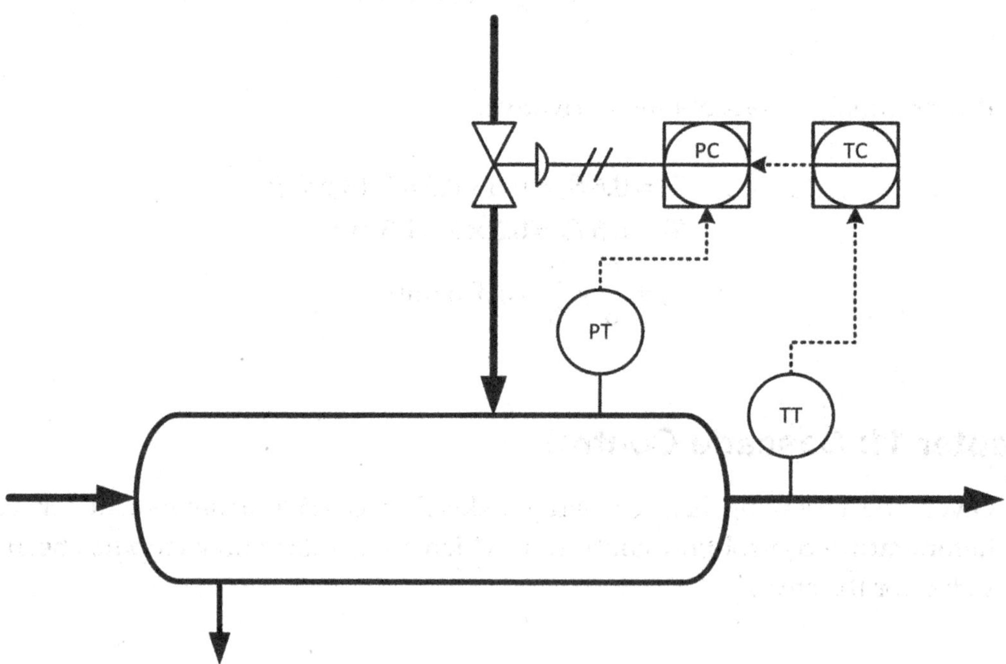

11.3 Compare the results of Exercises 11-1 and 11-2 and determine which gives a better initial response to disturbances in inlet process temperature (T_I).

Answers:

The tubes will be practically at the temperature that corresponds to the steam-side pressure because the liquid film offers a much greater resistance to heat transfer.

Chapter 12: Feedforward Control

12.1 Refer to the following figure. Design a steady-state feedforward control system for this unit.

Answers:

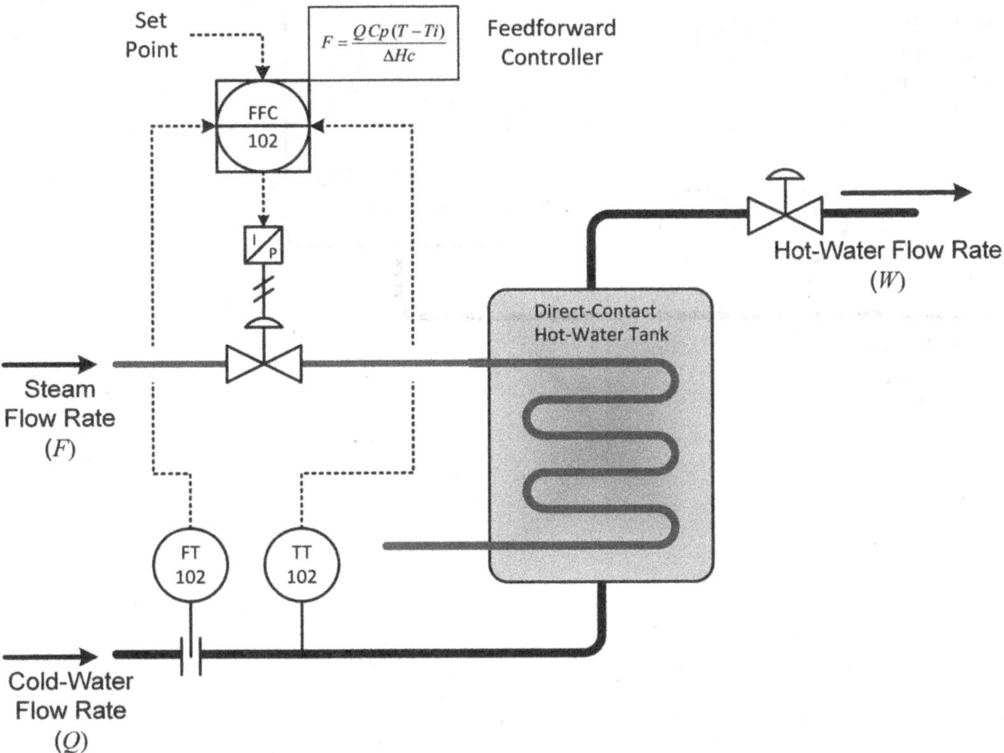

12.2 To Exercise 12-1, add a lag-lead dynamic compensation of the following form:

$$\frac{1+\tau 1s}{1+\tau 2s}$$

where the ratio of $\tau 1/\tau 2$ is adjustable (tunable).

Answers:

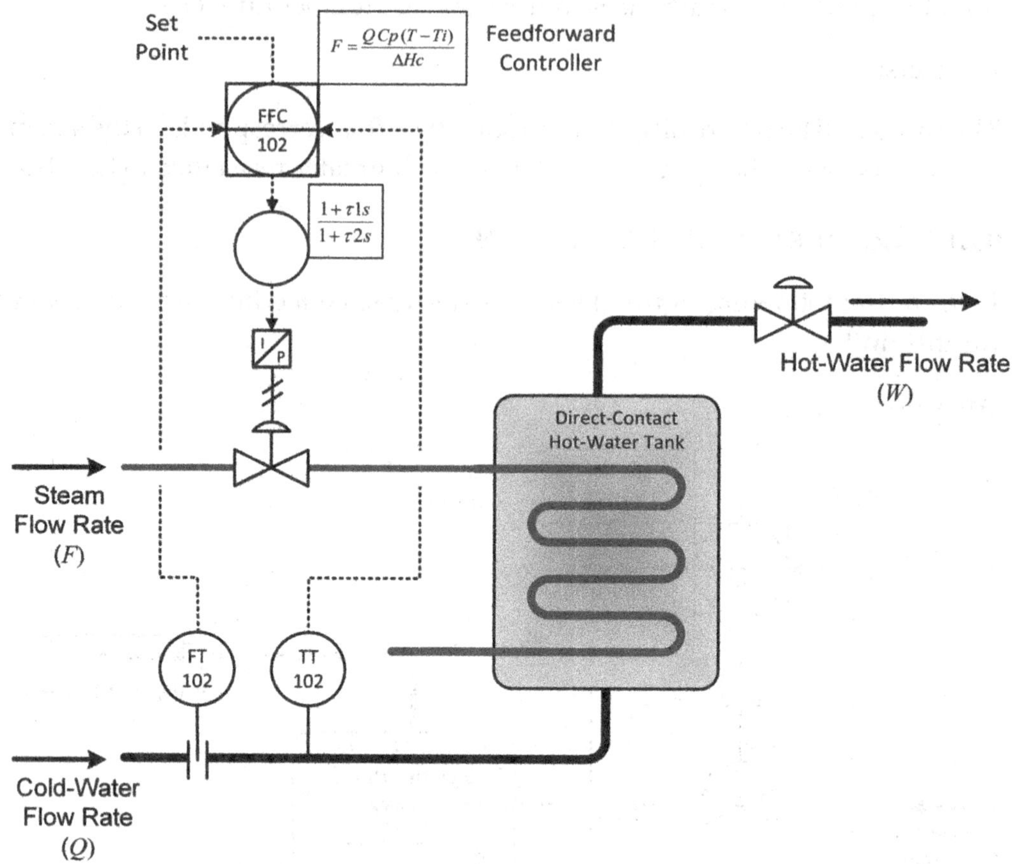

Set Point
$F = \frac{QCp(T - Ti)}{\Delta Hc}$
Feedforward Controller
FFC 102
$\frac{1 + \tau 1s}{1 + \tau 2s}$
I P
Hot-Water Flow Rate (W)
Direct-Contact Hot-Water Tank
Steam Flow Rate (F)
FT 102
TT 102
Cold-Water Flow Rate (Q)

12.3 Add cascade control of the manipulated steam flow to the diagram in Exercise 12-2.

Answers:

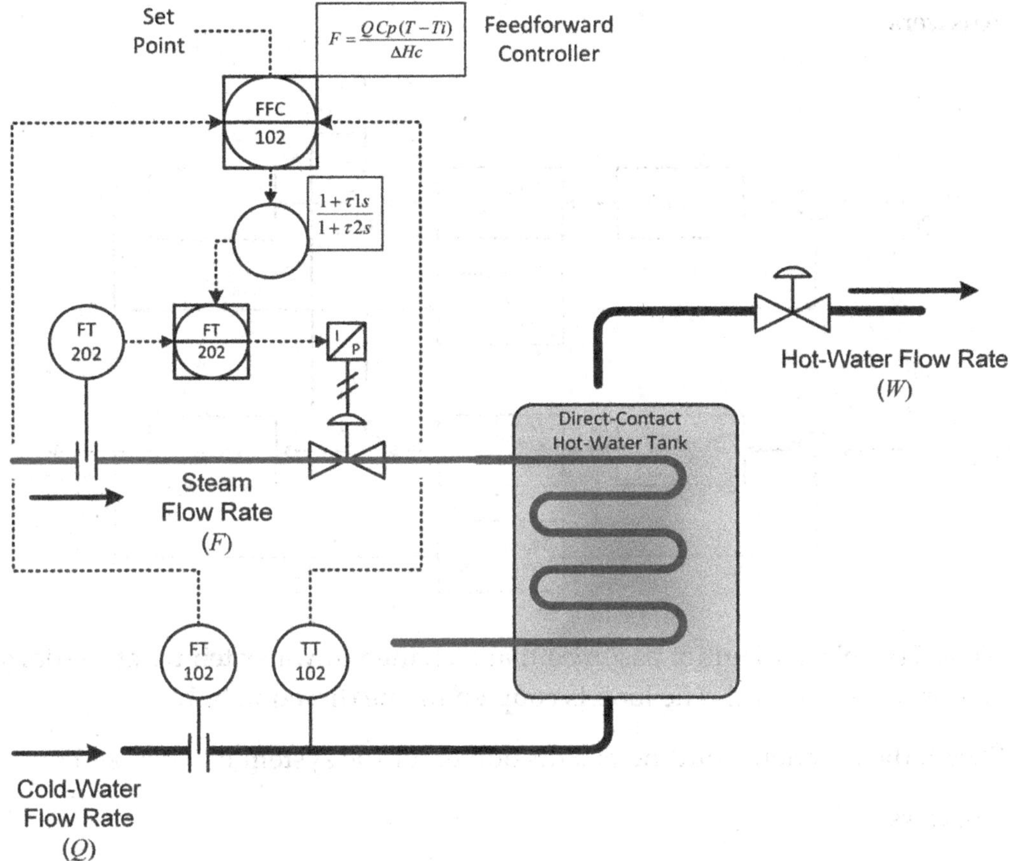

Chapter 13: Multivariable Control

13.1 Given the following 2 × 2 control system, draw the functional block diagrams for these two loops.

Answers:

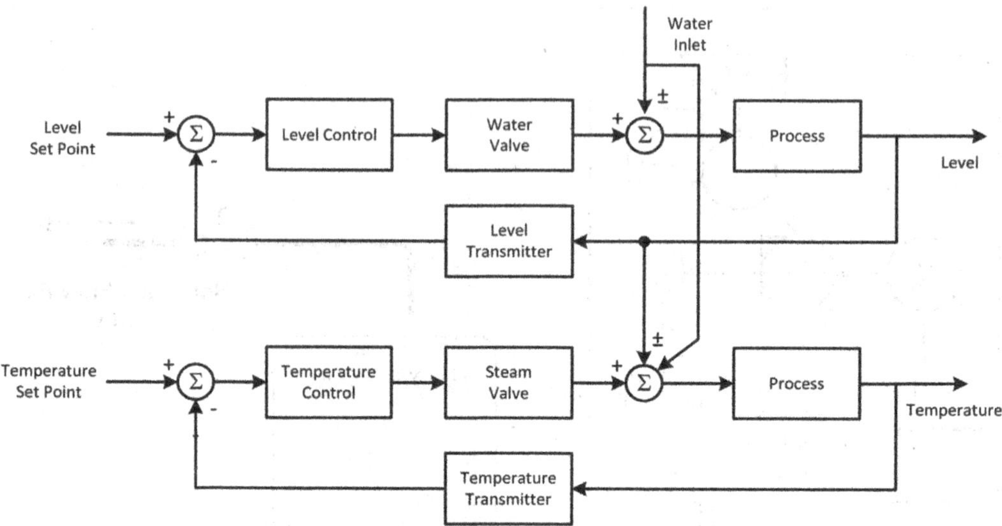

Note: This block diagram assumes that a change in water temperature does not change its water level. The loop is coupled in one direction only.

13.2 Sketch the functional outline of a decoupler for the system in Exercise 13.1.

Answers:

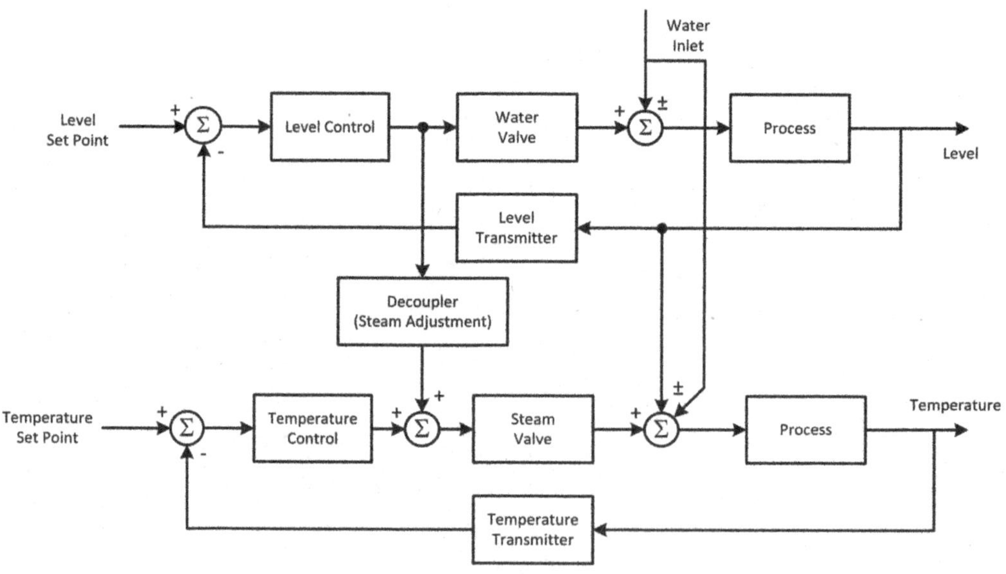

Chapter 14: Ratio Control

14.1 Diesel oil is injected into the suction of a fuel oil pump. (The impeller does the mixing to control viscosity in fuel oil.)

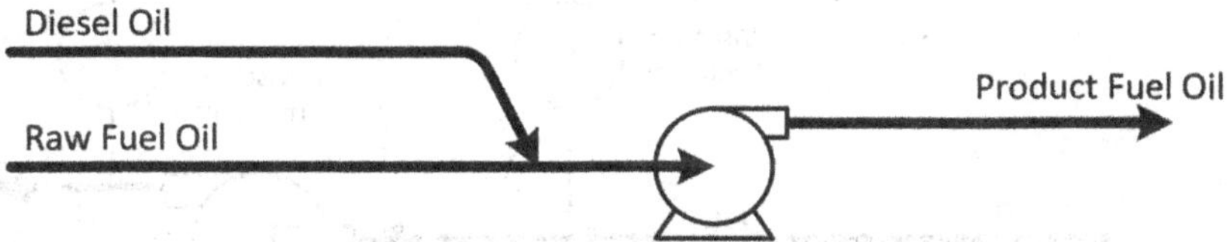

Design a ratio control system for this pump.

Answers:

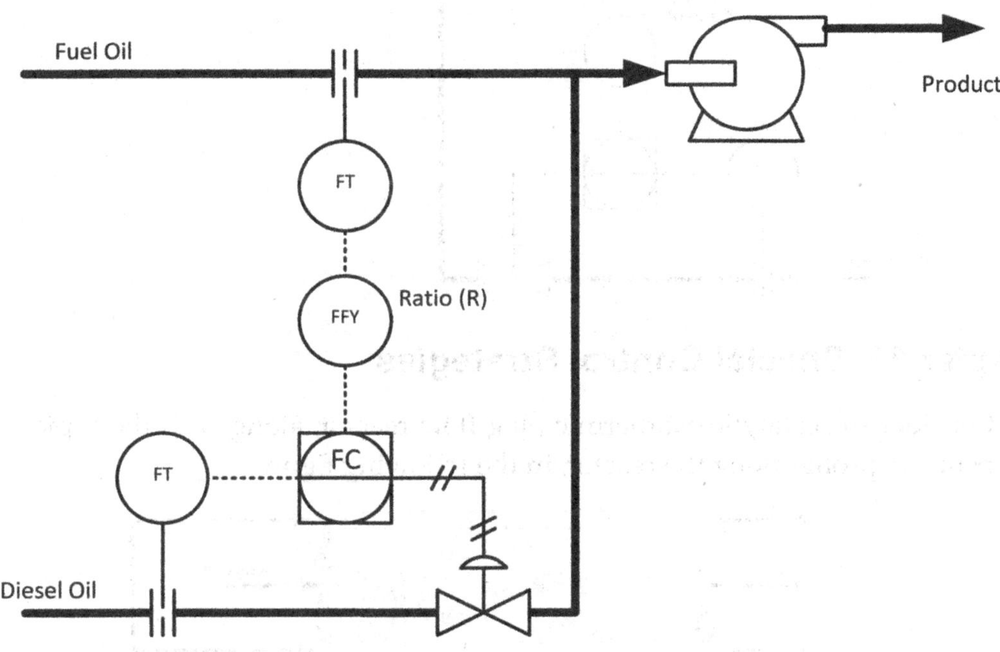

14.2 Install a viscosity meter as a sensor on the product fuel oil line of Exercise 14.1 and use it to adjust the ratio between the two streams.

Answers:

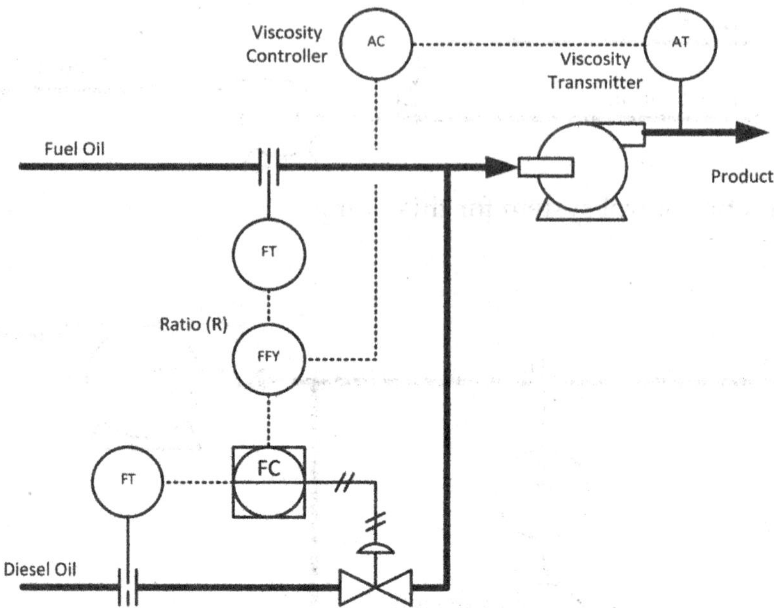

Chapter 15: Special Control Strategies

15.1 Consider the catalytic exothermic plug flow reactor, along with the typical temperature profile along the reactor, in the following figure.

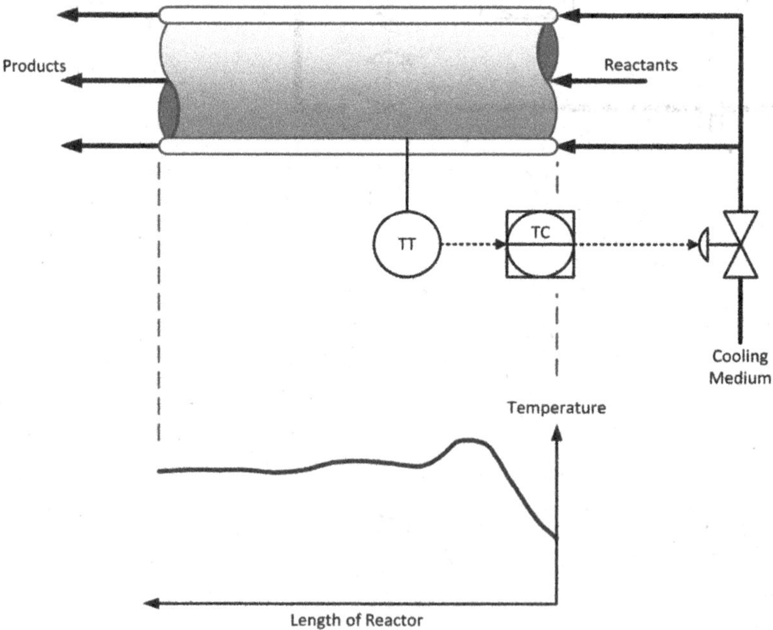

As conditions change and as the catalyst ages, the hot spot moves. Design a selective control scheme so that the measured variable moves as the hot spot moves. Use multiple temperature measurements with a high selector.

Answers:

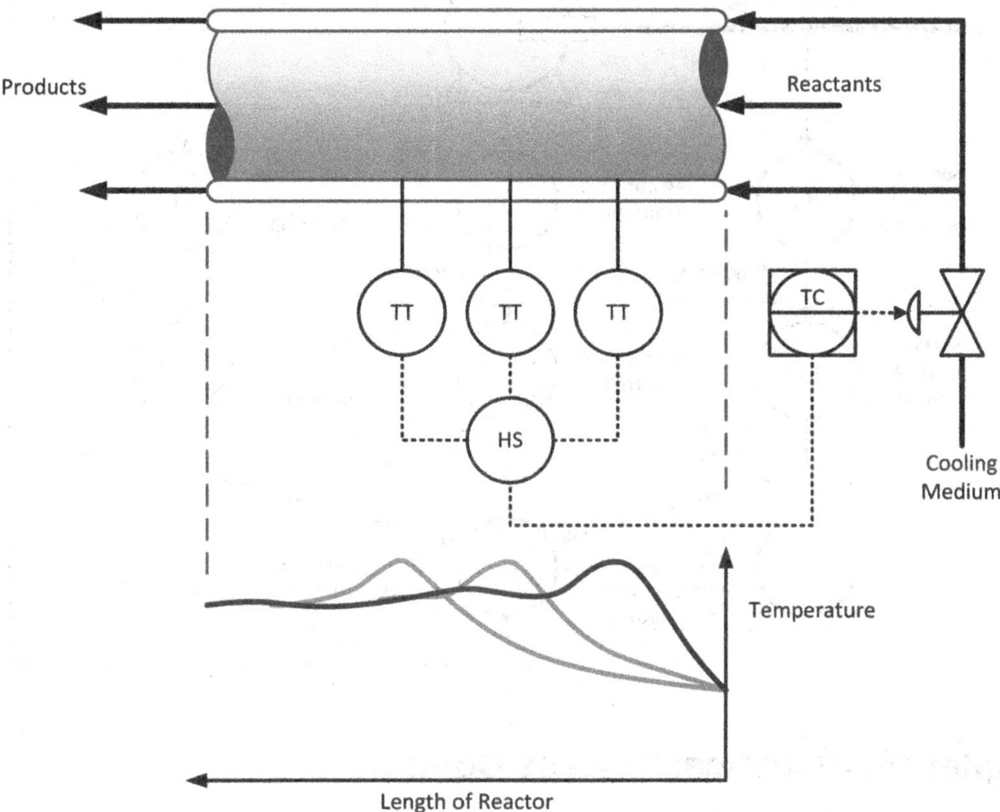

The high selector (HS) always selects the transmitter with the highest output, and, in so doing, the controller process variable is always the highest temperature.

15.2 To operate a pumping station on a pipeline efficiently, it is best to operate the control valve wide open on all stations but one. This station becomes the throttling unit. It is necessary to cut back on the control valve position if the:

- o Suction pressure drops too low
- o Motor load rises too high
- o Discharge pressure rises too high

Sketch an auto-select/cutback control system to handle this.

Answers:

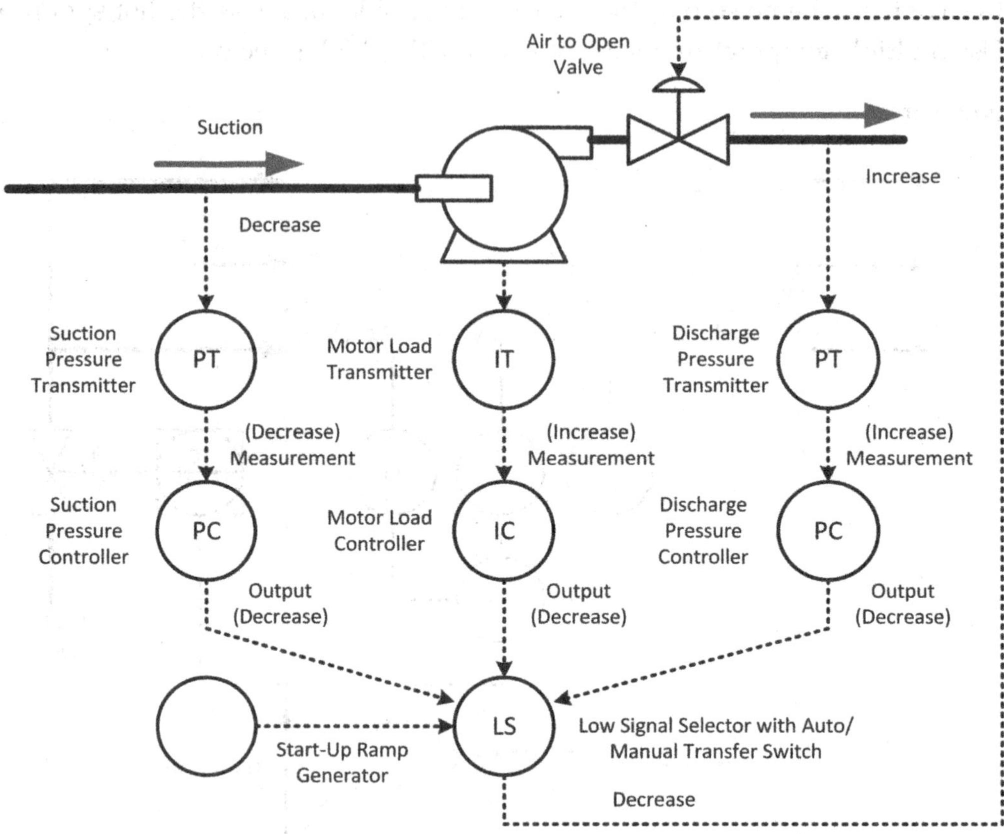

Chapter 16: Statistical Process Control

16.1 Raw material used in a chemical process contains some impurities. To ensure that the impurities are below the requirement specified, three samples are tested at regular intervals. When the process is running satisfactorily, the mean percentage of impurities was found to be 16.7 with a standard deviation of 3.4.

Set up a control chart for means and for standard deviations for the following set of samples.

Sample 1	Sample 2	Sample 3	Mean	Standard Deviation
16.9	19.3	20.2		
14.2	19.1	5.2		
9.3	12.2	8.1		
24.2	25.6	22.0		
22.7	19.3	23.1		

Answers:

Sample 1	Sample 2	Sample 3	Mean	Standard Deviation
16.9	19.3	20.2	18.80	1.39
14.2	19.1	5.2	12.83	5.76
9.3	12.2	8.1	9.87	1.72
24.2	25.6	22	23.93	1.48
22.7	19.3	23.1	21.70	1.70

16.2 In the production of widgets, four samples are taken and measured at regular time intervals. The following table shows the measurement of the four samples at eight sequential time intervals. The target measurement value is 9 mm.

Sample	Time interval							
	1	2	3	4	5	6	7	8
1	9.0	10.4	8.2	7.9	8.2	8.4	7.4	7.6
2	8.1	9.0	9.2	7.7	9.0	8.1	8.0	8.5
3	8.7	7.9	7.9	9.7	7.4	8.4	8.9	8.1
4	7.5	7.2	7.7	9.3	8.6	8.7	9.8	8.8

A. Calculate the mean sample range and the standard deviation of the process.

B. Using the result of A., draw a control chart for means showing 95% upper and lower limits, and plot the eight samples.

C. Briefly discuss your observations of the patterns of the chart.

Answers:

Mean

$$\bar{x} = \frac{1}{n}\sum_{i=1}^{n} x_i = 8.4$$

First calculate the variance (s^2)

$$s^2 = \frac{1}{n-1}\sum_{i=1}^{n}(x_i - \bar{x})^2 = 0.02$$

Then calculate the standard deviation (σ)

$$\sigma = \sqrt{s^2} = o.1414$$

Chapter 20: Nonlinear and Adaptive Control

20.1 Given Equation 20-1, what is the steady-state process gain?
How does this vary with flow?

Answers:

Differentiating controlled variable T_o with respect to manipulated variable W gives the following:

$$\frac{dT_o}{dF} = \frac{\Delta H}{WC_P}$$

The gain varies inversely with liquid flow.

20.2 Given the following installed characteristic curve for a butterfly valve, how could you compensate for the nonlinear character of the valve?

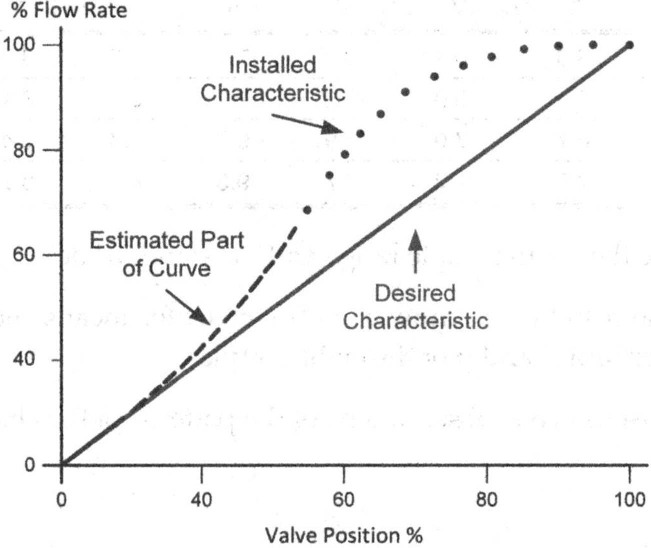

Answers:

Given the flow characteristic curve for the valve, a compensator is needed whose characteristics are the reverse of the valve's. For example, a compensator was available in the chosen controller that approximated the curve by five straight-line segments. The segments need not be of equal length, so most of them can be in that portion of the curve where you expect to operate.

Bibliography

Amy, Lawrence T. *Automation Systems for Control and Data Acquisition.* Research Triangle Park, NC: ISA (International Society of Automation), 1992.

ANSI/ISA-88.00.01-2010. *Batch Control – Part 1: Models and Terminology.* Research Triangle Park, NC: ISA (International Society of Automation).

Aström, Karl J., and Tore Hägglund. *PID Controllers: Theory, Design, and Tuning.* 2nd ed. Research Triangle Park, NC: ISA (International Society of Automation), 1995.

Bateson, Robert N. *Introduction to Control Technology.* 6th ed. Upper Saddle River, NJ: Prentice-Hall, 1999.

Bauman, Hans D. *Control Valve Primer: A User's Guide.* 4th ed. Research Triangle Park, NC: ISA (International Society of Automation), 2009.

Blevins, Terrence L., Gregory K. McMillan, Willy K. Wojsznis, and Michael W. Brown. *Advanced Control Unleashed: Plant Performance Management for Optimum Benefit.* Research Triangle Park, NC: ISA (International Society of Automation), 2003.

Borden, Guy Jr. *Control Valves: Practical Guide for Measurement and Control.* Research Triangle Park, NC: ISA (International Society of Automation), 1998.

Bucek, Victor J. *Control Systems: Continuous and Discrete.* Englewood Cliffs, NJ, Prentice-Hall, 1989.

Cogan, D. A., ed. *Fundamentals of Industrial Control.* 2nd ed. Research Triangle Park, NC: ISA (International Society of Automation), 2005.

Connell, Bob. *Basic Math for Process Control.* Research Triangle Park, NC: ISA (International Society of Automation), 2003.

Corripio, A. P. *Tuning of Industrial Control Systems.* 2nd ed. Research Triangle Park, NC: ISA (International Society of Automation), 2001.

Dorf, Richard C. *Modern Control Systems.* 2nd ed. Don Mills, ON: Addison-Wesley, 1974.

Driels, Morris. *Linear Control Systems Engineering.* New York: McGraw-Hill, 1996.

Friedmann, Paul G. *Automation and Control Systems Economics.* 2nd ed. Research Triangle Park, NC: ISA (International Society of Automation), 2006.

Friedmann, Paul G. *The Automation, Systems, and Instrumentation Dictionary.* 4th ed. Research Triangle Park, NC: ISA (International Society of Automation), 2005.

Harrington, John. *Automated Process Control Electronics.* Albany, NY: Delmar, 1989.

Herb, Samuel M. *Understanding Distributed Processor Systems for Control.* Research Triangle Park, NC: ISA (International Society of Automation), 1999.

Hodson, William R. "How Fieldbus Will Affect DCS Architecture." *InTech* 43, no. 11 (November 1996): 50–53.

Hughes, Thomas A. *Measurement and Control Basics.* 5th ed. Research Triangle Park, NC: ISA (International Society of Automation), 2015.

ISA-5.1-2009. *Instrumentation Symbols and Identification.* Research Triangle Park, NC: ISA (International Society of Automation).

ISA-51.1-1979 (R 1993). *Process Instrumentation Terminology.* Research Triangle Park, NC: ISA (International Society of Automation).

ISA-5.3-1983. *Graphic Symbols for Distributed Control/Shared Display Instrumentation, Logic and Computer Systems.* Research Triangle Park, NC: ISA (International Society of Automation).

ISA-5.4-1991. *Instrument Loop Diagrams.* Research Triangle Park, NC: ISA (International Society of Automation).

ISA-5.5-1985. *Graphic Symbols for Process Displays.* Research Triangle Park, NC: ISA (International Society of Automation).

Kilian, Christopher T. *Modern Control Technology: Components and Systems.* 3rd ed. Clifton Park, NY: Delmar Thomason Learning, 2006.

Kleiber, Michael. *Process Engineering.* Berlin/Boston: Walter de Gruyter GmbH, 2016.

Macia, Narciso F., and George J. Thaler. *Modeling and Control of Dynamic Systems.* New York: Thomson Delmar Learning, 2005.

Murrill, Paul W. *Fundamentals of Process Control Theory.* 3rd ed. Research Triangle Park, NC: ISA (International Society of Automation), 2000.

O'Conner, Dan. *A Process Control Primer.* Freeport, IL: Honeywell, 2000.

Ogata, Katsuhiko. *Modern Control Engineering.* 3rd ed. Upper Saddle River, NJ: Prentice-Hall, 1997.

Rohrs, Charles E., James L. Melsa, and Donald G. Schultz. *Linear Control Systems.* New York: McGraw Hill, 1993.

Sands, Nicholas P., and Ian Verhappen, eds. *A Guide to the Automation Body of Knowledge.* 3rd ed. Research Triangle Park, NC: ISA (International Society of Automation), 2018.

Sante, Daniel P. *Automatic Control System Technology.* Upper Saddle River, NJ: Prentice-Hall, 1980.

Seal, A. M. *Practical Process Control.* Burlington, MA: Elsevier Butterworth-Heinemann, 1998.

Shewhart, Walter A., and W. Edwards Deming. *Statistical Method from the Viewpoint of Quality Control.* Washington, DC: Graduate School of the Department of Agriculture, 1939.

Smith, Carlos A., and Armando Corripio. *Principles and Practice of Automatic Process Control*. 3rd ed. Hoboken, NJ: John Wiley & Sons, 2006.

Smith, Cecil L. *Basic Process Measurements*. Hoboken, NJ: John Wiley & Sons, 2009.

Smith, Cecil L. *Practical Process Control: Tuning and Troubleshooting*. Hoboken, NJ: John Wiley & Sons, 2009.

Verhappen, Ian, and Augusto Pereira. *Foundation Fieldbus*. 4th ed. Research Triangle Park, NC: ISA (International Society of Automation), 2012.

Wade, Harold L. *Basic and Advanced Regulatory Control: System Design and Application*. Revised 3rd ed. Research Triangle Park, NC: ISA (International Society of Automation), 2017.

Weyrick, Robert C. *Fundamentals of Automatic Control*. New York: McGraw-Hill, 1975.

Williams, Theodore J. *The Use of Digital Computers in Process Control*. Research Triangle Park, NC: ISA (International Society of Automation), 1984.

Ziegler, J. G., and N. B. Nichols. "Optimum Settings for Automatic Controllers." *Transactions of the ASME* 64 (November 1942): 759–768.

Index

Note: "n" refer to the note.

C

D

G

H

N

O

P

Q

R

S